HUMBLE WORKS
FOR
HUMBLE PEOPLE

For Isaac

HUMBLE WORKS FOR HUMBLE PEOPLE

A History of the Fishery Piers of Co. Galway and North Clare, 1800–1922

NOËL P. WILKINS

IRISH ACADEMIC PRESS

First published in 2017 by
Irish Academic Press
10 George's Street
Newbridge
Co. Kildare
Ireland
www.iap.ie

9781911024910 (Cloth)
9781911024927 (Kindle)
9781911024934 (Epub)
9781911024941 (PDF)

British Library Cataloguing in Publication Data
An entry can be found on request

Library of Congress Cataloging in Publication Data
An entry can be found on request

Interior design by www.jminfotechindia.com
Typeset in Minion Pro 11/14

Jacket design by edit+ www.stuartcoughlan.com

Jacket front: Roundstone Harbour, Co. Galway (David Robertson/Alamy Stock Photo).

Jacket back: Cleggan pier (top): detail from NLI maps, 16 H 5 (5), courtesy of the National Library of Ireland; Rosroe harbour: detail from 20th Report of the OPW, BPP 1852–3 [1569]; Clifden harbour map: detail from OPW/8/84, courtesy of the Director, the National Archives of Ireland; Trellis pier, Clifden: detail from OPW/5/4076/5, courtesy of the Director, the National Archives of Ireland.

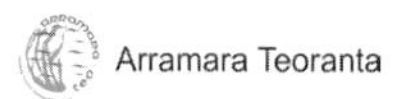

Contents

Brief Chronology

1819. The Act 59 Geo III C. 109, *an Act for the Further Encouragement and Improvement of the Irish Fisheries,* sets up the first Commission for Irish Fisheries.

1822. Famine in the West. The Commission instructs Alexander Nimmo to design small piers and harbours as public works largely for the relief of distress.

1824. The Act 5 Geo IV C. 64 makes the first legislative provision of funds for Irish fishery piers and quays.

1830. The Commission goes out of office having built 23 piers and quays in Galway and north Clare (65 commenced in all Ireland).

1831. The Act 1&2 Wm IV C. 33, A*n Act for the Extension and Promotion of Public Works in Ireland,* sets up the Irish Board of Public Works (OPW). The Board takes over responsibility for erecting fishery piers and harbours.

1835. Public Enquiry into the state of the Irish Fisheries.

1835 to 1841. OPW does little on piers and quays beyond general repairs and maintenance.

1842. The Act 5&6 Vict. C. 106, *An Act to regulate the Irish Fisheries,* creates entirely new legislation for Irish fisheries. The Commissioners of the OPW made Commissioners for Irish Fisheries and appoint two Inspectors of Irish Fisheries.

1842 to 1845. Having no funds for piers, the OPW does little to build new structures, but Ballyvaughan pier in north Clare built with existing resources.

1846 and 1847. First (9 Vict. C. 3) and Second (10&11 Vict. C. 75) *Piers and Harbours* Acts provide funds for building fishery piers and quays as relief works in response to the Great Famine.

1846 to 1853. Nine piers and quays built in County Galway (64 in all Ireland) under the Acts.

1854. The Act 16&17 Vict. C. 136 legislates for the fishery piers and quays to be vested in the relevant counties. 21 vested in Galway and north Clare. 93 vested in all Ireland.

1855 to 1865. OPW exhausted after the Great Famine and very little new pier building undertaken.

1866. New Piers and Harbours Act increases the sums available for individual piers from £5,000 to £7,500 and extends funding support to pier extensions and restorations. This stimulates renewed building. Nine works undertaken from 1867 to 1879 in Galway and north Clare, including *Spidéal* new pier.

1869. A new Irish Fishery Inspectorate set up under the Act 32&33 Vict. C. 92, *An Act to amend the Laws relating to the Fisheries of Ireland.*

1875 to 1879. Treasury suspends all funding for small piers and quays in Ireland to facilitate the funding of major harbours on the east and south coasts. (Ardglass, Arklow, Kinsale)

1879. Famine threatens again. Suspension of pier building lifted.

1880. $100,000 provided by the Government of Canada for the relief of distress in the threatened famine. $20,000 of this allocated to the construction of piers and quays, eight in Galway and 29 over all the west coast.

1881. UK Government provides £45,000 for piers under the Relief of Distress Acts, 1881. Fishery Piers Committee (FPC) set up to select the sites to benefit from these funds.

1881 to 1884. Twelve piers and quays built mainly as relief works in Galway (N=11) and north Clare (N=1) (39 in all Ireland).

1883. *The Sea Fisheries (Ireland) Act 1883*, 46 & 47 Vict. C. 26, grants £250,000 for Irish fishery piers and quays. Fishery Piers and Harbours Commission (FPHC) set up to determine the sites that were to benefit from the Sea Fisheries Fund. Piers were to be built by the OPW.

1883 to 1889. 14 piers and harbours built or greatly restored in Galway and north Clare under the Sea Fisheries (1883) Act (63 in all Ireland).

1886. The *Relief of Distressed Unions Act*, 49 Vict. C. 17, provides £20,000 to be administered by a Piers and Roads Commission (P&RC) for public relief works, including piers and quays, in counties Galway and Mayo. P&RC set up under C.T. Redington.

1886 to 1887. The P&RC completes 21 marine works in Galway and Mayo.

1887 to 1889. Col. T Fraser of the P&RC completes a further 13 works under the Public Works and Industries (Ireland) Special Grant.

1891. The Congested Districts Board for Ireland set up under the *Purchase of Land (Ireland) Act*, 54 & 55 Vict. C. 48, and takes over responsibility for piers and quays in the congested districts.

1892 to 1900. The CDB undertakes 21 small marine works in county Galway.

1898. The *Local Government (Ireland) Act* sets up County Councils in place of Grand Juries. Councills take over responsibility for further pier construction.

1899. *The Department of Agriculture and Technical Instruction Act, 1899* is passed. It takes over responsibility for fisheries and fishery piers outside the congested districts counties.

1902. *The Marine Works (Ireland) Act, 1902* provides funds for industrial developments, including fishery piers and harbours, within the congested districts counties.

1903. *The Ireland Development Grant Act, 1903* provides development funds for works including piers and quays, outside the congested districts counties.

1901 to 1911. CDB undertakes 11 small marine works in County Galway with its own resources. It also contributes to five large piers in County Galway (total 19 in all Ireland) funded mainly under the Marine Works (Ireland) Act.

1908. The Royal Commission on Congestion (the Dudley commission), set up in 1906, reports. As a result, major changes occur in the CDB and in its involvement with fishery piers.

1899 to 1914. Thirteen works in all Ireland (one in county Galway) were sanctioned and scheduled under the *Sea Fisheries (Ireland) Act 1883*, the *Ireland Development Grant Act 1903* with DATI and County Council support during this period. Some were never commenced due to the outbreak of World War 1.

1915. All funding for piers and quays suspended due to the war.

1918 to 1922. Little further action taken in connection with large piers. CDB builds two small piers in County Galway.

1923. Old OPW replaced by new OPW. CDB and DATI replaced by Department of Fisheries of Saorstát Éireann.

ACKNOWLEDGEMENTS

This work has been a long time in preparation and many persons have helped in a variety of ways, pointing me in fruitful directions and providing information and other practical assistance. They include: Seamus Breathnach; Micheál Corduff; Paul Duffy; Kevin Finn; Michael Gibbons; Paul Gosling; John S. Holmes; Jim Houghton; Catherine Jennings; Des and Tom Kenny; Marie Mannion; John Mercer; Seamus Ó Scannail; Mike Taylor; Kathleen Villiers-Tuthill; Fr Kieran Waldron; Brendan Wilkins; and very many others whose information, freely shared, went into the substance of the work and whose encouragement kept it all alive. Thank you all for making this work possible.

Jerry O'Sullivan, Pádraic de Bhaldraithe, Jim Gosling and Eoin McLoughlin read an early draft of the work which was shaped and greatly improved by their special comments and good advice. Their keen interest and constant encouragement were essential when my energy seemed to flag. From the project's conception through to the final draft, Dr Micheál Ó Cinnéide was a constant supporter, especially during the long research phase and in the final process leading to publication. Without his unflagging enthusiasm and belief in the worth of the project, there would never have been any successful outcome.

Special thanks are due to the staffs of the National Archives of Ireland, the National Library of Ireland, the staff of the Special Collections and Archives section of the James Hardiman Library, NUI Galway (especially Marie Boran, Kieran Hoare, Margaret Hughes, Margo Donohue and Geraldine Curtin) for their unfailing courtesy and professional attention.

Publication could not have gone to completion without the generous sponsorship of the Marine Institute; Ryan Hanley Consulting Engineers; *Údarás na Gaeltachta*; *Arramara Teo*; and NUI, Galway. I thank them most sincerely for their financial help, and also for the confidence they showed in what must have sounded like a very specialised if not esoteric work. Hopefully they will find some satisfaction in the outcome. Should the book fail to meet their expectations, they are in no way responsible; neither are they responsible for (and they may not even agree with) the opinions or approaches taken in this book. For these, I alone am entirely responsible and all errors are entirely my own.

For many years, NUI Galway has extended help with computing, library resources and in numerous other ways. Without that background help and the research ethos of the university, the research for this work would never have been sustainable; I can only hope that the result justifies that long-term support.

Conor Graham and his colleagues at Irish Academic Press have been helpful beyond anything I could have expected, for which I am most grateful.

As always, my wife and family put up with me and my needs throughout the long process, especially the difficult later stages when the burden of it all seemed too much to continue with. They saved it from failure.

Note on text and references

References to sources are made in the text using superscript numerals. The full reference data in each case is explained in the numbered notes at the end of each chapter.

Many of the sources are British Parliamentary Papers and these are referenced as follows: Paper title, BPP, Session Year, Paper Number. The title may be abbreviated, but it contains the relevant information regarding the type of material involved. The session year is the year the paper was presented in Parliament. The paper number refers to the actual paper in that session year, and is sometimes enclosed in square brackets, sometimes in round brackets and occasionally without brackets. The number may be preceded by C. or Cd. or Cmd. Examples are:

48th Annual Report of the OPW, BPP 1880 [C. 2646];

11th Report of the Commissioners of Fisheries, BPP 1830 (491);

16th Annual Report of the Congested Districts Board for Ireland, BPP 1908 [Cd. 3767].

The British Parliamentary Papers have been digitised and can be accessed on the internet through various search engines. When searching it is essential to use the exact BPP reference given, i.e. year and paper number exactly as indicated in the notes; omitting the brackets, where they are called for, or using the wrong brackets, or using an incorrect prefix, C. instead of Cd. (for example), may result in failure to locate the relevant paper. Sometimes it is possible to buy a copy – original or 'print on demand' – of parliamentary papers. In this case, relevant words from the title can be input into book search engines like Abebooks.com or Bookfinder.com, where the material may be shown for sale as the search engine indicates. (That is why the titles of papers are given here. They are not essential when locating material by session year and paper number; in searching for copies to buy, such keywords are essential and session dates and numbers usually of no value.)

Full data on a source is given in the notes even when that source is referred to multiple times. While this may seem repetitive, it reduces the generally confusing use of terms like *ibid.*, *loc. cit.* and so on.

Quotes in parentheses are taken unaltered from the sources shown in each case, except where intervening material is omitted, indicated by '...' within the quote. Square brackets *within* the quote are explanatory comments; round brackets *within* the quote are part of the original.

In the text, certain words are printed in italics for emphasis (very occasionally). Where the emphasis is added here, that is stated in square brackets, i.e. [emphasis added]; where the emphasis occurs in the original, it is followed by '[*sic*]'. Apparent misspellings of place names are given as they appear, followed by '[*sic*]' to indicate they are original. This can be important if the misspelling occurs in the original and may be essential in locating the item in the records.

The treatment of place names in Irish is explained in chapter 1, page 4. Generally Irish words and foreign words are given in italics.

Abbreviations

BPP	British Parliamentary Papers.
CDB	The Congested Districts Board, Ireland.
DATI	Department of Agriculture and Technical Instruction, Ireland.
ESB	The Electricity Supply Board of Ireland.
FPC	Fishery Piers Committee.
FPH (FPHC)	Fishery Piers and Harbours (Commission).
LWST	Low Water of Spring Tide.
MWI	The *Marine Works (Ireland) Act 1902.*
NAI	The National Archives of Ireland.
NLI	The National Library of Ireland.
OPW	The Office of Public Works, Ireland; The Commission for Public Works, Ireland.
OS	The Ordnance Survey of Ireland.
P&RC	Piers and Roads Commission.
RCCI	The Royal Commission on Congestion in Ireland.
SFI	*The Sea Fisheries Ireland Act (1883).*

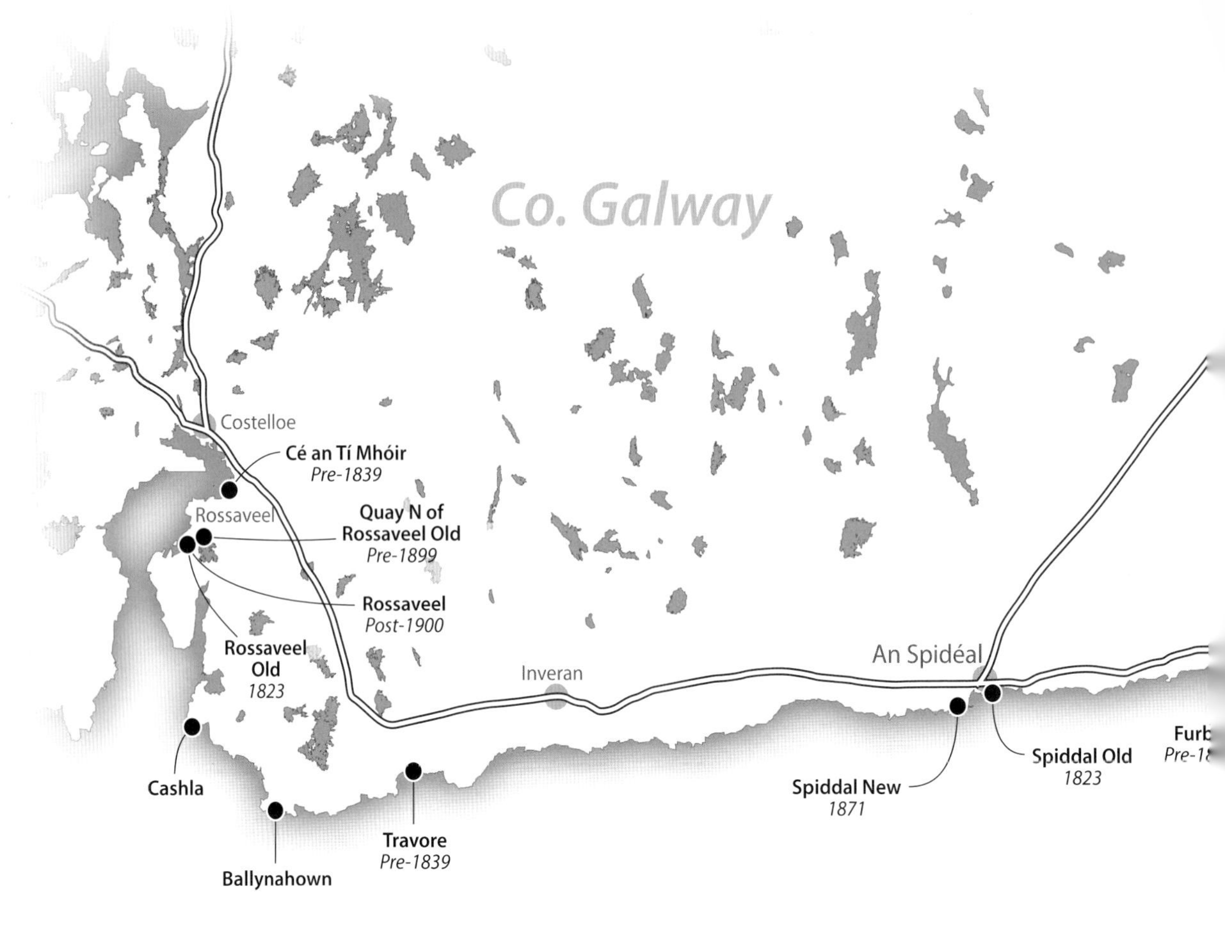
Co. Galway
Costelloe
Cé an Tí Mhóir
Pre-1839
Rossaveel
Quay N of
Rossaveel Old
Pre-1899
Rossaveel
Post-1900
Rossaveel
Old
1823
Inveran
An Spidéal
Spiddal Old
1823
Spiddal New
1871
Cashla
Ballynahown
Travore
Pre-1839
GALWAY BAY

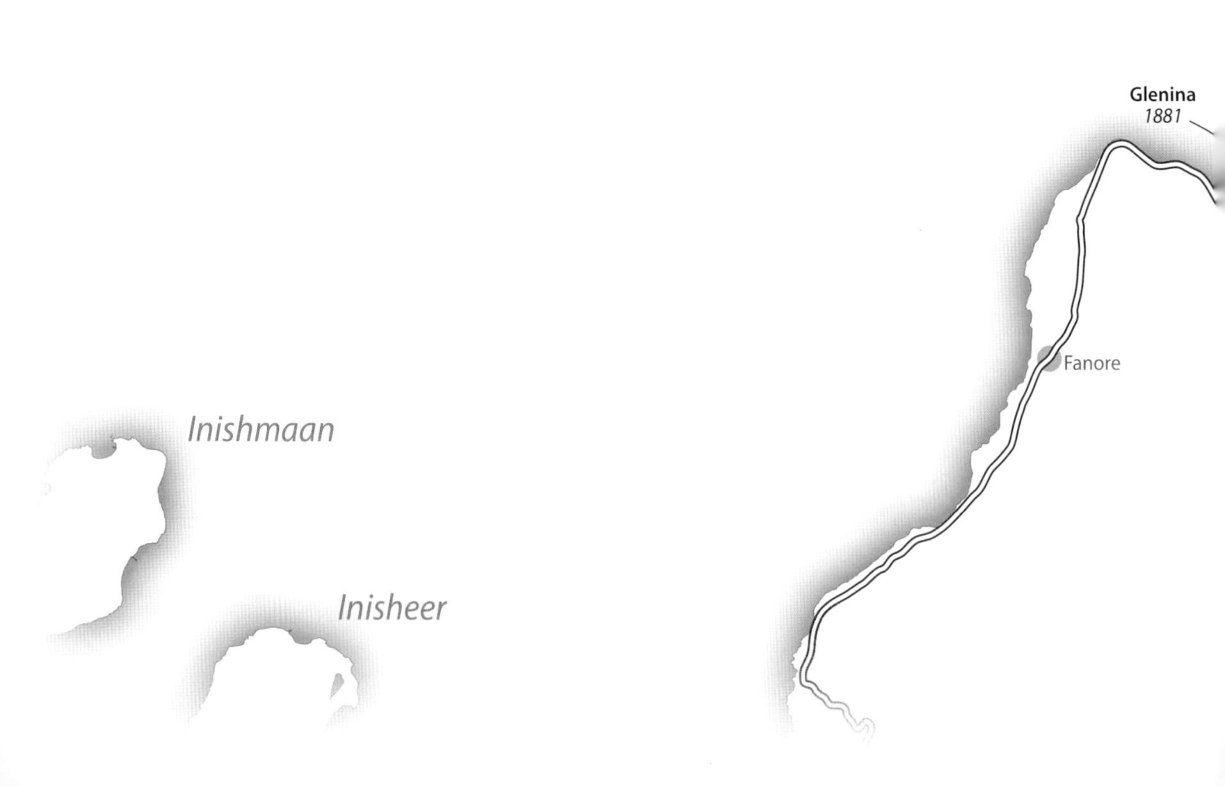
Glenina
1881
Fanore
Inishmaan
Inisheer

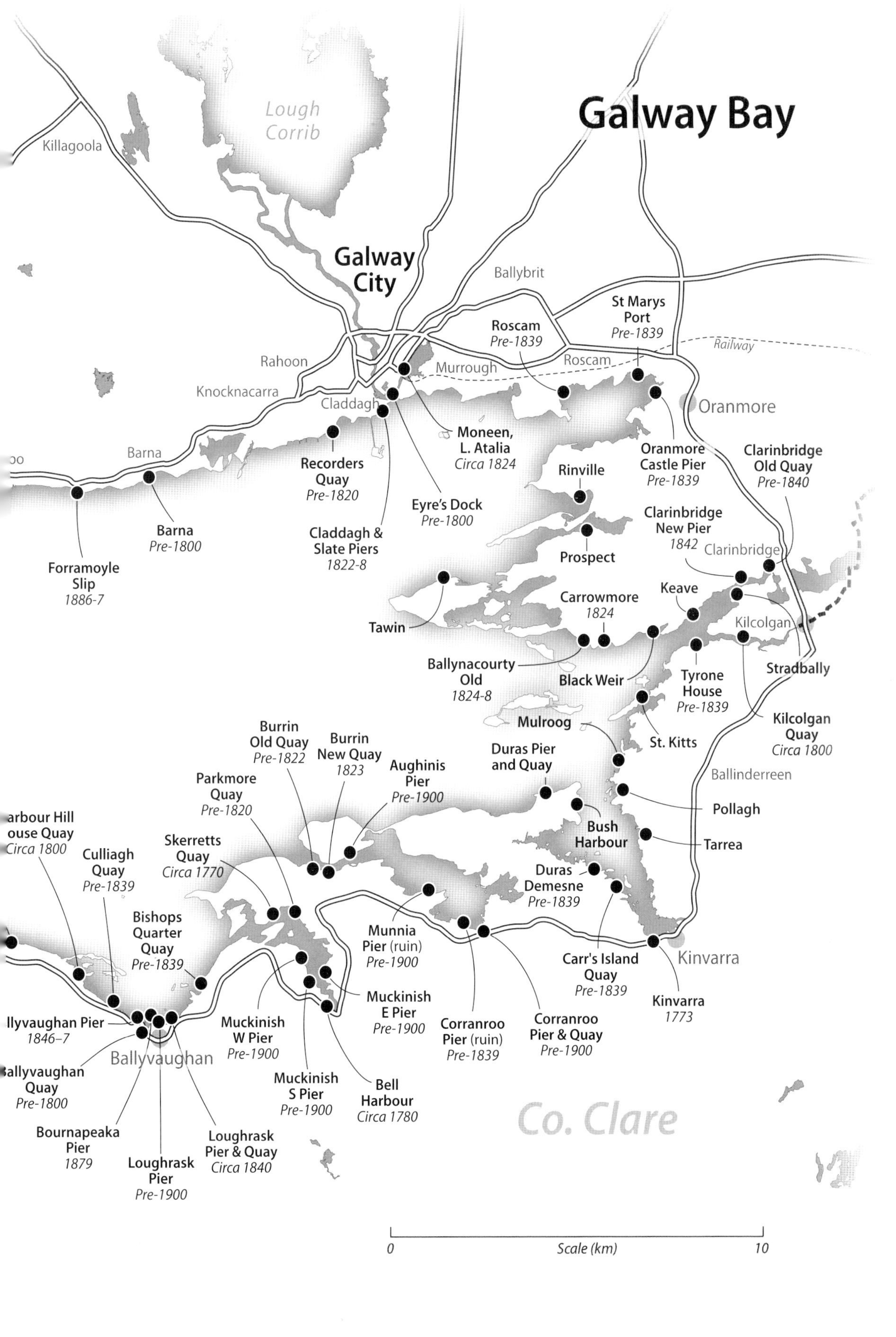

Galway Bay
Lough Corrib
Killagoola
Galway City
Ballybrit
Rahoon
Knocknacarra
Claddagh
Murrough
Roscam
Railway
Oranmore
Barna
St Marys Port Pre-1839
Roscam Pre-1839
Moneen, L. Atalia Circa 1824
Recorders Quay Pre-1820
Eyre's Dock Pre-1800
Claddagh & Slate Piers 1822-8
Barna Pre-1800
Forramoyle Slip 1886-7
Oranmore Castle Pier Pre-1839
Clarinbridge Old Quay Pre-1840
Rinville
Prospect
Clarinbridge New Pier 1842
Clarinbridge
Keave
Carrowmore 1824
Kilcolgan
Stradbally
Tawin
Ballynacourty Old 1824-8
Black Weir
Tyrone House Pre-1839
Kilcolgan Quay Circa 1800
Mulroog
St. Kitts
Burrin Old Quay Pre-1822
Burrin New Quay 1823
Aughinis Pier Pre-1900
Duras Pier and Quay
Ballinderreen
Parkmore Quay Pre-1820
Pollagh
Bush Harbour
Tarrea
Culliagh Quay Pre-1839
Skerretts Quay Circa 1770
Duras Demesne Pre-1839
Bishops Quarter Quay Pre-1839
Munnia Pier (ruin) Pre-1900
Carr's Island Quay Pre-1839
Kinvarra
Kinvarra 1773
Muckinish E Pier Pre-1900
Muckinish W Pier Pre-1900
Corranroo Pier (ruin) Pre-1839
Corranroo Pier & Quay Pre-1900
Ballyvaughan
Muckinish S Pier Pre-1900
Bell Harbour Circa 1780
Co. Clare
Bournapeaka Pier 1879
Loughrask Pier Pre-1900
Loughrask Pier & Quay Circa 1840
0
Scale (km)
10

Inishbofin
East End Inishbofin
Main Harbour Inishbofin
Main Harbour Inishbofin
Rusheenduff
Renvyle
Gurteen
Mullaghglass
Glassillaun
Little Killary 2
Letter
Derryinver
Dawros 2
Dawros 1
Letterfrack
Dooneen
Rossadillisk
Aughrusbeg
Cleggan
Aughrusmor
Claddaghduff
Kingstown Slip
Cul an Chlaí
Inishturk
Inishturbot
Fahy
Clifden Beach Rd
Clifden
Clifden Beach
Drimeen
Faul
ATLANTIC OCEAN
Curhownagh
Derryginlagh
Derryadd West
Dunloughan
Pollrevagh
Cuan na Luinge
Slackport
King's and Joyce's
Aillencally
Aillebrack
Bunowen
Ballyconnely
Doohulla
Emlaghmore
Paul's Quay
Callagh
Murvey
Dolan
Ervallagh
Inishlacken
Roundstone
Monastery 1
Monastery 2
Roundstone
Rosro
Canow
Bealcarra
Cuan Caorthain
Dawros
Glean tSrut
Moyrus 1
Moyrus 2
Corradan
Cuilee
Caladh an Bhaid
Oilean an Bhromaigh
Carna
Half Mace
Mace
Ard West
Ard East
Inis Macdara
Masson Island
Portach Mweenish
Meall Rua Mweenish
Crumpaun Carna
Aill an Eachra
Bungabhla
Kilmurvey (Main Pier)
Kilmurvey (Small Pier)
Kilronan (Jetty)
Kilronan (Slip)
Kilronan (Main Pier)
Kileany
Gort na gCapall
Inishmore
Port na gCurrach
Caladh Mór
Cora Point
Inishmaan
Inisheer (Main Pier)
Tra Teach
Port na Cille
Poll na gCaorach
Corrait
Lighthouse (Jetty)
Inisheer

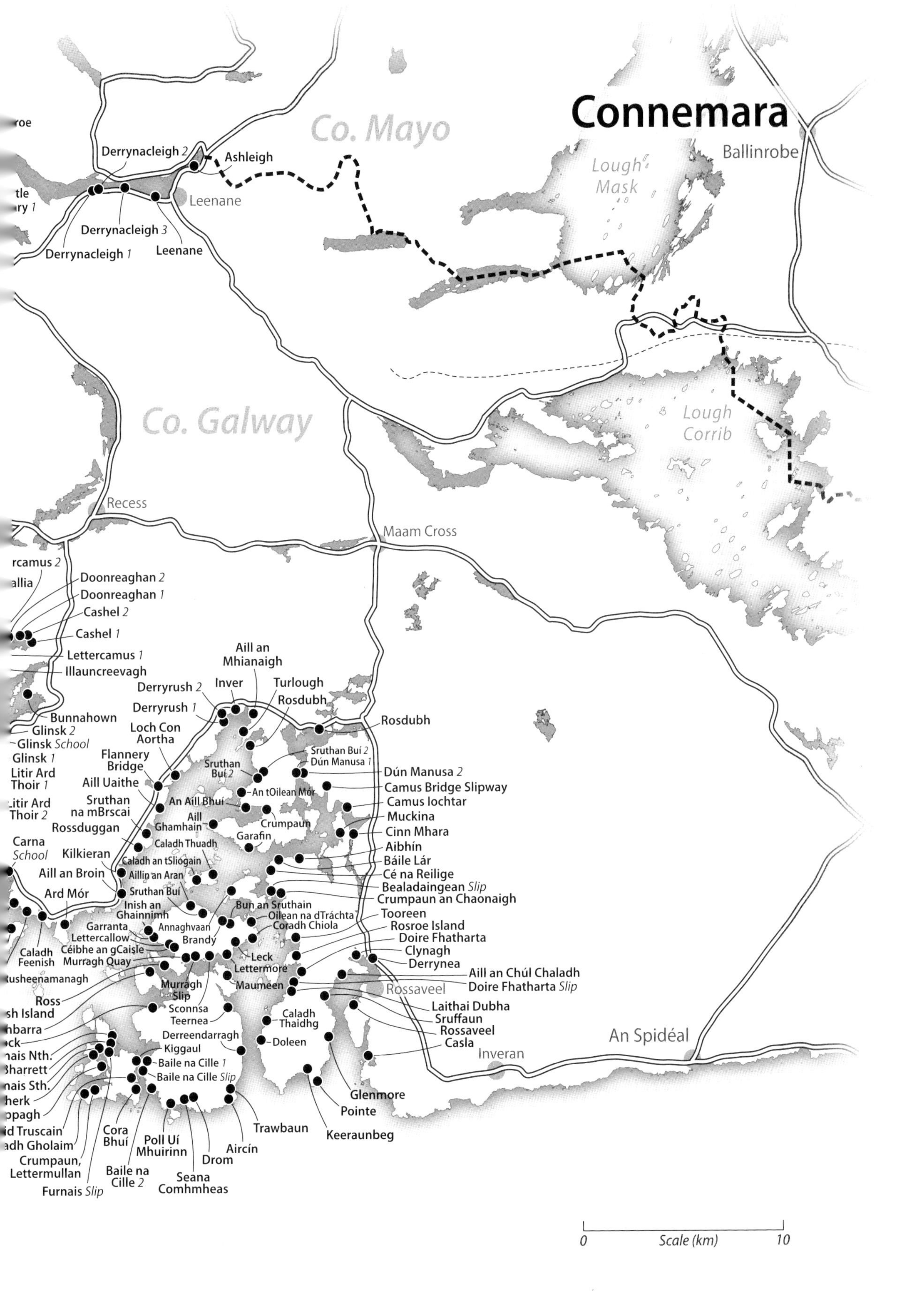

Connemara
Co. Mayo
Co. Galway
Lough Mask
Lough Corrib
Ballinrobe
Leenane
Recess
Maam Cross
Rossaveel
Inveran
An Spidéal
Derrynacleigh 2
Ashleigh
Derrynacleigh 3
Derrynacleigh 1
Leenane
Doonreaghan 2
Doonreaghan 1
Cashel 2
Cashel 1
Lettercamus 1
Illauncreevagh
Bunnahown
Glinsk 2
Glinsk School
Glinsk 1
Litir Ard Thoir 1
Litir Ard Thoir 2
Carna School
Aill an Mhianaigh
Inver
Turlough
Derryrush 2
Derryrush 1
Rosdubh
Loch Con Aortha
Flannery Bridge
Aill Uaithe
Sruthan na mBrscai
Rossduggan
Kilkieran
Aill an Broin
Ard Mór
Caladh Feenish
Sruthan Buí 2
An tOilean Mór
An Aill Bhuí
Aill Ghamhain
Crumpaun
Garafin
Caladh Thuadh
Caladh an tSliogain
Aillin an Aran
Sruthan Buí
Inish an Ghainnimh
Garranta
Lettercallow
Céibhe an gCaisle
Murragh Quay
Annaghvaan
Brandy
Bun an Sruthain
Oilean na dTráchta
Coradh Chiola
Leck
Lettermore
Maumeen
Murragh Slip
Ross
Sconnsa
Teernea
Derreendarragh
Kiggaul
Caladh Thaidhg
Doleen
Baile na Cille 1
Baile na Cille Slip
Cora Bhuí
Poll Uí Mhuirinn
Baile na Cille 2
Seana Comhmheas
Drom
Aircín
Trawbaun
Keeraunbeg
Pointe
Glenmore
Crumpaun, Lettermullan
Furnais Slip
Sruthan Buí 2
Dún Manusa 1
Dún Manusa 2
Camus Bridge Slipway
Camus Iochtar
Muckina
Cinn Mhara
Aibhín
Báile Lár
Cé na Reilige
Bealadaingean Slip
Crumpaun an Chaonaigh
Tooreen
Rosroe Island
Doire Fhatharta
Clynagh
Derrynea
Aill an Chúl Chaladh
Doire Fhatharta Slip
Laithai Dubha
Sruffaun
Rossaveel
Casla
0
Scale (km)
10

CHAPTER 1

INTRODUCTION: SETTING THE SCENE AND PREVIEWING THE WAY AHEAD

An Abundance of Natural Landing Places

Early in the research for this book I asked a resident of Carraroe, Connemara, whether a pier, quay or landing place existed at the end of a particularly long *bóithrín*. He replied that in Connemara there is a landing place at the end of every *bóithrín* leading to the shore. In the face of this declared abundance, any attempt to record all of the landing places in Co. Galway and north Clare would be a task so demanding as to be virtually impossible – demanding not only because of the intimate perambulation of the coast that it would involve, but because every piece of shingly or sandy shore is a potential landing place, even if not used as such today. Nor would it be very instructive to attempt to record them all: their use can be casual, and unless there is clear evidence of significant manmade structures – piers, jetties, quays, wharves or breakwaters – they really give us little specific insight into the nature and history of maritime activity in this or any other region. That is not to say that they are of no interest at all.

Indeed, the Carraroe man might well have added that every landing place has its own *scéal* – its own unique story – to tell about its origin and its importance in the lives of the persons who used it. The problem is simply that, in the complete absence of manmade structures, it is almost impossible to retrace those lives and to recount the role of these places in them. Many, but not all, piers and quays started out as natural landing places and were transformed through human intervention into the structures we see today. This evolution was driven by the ordinary needs and deeds of many generations of coastal people whose occupations included seaweed harvesting, fishing, kelp burning, turf distribution, oyster dredging, inter-island communication and, from the early nineteenth century in particular, the drain and pain of emigration. These, we will see, were the drivers of the development of the fishery piers of the West. This book aims to open the history of the public piers and quays of Co. Galway and north Clare – essentially the coast from Black Head in Co. Clare to Killary Harbour on the Galway/Mayo border – in the hope that, in due course, others may be able to tease out the hidden human stories, all the *scéalta*, that underlie their historical use.

How Can We Find Out about Piers and Quays?

If we do not know precisely the number of natural landing places that were regularly used in the past, we can at least estimate reasonably well the number of manmade piers and quays that were constructed from around the beginning of the nineteenth century onwards. Recording these, and telling part of their story, are feasible tasks that do not appear to have been comprehensively addressed before, for this or any other part of the Irish coast.

We can glean information from public documentary sources not relating specifically to piers and quays, and from other public records that were drawn up with such structures particularly in mind. The absence of manmade piers and quays is remarkable in detailed sea charts of the region made by French hydrographers in the late seventeenth century, which otherwise record many good anchorages and even places for careening ships.[1] William Larkin's later (1819) *Map of County Galway*,[2] gives no information on landing sites, although it does record some anchorages and displays the word 'quay' at a very small number of locations. None of these cartographic sources indicate the existence of any manmade structures at the sites in question.

As is so often the case, the best sources of information are the Ordnance Survey (OS) maps, especially the first and second editions of the six-inch series (1:10,560) surveyed in 1838–9[3] and 1898–9[4] respectively. In modern times, and for most ordinary purposes, we can use the current Ordnance Survey maps, for instance the Discovery Series (1:50,000), which are the maps that visitors and others who are interested are most likely to avail of. For many people, Google Earth images and other online sources will be invaluable, although the coverage of much of the Galway coast is not yet available at sufficient magnification, or with sufficient clarity, to enable all the structures to be readily identified.

No foray into Connemara, the Aran Islands or north Clare can be complete without the maps of Tim Robinson: they are not simply maps, but comprehensive visual essays on the topography, history, heritage and culture of these places.[5] They are amplified by his books, which are also invaluable sources of reference.[6] In all these, Robinson does not focus specifically on piers and quays, so we should not expect him to have recorded every individual structure, although he did cover a very large proportion of them.

In the millennial year 2000, Galway County Council commissioned a survey entitled *Assessment of Piers, Harbours and Landing Places in County Galway*,[7] carried out by Ryan Hanley Consulting Engineers, Galway, the results of which are available on the Galway County Council website.[8] That survey was completed prior to a blitz of pier 'restoration' by the Council that has obliterated many of the old structures by encasing them indiscriminately in concrete. It is therefore an irreplaceable, thorough and professional survey and review of the piers and harbour structures as they existed at the start of this century. Profusely illustrated with photographs of almost 190 piers and quays, the surveyors were not required to make any study of, or comment on, the date of origin or the history of any of the structures recorded. This book now provides as much of that information as could be gleaned from public records and site examinations made from 2010 onwards.

Another very useful source is the equivalent survey of the Co. Clare coast of Galway Bay, the *Clare Coastal and Architectural Heritage Survey*, compiled by Sarah Halpin and Grainne O'Connor in 2006–07,[9] available on the Clare County Library website.[10]

The Annual Reports of the Board of Public Works, Ireland (the OPW) are an absolute mine of information on fishery piers and quays. The Board, after all, had responsibility for designing, costing and erecting most of them, and its Annual Reports are the published record of that activity from 1832 to 1922. Behind these published documents are the printed and manuscript reports, memorials, correspondence, designs, specifications, maps and so on, connected with them. Fortunately, many of these important primary sources, although very far from complete, are conserved in the National Archives of Ireland, catalogued under the Office of Public Works. A brief, useful and attractive introduction to the piers and harbours section of that archive is given in the *Guide to the Archives of the Office of Public Works* by Rena Lohan, published in 1994.[11] It is essential to read that guide before approaching the original material.

The Research Needs Care and Focus

These then, along with a multitude of British parliamentary and other official papers of the period, are the main archival sources used here to document, as far as possible, the history of the fishery piers and quays of the region. One must caution, however, that it is impossible to know with complete accuracy and lack of ambiguity the total number of manmade marine structures that may have existed at one time or another since 1800. How, for instance, should we record the harbour at *Clais na nUan* (Teeranea Pier and Quay) (No. 120), where the breakwater that transforms the site of an isolated pier into a small harbour was not built until many years after the pier itself? Or what about Pollrevagh (No. 027), where there are two quite distinct piers of differing age which are spatially well separated, with a slip and a number of distinct laying-up places for currachs within the site? At Ballyvaughan (CS 031) the existing pier is located close to the site of an older quay no longer extant, but completely separate from it. Should we record one or two distinct structures at places like these, where there is only one structure today? At Oranmore (No. 147), the existing cut-stone pier was built in 1881 beside the castle, replacing a mediaeval quay on the same site. Here, too, where only one pier exists today, should we record one or two structures?

Clearly, just as the true length of the coast cannot be accurately measured without being fractal or fractious,[12] the exact number of piers, quays and landing places is also fractal: the closer we look, the more individual structures, often of differing dates, we can identify even at a single site. Since the research for this book was completed, for instance, evidence has come to light of quays and piers in Killary Harbour and Ballinakill Bay that have not been recorded publicly to date. For these reasons the number of structures recorded will differ from one commentator to another, and the figures should be treated with appropriate caution. Not everyone will agree with the approach taken here: the simple landing places that entirely lack manmade structures are generally ignored and the complex places *Clais na nUan*, Leenaun (No. 002) and Roundstone (No. 040), for example, are treated as 'dual' or two-part structures, each part being recorded and treated as a separate entity.

Finally, the names of places and sites in official public documents do not always correspond exactly with the names in present-day use. This discrepancy is particularly obvious in spelling, and certain place names are spelt in different ways at different times, due mainly to transcriptional and translational errors. Sometimes a single place may be recorded under different names at different times. On the other hand, the recurrence or continuance of a name in the records does not always signal the exact same site or structure. Cashla pier, for instance, refers to one particular structure at one time, but to a completely different place and structure later on. The same happens with *Spidéal* pier where the public records refer to one particular site and structure before 1869 (now called *an tSeancéibh*) and to a completely different site and structure (called Spiddal pier) after that date.

Place names, too, have changed over time, and names in use in the nineteenth century may be lost or replaced today. These discrepancies may be minor, but their effect can be critical in computer-based searches of the records. In this work, the alternative names of sites are given where these are known, even where they represent common misspellings. Special care is needed with place names in Connemara, a Gaeltacht (Irish-speaking region). Today the only official names of places in all such regions are the Irish-language versions as laid down in the Statutory Instrument No. 599 of 2011.[13] However, in the published records, the piers and quays are recorded only under their English names and the old records cannot be accessed properly without using these. Here, the English names alone are used where places are commonly known and recognised – like Carraroe, or Kilronan, or Kilkieran. Less well-known places like *Sruthán Bhuí* or *Trácht Each* are named in Irish (in italic script), accompanied by the English version of the name. That acknowledges the only officially correct names for them today, while allowing them to be traced in the published records.

One final caveat should be mentioned here: as with all research, what one sees depends on the lens through which one makes observations. Looking at piers and quays through the lens of language, especially the Irish language in this case, one meets with established place names like *Caladh Mhuiris* (Morris's quay), *Port Bád* (Boat quay), *Caladh na Loinge* (the ship's quay), *Port Mór* (Large port) and so on, all indicative or suggestive of natural landing places and vernacular ports and harbours. Unless documentary records exist, or physical on-site evidence of manmade structures is evident in such places (like, for instance, at *Caladh Thaidhg* or *Caladh Feenish*), they will not, generally speaking, be referred to here, where the lens is one of documentary and on-site structural evidence.

So How Many Piers and Quays Are There?

Bearing these caveats in mind, we can be confident from the Ryan Hanley survey and the Clare Coastal Heritage survey that approximately 300 identifiable, manmade piers and quays were erected between Black Head and Killary Harbour since around 1800 (274 in Co. Galway and twenty-six on the north Clare coast). Most of them, except those on a few of the islands, have been visited in the course of a personal perambulation of the localities (over 250 sites actually visited), supplemented by a detailed study of the various maps and documents listed in the endnotes and the bibliography. Research for this work has increased the total number of identified manmade structures in the region to about 320, representing an average of one for every two to three kilometres of coast.

The various maps and other sources differ in the actual number each one records: the OS Discovery Series maps show structures at 144 of the 320 sites recorded here (about 45 per cent); Robinson mentions in text, or records in maps, 170 at the 250 locations common to this study that he surveyed (68 per cent); and the Ryan Hanley survey records 193 at the 234 (83 per cent) locations common to it and this study. A small number have been overlooked by all previous sources, and are identified here for the first time. A very approximate date of origin of many can be estimated from the appropriate sheets of the first and second editions of the six-inch OS maps. For example, fifty-nine are recorded in the first 1839 edition, so they were already in existence at that date. From other documentary records,[14] the origin of twenty of these pre-1839 piers and harbours can be dated accurately to the 1820s; the remaining thirty-nine may date from the start of the nineteenth century or even before then, although only a handful can be attributed with any confidence to a specific date before 1800. The second 1898–99 edition of the same sheets of the six-inch OS maps records 149 piers and quays. The difference between the two editions is, therefore, ninety and this figure might be taken, at a first approximation, to represent the net number of new piers and quays that were built in the region between 1839 and 1899, an average of more than one each year. However, fifteen of the piers in the 1839 edition are not recorded in the 1898 edition, so the real increase in recorded new piers between the two surveys is, more accurately, 105.

The Congested Districts Board and the OPW erected about thirty new piers, quays and slips in Co. Galway between 1900 and 1922, so that the minimum number in existence in the county at the foundation of Saorstát Éireann was not less than about 194 (149+15+30). The north Clare piers were all erected before 1922, so that the minimum number erected in the whole region before the foundation of Saorstát Éireann was over 200. Their distribution in the four sub-regions of Galway and north Clare, viz. Galway Bay, south Connemara, west Connemara and the offshore islands, is given in Table 1.1.

For comparison, the Inspectors of Irish Fisheries recorded 160 fishery piers and quays in the whole island of Ireland in 1890 (sixty-six of them in Galway and north Clare) and the Department of Fisheries of Saorstát Éireann estimated there were approximately 347 fishery piers and quays in the entire twenty-six counties in 1823.[15] Of the pre-1839 piers, most of those in Galway and north Clare are still extant, although some were greatly modified during the later nineteenth and the twentieth centuries, as one might expect, and others are mere ruins.

TABLE 1.1

Number of fishery piers and quays recorded between Black Head and the Galway shore of Killary Harbour at differing periods.

Sub-region	**Pre-1820**[1]	**End 19th century**[2]	**Today**[3]
Galway Bay	28	46	64
South Connemara	6	63	169
West Connemara	4	31	57
Islands	1	9	25
Totals	39	149	315

[1] From the 1st edition of the six-inch OS maps, with the piers and quays known to have been built in the 1820s excluded. [2] From the 2nd edition of the six-inch OS maps. [3] From the Ryan Hanley survey (ref. 8) the Clare Coastal survey (ref. 10) and newly recorded in this work.

These are just the very broad features of thc data. Mention is made in the nineteenth-century official records of about 130 piers and quays in the region that were built or modified with public money under Government auspices, mostly in the course of four distinct periods: the 1820s; the Famine and immediate post-Famine period, 1846 to 1855; the great decade of fishery pier building from 1879 to 1889; and, finally, the period 1891 to 1922 during the time of the Congested Districts Board. The available records permit many of them to be dated to the precise year of their construction, and other sources, such as family papers and estate records, may add to the total in future research.

It must be emphasised that the information here is based almost entirely on official public documents and therefore refers overwhelmingly to publicly funded piers and quays. These comprise the most substantial structures that were built. Private piers were generally of less impressive mass and style, and such piers were erected with declining frequency as the nineteenth century progressed. This book lays no claim to be definitive as to the exact total number of structures erected, although it is the closest it is reasonably practical to get today to the true number. For a very large number of them, there are structural, but no documentary, records. Most if not all of these were built by private charitable and humanitarian concerns. There are, no doubt, many small piers and quays still to be documented, and many more where time, tide, weather and thoughtless redevelopment have wreaked their familiar destruction.

When and Why Were the Piers Built?

The vast majority of the piers and quays built in Ireland and the UK in the nineteenth century were built ostensibly for fishing purposes, constructed against the wider background of the progressive exploitation of fisheries as the century progressed. Technological evolution resulted in many new ports and harbours being built, mostly in England and Scotland, but in Ireland also. Advances in boat and net design and propulsion – largely the replacement of sail by steam, and then of steam by motorised vessels – and the enormous growth of the fishing trade as it industrialised, demanded changes in piers, quays and wharves that increased their size and sophistication. This wider UK context is important when the fishery piers of the whole of Ireland, particularly those of the east, northeast and south coasts, are being considered. But the British situation is a separate, major study in itself, that has already been treated elsewhere.[16] It will be considered here only to the extent that it sheds light on developments in Ireland, and in Galway and north Clare in particular.

The known construction dates of the various Galway and north Clare structures show that 160 of the approximately 200 built with public money between 1820 and 1920 were first erected during the main periods mentioned above. Many of those were times or episodes of famine and severe distress. So, although built overtly for the fisheries, almost all the piers built in Co. Galway up to 1891 were built as relief works, designed to give paid work to the destitute, rather than as properly planned infrastructural elements of a growing fishing industry.

Many were aided by voluntary contributions from domestic and international charities; most were built at sites that were not ideal for fishery purposes, but nevertheless they came to serve the coastal communities legitimately and well by fostering trade and communication between isolated and remote districts. Almost 60 per cent were built or renovated in the last quarter of the century and about 30 per cent date from before the Great Famine: generally speaking, the stock of publicly funded piers is not as old as some might think. On the other hand, many persons in Galway incorrectly attribute the majority of the piers, especially the larger ones, to the Congested Districts Board when, in fact, they date from many decades before that Board was operating.

The Challenge of Different Ports for Different Boats

For most of the century there was a tension inherent in the precise positioning and design of piers, ignoring entirely the wishes of landowners and other personal and legislative requirements. The challenge was to balance the requirement for a place where boats could lie safely in shelter, with the need for freedom to come and go at all phases of the tide, and to do this within budget at locations that suited the size of boats in common use, while still being convenient to the homes of the boatmen. In the era of currachs and small rowboats, boats could be drawn up above high water mark, either by manual effort as in the Aran Islands and elsewhere, or by a crab-and-chain mechanism at manmade slips, as at Inishshark, where landing conditions were extremely difficult.[17]

Where piers and quays were lacking, boats could be beached and tied down above high water, generally safe in inclement weather. However, beaching at or above high tide often meant that boats could not be conveniently launched again until the tide was sufficiently full to make easy the reverse operation. The tide recedes a long way in many of the shallow bays and inlets of the Galway and north Clare coasts, so that carrying even a relatively light currach from high to low tide mark can be demanding. Waiting to launch later, when the tide had filled sufficiently, could often mean a delayed departure. When the destination was a pier or landing place some distance from home, the boats risked arriving there when the tide was already well into the ebb, necessitating a further wait in the destination offing. Alternatively, if they arrived at high tide or early in the ebb, it could necessitate a rushed unloading to avoid becoming stranded. This problem was more serious for larger, heavier boats like hookers: they could not be conveniently carried or drawn above high water and, in the absence of a suitable quay, had to lie beached in the intertidal zone until the tide filled sufficiently to refloat them. At Conroy's shop beside Garafin harbour (No. 088) in Rosmuc, for example, customers arriving by boat were given precedence in service before those arriving on foot, because of their need to depart the harbour before the tide ebbed and stranded them for six or more hours; this precedence practice survived into living memory.[18]

One solution to the tidal difficulty was to extend a pier or quay down below low water of Spring tide, so as to be accessible for boats at the lowest ebb. Another was to excavate inwards, as was done at *Trácht Each* (No. 185) on *Inis Oirr*, so that boats could lie afloat in relative shelter yet be ready to come and go irrespective of the stage of the tide (both these types of quay were termed 'floating docks'). The former solution required construction work to be carried out underwater, necessitating the use of coffer dams, or of the diving bell, both of which demanded specialist workers not that readily available in remote places. Underwater construction requiring skilled workers and specialised equipment was costly beyond the financial capacity of most fishery grants at the time. Considerations like these should be borne in mind when accounting for the small size and structural simplicity of the piers that were built, up to and well beyond the Great Famine. One further important consideration was the topographical, commercial and social convenience of the site: where piers were located in remote places, fish and other cargoes needed to be carried to or from them to markets, homes or centres of population, a difficult task in a region where there were no proper roads. For most of the nineteenth century, transport by sea rather than by road was the principal, oftentimes the only, means of communication in most of Co. Galway and north Clare.

Piers and Quays in Galway in Early Times – The Lordship of the O'Flahertys

We know almost nothing about manmade piers and quays in the region before 1820, but it seems there were few, if any, of major significance. This apparent absence, however, is not evidence for the absence of trade or fishing outside the city. The lands of *Iar Connacht* in the west of Co. Galway were in O'Flaherty hands for much of the time between the fifteenth and eighteenth centuries, and the lands of south Mayo in the hands of their in-laws and sometime rivals, the O'Malley's, of whom perhaps the most notorious was Grace O'Malley, or *Gráinne Uí Mháille* (Granuaile) in Irish. O'Flaherty castles were dotted throughout the west and northwest of the county, one of their principal centres being the district later called the Barony of Ballinahinch, synonymous with the very heart of Connemara. During the O'Flaherty lordship, Ballinahinch was an important place, both ecclesiastically and civilly. It had good access to the open sea through Roundstone Bay and Ballyconneely Bay. Members of the sept engaged in robust clandestine maritime activity, although many commentators take a jaundiced view of the nature and aims of that engagement.

Ballinahinch is located close to the head of Roundstone Bay, deep inside Bertraghboy Bay, with anchorages well surrounded by land. It is a safe and sheltered, almost hidden, place where any O'Flaherty ship could lie beached or at anchor in perfect safety. There was no need for shelter piers or stone wharves in such a place: it was a foolhardy person who would follow a ship into this lair of the ferocious O'Flahertys and hope to emerge unscathed, or even to emerge at all.

The lower, seaward, section of Roundstone Bay, near the present village, was a convenient anchorage in which ships would sometimes shelter (perhaps unaware of the attendant potential risk) while awaiting suitable winds before rounding Slyne Head or making passage southwards. Another O'Flaherty castle was located at Bunowen, near to, and overlooking, Slyne Head. It belonged to *Dónal an Chogaidh* O'Flaherty and his wife *Gráinne Uí Mháille* in the mid-sixteenth century.[19] There was little shipping at the entrance to Galway Bay that that family could not observe at leisure from Bunowen and act upon as it wished. It is said of the sixteenth century that 'the mariner or merchant who left Galway in pursuit of trade seriously risked his life in more ways than one'.[20] Among the greatest risks was piracy, not least from the entirely home-grown O'Flahertys, male and female, who were expert exponents of that nefarious art, and no respecters of the Galway city merchants. Because of the nature of their activities, artificial stone piers and wharves would have been of little value to them or to any other pirates; they would only have served to reveal to their pursuers the exact whereabouts of their lairs.

Piracy, Smuggling and Skullduggery

By the late seventeenth century, the power of the O'Flahertys was waning considerably and that of the Galway city merchants had increased. The latter had purchased, or obtained by forfeiture, many of the lands previously in the ownership of the O'Flahertys. City merchant family names like Blake, Martin, Bodkin, and Skerritt became common in the Galway and Clare countryside, where they built manors, castles and fine houses. The O'Flaherty estate at Ballinahinch came into the hands of the Martins in the mid-eighteenth century and piracy declined throughout Galway Bay in consequence. However, it was only to be replaced by another clandestine maritime activity: smuggling.

The 'native Irish' smugglers, men with names like McDonagh and O'Malley, were joined by some of the most distinguished (one hesitates to say 'reputable') Galway merchants, some of whom profited from long careers in smuggling Valentine Browne, for example, owned one of the most glaring, if not daring, surreptitious entry points through Galway city walls, according to Hardiman:[21] 'By the marsh, a hole broke through at Val Browne's house, shut up and opened as often as he has occasion to bring ankers of brandy into town' (an anker was a liquid measure, used mainly at Amsterdam, about eight and a half imperial gallons).[22] The greater portion of the city walls was levelled at the start of the nineteenth century,[23] but that did not stop the smuggling activity, further anecdotes of which are given by Robinson.[24]

That such activities did actually transpire is shown conclusively in the public records of London and Nantes, analysed in detail by Louis Cullen.[25] Cullen concentrates on the decade of 1730 to 1740, when wool smuggling (known by the delightful term 'owling') was rampant, and accompanied by extortion, kidnapping and paid informers. The smugglers' ships would lie to in Roundstone Bay and the wool was delivered to them by small boats moving along the coast from Oranmore, Rinville and Tyrone in inner Galway Bay. In this way, it was usually possible to slip quietly by the naval patrol vessel, aptly named HMS *Spy*, which the Government had stationed in the Galway roadstead specifically to prevent smuggling.

Roundstone Bay was only one smuggling centre in the county during the early eighteenth century involving both inbound and outbound goods; Casheen Bay, Costello Bay and Killary harbour are also mentioned in the records.[26] Cullen concluded: 'the smuggling merchants of Galway were a small and fairly compact group and the interests of the continental trade bound them and some of the land-owning gentry closely together. Many, if not most, of the Galway merchants of the 1730s had some share in the "clandestine" trade'.[27] In light of this, it seems probable that many parts of the coast hosted some illicit landings at one time or another. Unlike piracy, which was outright theft, smuggling involved evasion of revenue and the circumvention of commercial embargos; therefore, just as with pirates, artificial piers and quays, had they been erected, would only have given away the location of the smugglers' trysts and illicit transhipments, so they really were 'surplus to requirements'.

Inshore Fishing

The above digression serves to illustrate that the region from Black Head to Killary Harbour did not have the topography, the infrastructure, the social, commercial, demographic or other features that would justify the creation of large centres of legitimate trade separate from the legal quays of the administrative and revenue centre that Galway city represented. Despite the absences of these characteristics, which remained largely unchanged during the nineteenth century, it may be wondered how it came about that so many piers and quays were built in that century, and how it is that around 320 came to exist in a region with a coastline of about 700 kilometres?

One necessary, but partial, answer is fishing. There is not a coastal population anywhere that does not avail, to some extent, of the resource freely presented by inshore fishing; the communities of the west of Ireland are no different in this regard. Fishing close inshore in currachs or small rowboats, working out of local natural landing places, was common here from the earliest times. Fish, including shellfish, were taken for family needs and were rarely traded. Until the end of the nineteenth century, the opinion was widely held that offshore fish stocks were inexhaustible and that the Atlantic Ocean held an untapped bonanza for fishermen, if only they would venture out far enough. The west of Ireland in particular, it was held, could and should benefit enormously from this abundance. As late as 1883, the Select Committee on Harbour Accommodation declared how impressed it was that 'on the Atlantic seaboard of Ireland, fishing grounds teeming with fish abound'.[28] Historical accounts of Spanish, Portuguese and other foreign fleets paying levies to the English monarch for permission to fish off the Irish coast appear to have fuelled this impression.[29]

Few commentators seem to appreciate that the levies those fleets were charged may have had more to do with contested political concepts of 'sovereignty of the seas' and 'maritime dominion' claimed by the Stuart and later monarchs, than with any firm expectation or anticipation of abundant catches.[30] That some fishing did take place off the Atlantic coasts of Ireland, especially by foreign boats, is beyond serious doubt, but its extent and location are based only on skimpy documentary records. Before the nineteenth century, fishing boats did not have the gear, the propulsive power or the navigational and other aids needed to fish profitably in deep water far from their home ports.

Closer to shore, it is certain that from the late eighteenth century on, some boats from Skerries, and Scottish fishermen in their herring busses (these were shallow-keeled, two or three-masted sailing vessels used to catch herrings), were coming as far as the Donegal coast in the appropriate season and making good profits from their long-distance ventures. They were undoubtedly far from home, but their catches were normally made by local Irish fishermen working within the shallow local bays and selling their catches to the visiting boats, which there and then salted and barrelled the fish.

Facing out into the unknown deep Atlantic Ocean, and sailing into the prevailing onshore southwesterly winds that could drive small boats on to an unforgivingly hard coastline should conditions turn stormy, was never an attractive proposition to those fishing the west coast of Ireland. They rarely attempted, or even considered, offshore deep-sea fishing, beyond the Aran Islands, for example. That would have required larger, decked vessels and these in turn would have required better landing facilities, accessible at all states of the tide, if they were ever to be really useful. Therefore the local fishermen generally fished from small boats close to the shore and in the bays and inlets to meet their own requirements.

The Piers of the Nineteenth Century

The earliest period of public pier building on a large scale for which there are reliable records was instigated by the first Commission for the Irish Fisheries, a body established in 1819, and that lasted for only eleven years. At the beginning, it had no specific funds earmarked for piers. When faced with the outbreak of famine in 1822, the Commissioners decided to apply their limited general financial resources to building small piers and quays as a means of providing paid work for the distressed poor, as well as providing facilities that would encourage more fishing. They undertook to offer free grants of one half of the estimated cost of any proposed pier or quay, provided the local applicants (who were normally the local landowners) made up the other half, and subject to a maximum expenditure of £500 per site.

The scheme did not work particularly well at first, because few landowners were willing, or able, to contribute their one half of the estimated costs; many had no interest at all in having quays or piers erected in their localities. The 'local' contributions were therefore made up by donations from charitable societies in Dublin and London, a humane reponse to the horrors of the 1822 famine. In 1824, the Government introduced legislation formalising the Commission's free grant scheme,[31] and by reducing the local contribution from one-half to one-fourth of the estimated cost, in recognition of the proven difficulty or inability of local sources to meet the requirement. This model, of free grants from the public purse of three fourths of the cost of construction, a local contribution of the residual one-fourth, and a cap on the maximum expenditure permitted for any one structure, became the model for public funding of fishery piers and quays that would last for most of the century.

For the next sixty years, almost all public piers and quays would be erected mainly as public works for the relief of distress, rather than primarily for the benefit of the fisheries. When the first Acts that provided finance for piers were passed as relief measures during the Great Famine in 1846 and 1847, the funds granted were managed exclusively by the OPW. These Great Famine piers, together with all the older piers and quays that had been built with public money from the start of the century, were vested first in the OPW and later, in 1853, transferred to the control of the County Grand Juries who administered the counties before County Councils were created later in the nineteenth century. After 1853, very few new piers were built anywhere in Ireland for a number of years. New, independent Fishery Inspectors were appointed in 1869 but the construction of new fishery piers remained the exclusive responsibility of the OPW. Relations between the Inspectors and the OPW started to deteriorate straight away.

When new relief funds were granted by further Acts of Parliament in 1880, the two bodies, the Inspectors and the OPW, were barely corresponding civilly. By the time the 'Sea Fisheries Fund' was set up in 1883 – the greatest single allocation of funds ever made to build piers in Ireland exclusively as facilities for the fishing industry – relations between them had soured dramatically, to such an extent that the practical application of the enormous Sea Fisheries Fund (£250,000) never fulfilled the promise of its full potential. It did, however, result in the emergence of a national strategy for Irish fishery ports that would later be matured by new official bodies such as the Congested Districts Board (CDB) and the Department of Agriculture and Technical Instruction (DATI).

With substantial new funds made available to the new bodies by the *Marine Works (Ireland) Act* in 1902 and the *Development Grant (Ireland) Act* in 1903, the future of public fishery pier building, at the start of the twentieth century, seemed set on a prosperous road. Little enough of that prosperity would fall to Co. Galway, where efforts by the CDB to develop an offshore fishery came to nothing; north Clare came within the remit of the CBD too late to benefit in any serious way from that Board's activity. A few more years into the twentieth century, the turmoil that convulsed the whole of Europe and the world, and the events that led to independence in Ireland, put an end to serious fishery pier development.

This book, then, is the story broadly of the public fishery piers and quays of all Ireland, and, in detail, of those of Co. Galway and north Clare during the one hundred years from 1820 to national independence. It discusses the legislation, the social environment, the characters, commissions and personalities who dominated the fishery piers and harbours scene, the successes and failures, and the eventual outcome of it all.

The public documentary records for Galway and north Clare mention, however briefly, about 150 structures. It has been possible to date the origin of 130 of these (eighty-five precisely and forty-five broadly), about eighty of which are treated here in some detail. The narrative is broadly chronological and the piers are discussed as they first occur in the records, except in cases where particular structures re-occur many times; in their case (as, for example, Claddagh (No. 145A), Roundstone (No. 040), and Cleggan (No. 014)), the full information is given in the most appropriate chapter, even if out of chronological sequence.

The book identifies, for the first time, the piers and quays that were erected during the Great Famine of 1846–48; those of the 1879–80 famine; those of the Sea Fisheries Act and Fund 1883–88; those built in 1886–88 by the Piers and Roads Commission; and those erected by the CDB and DATI under the *Irish Land Act* 1891, the *Marine Works (Ireland) Act*, 1902 and the *Development Acts* of 1909 and 1914. These, along with the Galway and north Clare piers of the 1822 famine previously identified,[32] make up the principal piers and quays of the region, and this is their story in broad outline.

Anyone interested in pursuing the detailed story of the public piers of other counties will find the information given here on the legislative and administrative background to be essential and sufficiently comprehensive to make a valuable starting point for their own study. The appendices name the piers and quays built in all the other counties of Ireland during the different phases and under the various Acts passed during the century, extracted from the public records. This will help others who wish to commence similar investigations of the fishery piers in their own counties.

Many structures were built privately and are not mentioned in the public records. For these we have no precise documentary evidence at all, but the structural on-site evidence of them is incontrovertible. In some of these cases we can suggest, if we choose, possible dates of origin by extrapolation from other data. Many were built by local users funded by private interests or with the help of charities – although it is hard to see why a private landowner would willingly undertake the full expense when they might have received three-fourths of their costs by applying for public aid. On the whole, landlords did not see the need to provide piers for their tenants and others, so that many 'unofficial' piers and quays are small, vernacular structures, erected and improved by the users themselves, and therefore never received any official recognition in the public records. The very existence of these humble works is acknowledgement in itself of their importance to the humble people who built and used them.

The Twentieth Century

Over time, many of the old fishery piers and quays would fall into disuse and some into severe decay. Others were saved and modified by conversion to alternative functions, although they all still service inshore fishery interests in some way: Cleggan is now an important ferry port for Inishbofin; Roundstone is a major marine tourism destination; Rinville (New Harbour, No. 128) is a yachting centre; and many other piers support local aquaculture ventures. The main fishery port of the county is a modern development at Rosaveel (No. 137) beside, but not fully incorporating until very recently, Nimmo's original Costelo (Cashla) pier (No. 211) of 1823. The latter is now part of the ferry terminal for the Aran Islands, whose main piers see an impressive annual visitor throughput, never envisaged when they were first erected.

On the Galway Bay coast, Kinvara pier (No. 160) maintains its links with the past, largely through the annual *Cruinniú na mBád* festival, which commemorates the once significant turf and seaweed trades that linked the Aran islands, north Clare, Connemara and Galway city. Although the water-borne turf trade is now extinct, seaweed gathering has continued in a number of places. Small-scale inshore fishing (particularly for shellfish and lobsters) has not entirely died out; and many coastal communities endeavour to maintain their traditional maritime interests and activities with whatever vibrancy and vigour they can muster. But overall, the working life of most of the older piers and harbours has generally disappeared from public awareness and the number of boats using them for serious maritime work, rather than for leisure purposes, has declined greatly.

The Ryan Hanley survey[33] examined most of the piers and quays in detail, recording their deficiencies and decayed condition and indicating improvements, modifications and repairs necessary or appropriate to bring them once again to a safer, more useful state. Their suitability for land-based and heritage-oriented tourism and leisure was not generally prominent in the definition of 'useful state', although the heritage value of each pier was assessed realistically and recorded appropriately.

Paradoxically, the decay of many piers and quays was hastened, if not compounded, by the temporary wealth bubble that inflated nationally in the early years of the new millennium. Large sums were expended by Galway County Council in addressing the renovations necessary to make the structures safe. One consequence of this was the partial and, in some cases, total, entombment in concrete of many old stone piers, quays and jetties; the replacement of old stone steps; the addition of modern guard rails and foot rails; the removal of old bollards and other such 'improvements'. This was done with no clear knowledge of their origin or history and with little enough appreciation of their wider value. In consequence, many old structures that evoke and commemorate times long past – times blighted by hardship and hunger, works done by the hands of starving men – today stand like concrete mausolea, mute monuments of a recent developmental ethos that had different cultural and ethical values, and without much sympathy for the past.

The Ryan Hanley survey remains an invaluable and unique historical record of the maritime structural heritage of Galway and Galway Bay at the opening of the new millennium, two hundred years after the first flush of fishery pier and quay building. Hopefully, this book adds a little to that incomparable survey.

Endnotes

1 J. Conroy, 'Galway Bay, Louis XIV's Navy and the "Little Bougard"', *Journal of the Galway Archaeological and Historical Society*, 49 (1997), pp. 36–48.

2 W. Larkin, *Map of the County of Galway in the Province of Connaught in Ireland from Actual Survey by William Larkin* (London: Engraved by S.I. Neale and Son, 1819).

3 Ordnance Survey of Ireland. Six-inch Series, First Edition, 1839.

4 Ordnance Survey of Ireland. Six-inch Series, Second Edition, 1898.

5 T. Robinson, *Connemara. Part 1: Introduction and gazetteer. Part 2: A one-inch map* (Roundstone Co. Galway: Folding Landscapes, 1990); *Oileán Arann: A map of the Aran Islands with a companion to the map* (Roundstone, Co. Galway: Folding Landscapes, 1996); *The Burren. A map of the uplands of North-West Clare, Éire* (Roundstone, Co. Galway: Folding Landscapes, 1999).

6 T. Robinson, *Stones of Aran. Labyrinth* (Dublin: The Lilliput Press, 1995); *Connemara. Listening to the Wind* (Dublin: Penguin, 2006); *Connemara. The Last Pool of Darkness* (Dublin: Penguin, 2008); *Connemara. A Little Gaelic Kingdom* (Dublin: Penguin, 2011).

7 Colleran, J, and D.P. Hanley, *Assessment of Piers, Harbours and Landing Places in County Galway*, Vols 1–4. The actual surveys were carried out by Hugh Haughey, Damian Hanley, Ronan Waters and Sarah Neary.

8 http://data-galwaycoco.opendata.arcgis.com/datasets/, last accessed 2 March 2017.

9 *Clare Coastal and Architectural Heritage Survey, 2006-7.* The actual surveys were carried out by Sarah Halpin and Grainne O'Connor.

10 www.clarelibrary.ie. Last accessed 2 March 2017.

11 R. Lohan, *Guide to the Archives of the Office of Public Works* (Dublin: The Stationery Office, 1994).

12 T. Robinson, 'A Connemara Fractal' in T. Collins (ed.), *Decoding the Landscape* (Galway: Centre for Landscape Studies, 3rd edn, 2003), pp. 12–29.

13 *An tOrdú Logainmneacha (Ceantair Ghaeltachta)* I.R. Uimhir 599 de 2011 (Dublin: Office of the Attorney General, 2011).

14 Fourth Report of the Commissioners of the Irish Fisheries, BPP 1823 (383).

15 Report of the Department of Fisheries. Saorstát Éireann, 1927.

16 A. Jarvis, 'Dock and Harbour Provision for the Fishing Industry since the Eighteenth Century' in D.J. Starkey, C. Reid & N. Ashcroft (eds), *England's Sea Fisheries* (London: Chatham Publishing, 2000), pp. 146–56.

17 43rd Annual Report of the OPW, BPP 1875 [C. 1223], p. 53.

18 Jane Conroy, personal communication.

19 Robinson, *Connemara*, p. 62.
20 M.D. O'Sullivan, 'Glimpses of the Life of Galway Merchants and Mariners in the Early Seventeenth Century', *Journal of the Galway Archaeological and Historical Society*, 15, (1931–33), pp. 129–40.
21 J. Hardiman, *The History of the Town and the County of the Town of Galway.* (Dublin, 1820; reprinted by Galway Tribune Ltd., 1994), p. 174.
22 Samuel Johnson in his Dictionary delightfully defines an anker as follows: 'It is the fourth part of the awn, and contains two stekans; each stekan consists of sixteen mengles; the mengle being equal to two of our wine quarts.'
23 S. Lewis, *A Topographical Dictionary of Ireland* (London: S. Lewis and Co., 1837); Hardiman, *History*, 1994.
24 T. Robinson, *Connemara*, pp. 34, 51, 60, 111, 119, 123, 127, 132.
25 L.M. Cullen, 'The Galway Smuggling Trade in the 1730s', *Journal of the Galway Archaeological and Historical Society*, 30 (1962–63), pp. 7–40.
26 M. O'Malleyin Select Committee on Public Works in Ireland, BPP 1835 [573], queries 3323–24.
27 Cullen, *The Galway Smuggling Trade in the 1730s*, p. 22
28 1st Report of the Select Committee on Harbour Accommodation, BPP 1883 (255), p. 12..
29 A.E.J. Went, 1949. *Foreign Fishing Fleets along the Irish Coasts*, in Cork Hist. Arch. Soc. Jn., second series, 54, 197, pp. 17–24.
30 J. Selden, *Of the Dominion or Ownership of the Sea*, translated by M. Nedham (London: William Du Gard, 1652. Reprint edition by Arno Press, 1972); F. Hargrave, 'A Treatise in Three Parts from a Manuscript of Lord Chief Justice Hale' in *Collection of Tracts relative to the Law of England*, 1 (Dublin: E. Lynch et al., 1787).
31 5 Geo IV C. 64. (1824). *An Act to amend the several Acts for the Encouragement and Improvement of the British and Irish Fisheries.* Section 9.
32 N.P. Wilkins, *Alexander Nimmo, Master Engineer 1783–1832* (Dublin: Irish Academic Press, 2009).
33 Colleran and Hanley, *Assessment of Piers, Harbours and Landing.* The survey data, including maps and pictures, are in the public domain and can be accessed on the internet at the following website: http://data-galwaycoco.opendata.arcgis.com/datasets/. Click on 'Piers and Harbours of County Galway' in Recent Databases at that site. (Last accessed, 2 March 2017.)

CHAPTER 2

A Beginning in a Forgotten Famine

1800–1830

At several of the stations I have pointed out the sites where small piers would be useful ... The benefit of a small pier where none existed before, is sufficient to awaken a desire for future improvement; and, at present, it is my opinion that works of that kind should chiefly attract the attention of the Board.

Alexander Nimmo, 1821[1]

The Beginning of the Nineteenth Century

Prior to 1819 there were no Irish sea fishery authorities of any sort, or any other Irish public body with a dedicated role in fisheries or fishery piers and harbours. Experience in the closing years of the eighteenth century – with a failed French invasion at Bantry, Co. Cork (1798), and a small, successful French landing at Kilcummin, Co. Mayo, that same year – convinced the new United Kingdom Government that Irish harbours and landing places could present clandestine gateways into the kingdom for potential French invaders. Since so much of the west coast of Ireland was remote and difficult to defend, it was considered inadvisable to encourage the building of piers and landing places for fear of facilitating further incursions. Indeed, an Act 'for the Preservation of the Publick Harbours of the United Kingdom' was passed in 1806 requiring that no pier, quay or jetty be erected in any natural harbour anywhere in the United Kingdom without the proposer first giving at least one month's notice to the Admiralty.[2]

Nevertheless, when the first edition of the six-inch Ordnance Survey map of Ireland was published in 1839, it recorded sixty-one manmade piers and quays on the coasts of Co. Galway and north Clare. Twenty-three of these were built in the 1820s and can be accurately dated, leaving thirty-eight of uncertain, but early, date; the latter are listed in Table 2.1.

Considering the region in four notional sub-regions – Galway Bay (from Black Head in Co. Clare to Ballinahown in Co. Galway); south Connemara (Ballinahown to Roundstone);

TABLE 2.1

A list (N=38) of the manmade piers, quays and breakwaters of Co. Galway and north Clare recorded in the first edition of the six-inch OS maps, published in 1839. Piers known to have been erected between 1820 and 1830 (N=23) are omitted, but given separately in Table 2.3. The RH number is the number of the structure in the Ryan Hanley survey of piers and quays drawn up in 2001 (J. Colleran and D.P. Hanley, 2001). Numbers prefixed CS are those given in the County Clare Coastal Architectural Heritage Survey (S. Halpin and G. O'Connor, 2007).

Site	RH No.	Site	RH No.
Skeaghduff Quay	-	Clarinbridge (Tobernagloragh) Quay	155
Kingston Glebe Quay	-	Kilcolgan Quay	156
Doabeg West Wharf	-	Pollagh Quay	158
Salt Lake Pier	-	Mulroog Quay	-
Wallace's Pier	044	Kinvara Quay	160
Mainis Road Pier	-	Cushua Quay	-
Bunakill Pier	-	Duras Demense Quay	-
Lettermore Pier	123	Trawndaleen (*Inis Meáin*) Quay	184?
Sruffaun (*Sruthán*) Pier and Quay	133	Burrin old quay	-
Céibh Tí Mór	-	Callahvreedia Pier and Quay	CS023
Travore Bay Quay	-	Ballyvaughan Old Quay	CS 001
Furbo Quay	-	Corranroo Pier and Quay	CS028
Barna Pier	145	Harbour Hill House Quay	CS032
Galway Docks	145B	Culliagh Quay	-
Roscam Pier	146	Bell Harbour Quay	CS014
St Mary's Port	-	Parkmore Pier	CS009
Oranmore Castle Old Pier	147	Skerritt's Pier	CS008
North Quay	-	Muckinish Castle Pier	CS010
Prospect quay	149		
Tyrone House quay	-		

west Connemara (Roundstone to Killary Harbour); and the offshore islands of Aran, Inishark and Inishbofin – the numbers of piers and quays recorded before 1820 were as follows: Galway Bay, twenty-eight; south Connemara, six; west Connemara, four; and the offshore islands, one. That distribution suggests the main concentration of sea-borne commerce and communication was through and across Galway Bay, with Galway city at its apex. When the piers and quays on the outskirts of Galway city are excluded, the number in Galway Bay (twenty-four) still greatly exceeded the number in Connemara. Of the eleven piers on the north Clare coast of Galway Bay, seven were close to, and probably associated with, a so-called 'big house' in the locality. Four of the fifteen piers on the north shore of Galway Bay, and two of the ten in Connemara, were also close to a big house.

By the end of the nineteenth century the total number of recorded manmade piers and quays would be greatly increased and their distribution very different: Galway Bay would have about forty-six, south Connemara sixty-three, west Connemara thirty-one and the offshore islands nine, reflecting the extent of the construction of new piers (mainly in Connemara) as the century progressed.

This study records about 315 manmade piers and quays in the whole region, the vast majority of which are still extant. Although some are seriously dilapidated, physical on-site evidence remains even of those most decayed. In this book, there are sixty-four recorded in Galway Bay, 169 in south Connemara, fifty-seven in west Connemara and twenty-five on the offshore islands. Those of the south and west Connemara sub-regions combined make up over 70 per cent of the total in the whole region and almost 80 per cent of all the piers in Co. Galway alone. The vast majority are located in the division of south Connemara known as *Ceanntar na nOileán* (116 from Ballinahown to Ardmore) and in the Bertraghboy Bay and Carna division (53 from Ardmore to Roundstone). By and large, the funding for this massive increase in such remote and isolated coasts came from two principal sources: funds provided by Government legislation purportedly in support of piers and quays for use in the fisheries, and funds for the relief of distress in times of famine, provided in part by landowners, in part by charities and in part by Government. There was a considerable overlap in these sources as time went on.

About 40 per cent of all the piers built vested eventually in the relevant county, that is, legal ownership was transferred to the County Council (or to its predecessor, the County Grand Jury), which then maintained them on behalf of the public. These public structures constitute the vast bulk of all the piers and quays recorded in the region. A few others are owned privately, or their ownership is not clearly established; these are mainly small works built as famine relief works. The number of confirmed private piers is very small.

The First Commission for Irish Fisheries

Little had been done to develop the Irish fisheries after the Act of Union (1801). The existing fishery laws, made in the late Irish Parliament and dating from 1752, were repeatedly extended in operation until 1819 when, after much deliberation and a seminal report from the Select Committee on Disease and the Labouring Poor,[3] the Government finally enacted appropriate new fishery legislation. The new Act – *An Act for the Further Encouragement and Improvement of the Irish Fisheries, 1819* – set up a Commission for Irish Fisheries that was to stay in office initially for five years and from then to the end of the next session of Parliament.[4] This was an independent body whose main duty was to encourage and support the Irish sea fisheries mainly by introducing and managing a specially funded bounty scheme that financed in part the fitting-out of fishing boats and rewarded fishermen for landing fish. In addition, the Commissioners were granted a sum of £5,000 a year for use in the general administration of the coast fisheries. They were given no specific obligation or duty to construct, or maintain, any fishery piers. At the time they were appointed, the few piers that did exist had been privately built for the most part and, other than these and one or two public, general purpose trading piers built by the Grand Juries of both counties, there was little or nothing to facilitate fishermen and fishing activity in Co. Galway or in north Clare. The Commissioners themselves, men of the landed class who acted without salary or other emolument, were specifically precluded by the Act from having any personal involvement in the fishing industry. They decided, therefore, not to draw down any funds until 1823, by which time they would have gained some insights into, and practical knowledge of, the workings of the fisheries, such as they were.[5]

Indeed, their very first action was to appoint fishery inspectors, one reliable man in each province with knowledge of the local fisheries, charged with reporting the number of bays, harbours and creeks and the numbers of fishing boats and men belonging to each; the state of the fisheries and the markets supplied; the general accessibility of each district; whether the harbours were natural or artificial; and anything else of relevance to the development of fishery stations.[6]

The persons appointed, who were called Inspectors General of the Fisheries, were: Thomas Young of Derry, for Ulster; William Henry King of Galway, for Connacht; James Redmond Barry of Glandore, Co. Cork, for Munster; and John Madden of Balbriggan, Co. Dublin, for Leinster.[7] On a wider scale, they commissioned an engineer to make a general survey of the whole coast of Ireland 'to report the state of the harbours, the most advantageous sites for fishing stations and the most useful lines of communication between the principal harbours and the interior of the country through the mountainous districts'.[8] That post was given to a Scottish engineer, Alexander Nimmo (1783–1832).

Financing Piers and the Contentious Local Contribution

The Commissioners were soon apprised of the dearth of suitable landing facilities and the retarding effect that had on fishing activity. They resolved therefore to apply their financial resources to the erection of small piers and breakwaters where they were most lacking. Although not called for in the legislation, this was the very first official commitment to the construction of Irish fishery piers at public expense, and it established a funding protocol that would remain in place for the next six decades at least. The Commissioners held the view that their financial support should be on terms as advantageous to the public as possible. Therefore, they considered it reasonable to look to the landowners and inhabitants, the likely beneficiaries, for contributions towards the construction costs. This 'local contribution' could be given in cash or kind – that is, by way of the free provision of land, labour or materials – and the Commissioners expressed confidence in their expectation of success with their pier-building intentions.[9] In this they were initially disappointed, largely through their own excess of zeal: they had 'conceived it their duty to furnish but a moiety [one half] of the expense attending their [piers'] construction and in no instance to depart from a strict adherence to that principle'.[10]

In practice, there proved to be few persons willing to contribute the required local half of the cost from their own private resources. It was an onerous condition, to be sure: local inhabitants who were generally poor tenants could not, and many landowners either could not or would not, make up the necessary one-half contribution, an aspect of the scheme that would remain an obstacle for many years. In the few cases where landowners did contribute, the result was not always what the authorities expected. Landowning contributors often insisted on specifying the precise sites of the intended piers in accordance with their own private interests, not those of the fishermen, so that the piers and quays were not always built at the most useful or judicious places. This, too, would become a recurring criticism of public piers and quays as the century progressed. Henry Townsend, Secretary to the Commission, explained the requirement for so large a local contribution by 'the smallness of their [the Commissioners'] funds, and to render as diffusive as possible the benefits of their application'.[11] But it almost certainly had its roots in the principles of *laissez-faire* political economy then spreading in Britain.

The local contributions would not be revoked, not even temporarily during the height of the Great Famine, until more than sixty years later. Neither was this an exclusively Irish phenomenon. It mirrored the situation prevailing in Scotland, where sums taken from the accounts of estates forfeited to the Crown were granted for the construction of fishery piers, subject in each case to a local contribution equal to the sum granted.[12]

In Ireland, the Commissioners decided additionally that their moiety should not exceed £500 for any individual work. That certainly helped with diffusing their limited funds to a greater number of works, albeit ones necessarily of smaller scale and less sophistication, restricting as it did the total expenditure (the free grant plus the local contribution) to £1,000 at any one site.

The Famine of 1822

Despite the slowness with which it was initially taken up, the Commissioners' pier scheme was driven forward by circumstances that were both unforeseen and unwelcome. Early in 1822, famine and the distress it brought in its wake struck the west and south coasts of Ireland. There was no general or absolute shortage of food: the problem was that the indigenous poor had absolutely no money, nor any opportunity to earn money with which to buy food when their own crops failed. Disease came hand-in-hand with famine, as amply and pitifully recorded in letters to the London Tavern Committee[13] and in the contemporary correspondence of Henry Stratford Persse of Galway with his sons in America.[14] McNamara has succinctly reviewed the wider social causes of the 1822 famine and its detrimental consequences in the South and West.[15]

Public relief measures were immediately initiated by Government, managed by an official Relief Committee in Dublin Castle and a charitable Relief Committee in the

Mansion House, Dublin. Among other actions, these Committees considered various ways of providing relief work for the destitute. The Commissioners of Fisheries were quick to spot an opportunity. In their own words:

> On the commencement of the distress which during the summer of 1822 reduced the people of the Southern and Western districts of Ireland to a state of unparalleled misery, the Commissioners felt that such a period would be the most eligible for carrying their plans of safety harbours into effect, so far at least as regarded the districts in question. Could their construction be followed up consistently with the principle laid down by the Board (of only furnishing half the estimated expense) a double benefit would result from their immediate commencement; namely, the extension of the fisheries, by affording protection to suitable craft, and a supply of food to a large mass of starving population, through the medium of labour.[16]

Alexander Nimmo Called In

In 1822 the Commissioners of Fisheries wrote to the authorities in Dublin Castle seeking permission to remove the impediment of the local contribution so as to 'enable them at once to put great numbers into active employment'.[17] As it transpired, they did not need to abandon entirely their avowed principle of providing only one half of the costs; the authorities ruled that the local moiety could be provided by the Dublin Castle Relief Committee (giving one-fourth of the total cost) and the Mansion House Relief Fund (giving another one-fourth), so that the entire so-called local moiety could come from these combined sources. Allowing charity to provide the local contributions in this way gave the lie to the term 'local'; clearly, the authorities were content to permit external bodies, wherever they might be and for whatever reason they might choose, to provide the requisite contribution.

At the same time, the Commissioners recalled their engineer, Alexander Nimmo, from his coastal survey underway on the east coast, directing him to Galway with express instructions 'to furnish gross estimates of the expenses likely to be incurred in the erection of such small piers on the south and west of Ireland, where small sums so applied would be useful; and to have the same ready for consideration with the least possible delay'. Nimmo had already acquainted the Commissioners as early as January 1821 of his preference for small piers over larger ones: 'The benefit of a small pier where none existed before, is sufficient to awaken a desire for future improvement; and, at present, it is my opinion that works of that kind should chiefly attract the attention of the Board.'[18]

Within weeks Nimmo had produced a list of thirty-three sites distributed along the west and south coasts where small piers could be built for an estimated total cost of £12,865. A delegation of the Commissioners visited the West and selected twelve of these (three in Co. Galway) for immediate construction to designs supplied by Nimmo, to be built by day labour rather than by contractors. They forwarded their selection to the Relief Committees in Dublin who part-funded them as agreed, with the Commission contributing its moiety. The remaining twenty-one works on Nimmo's list were brought to the attention of the London Tavern Committee, which agreed to fund one-fourth of the cost of each of these, up to a total allocation of £2,000. The Dublin Castle Relief Committee agreed to add a further one-fourth and the Fishery Commissioners one half, so that all thirty-three works could go ahead without any delay.

Five more sites were added later to the list, four of them in Galway (Killeany, Duras, Clifden and Claddagh), making a final total of thirty-eight. They are listed in Table 2.2. The London Tavern Committee was a body of concerned persons who met in London in May 1822 and raised a subscription for the relief of the people in the famine-stricken counties. They were joined by others from various British cities and, from much further abroad, by like-minded persons in Calcutta and other dependencies of India.[19] By the time the London Tavern Fund was fully allocated and wound down in August 1822, it had given over £300,000 in charitable relief funds to Ireland, a largesse now almost completely forgotten. The residue of the fund, as we shall see, survived and was utilised up until the very end of the century.

Of the first twelve piers on Nimmo's list, funded by the Fisheries Commission, the Dublin Castle Committee and the Mansion House fund, three were in Co. Galway: Costello, Roundstone and Cleggan; there was none in north Clare. Of the next twenty-one piers on the initial list that were funded by the Fisheries Commissioners, the Dublin Castle Committee and the London Tavern Committee, five – Barna, *Spidéal*, Rinville (New Harbour), Ballinacourty and St Kitts (Killeenaran) – were in Co. Galway and one, Burrin New Quay, was on the north Clare coast. The final five were added later, four of them in Co. Galway: Killeany (*Inis Mór*), Duras, Clifden and Claddagh. The total number funded in Co. Galway and north Clare was therefore thirteen, over one-third of the total.

The Commissioners were fortunate in having Nimmo as their engineer: he had surveyed Connemara in 1813 on behalf of the Bogs Commission and was thoroughly familiar with the area.[20] Even at the latter early date, he had identified places where fishing villages

TABLE 2.2

List of the sites in all Ireland recommended by the first Commission for Irish Fisheries as locations for new fishery piers, together with the estimated cost of each. Nos 1 to 12 were approved immediately; Nos 13 to 33 were added later; the final 5 were the last to be added and approved.

	County	Location	Estimated cost
1	**Co. Clare**	Kilbaha	£500
2		Seafield	£500
3		Liscannor	£500

TABLE 2.2 ***Continued.***

	County	Location	Estimated cost
4	**Co. Galway**	Costello	£350
5		Roundstone	£400
6		Cleggan	£500
7	**Co. Kerry**	Bana	£250
8		Dingle	£400
9		Valentia	£495
10	**Co. Mayo**	Old Head	£500
11		Achill sound	£400
12		Ely Erris	£300
13	**Co. Clare**	Burrin	£200
14		Carrigaholt	£400
15		Dunbeg	£400
16	**Co. Cork**	Cullagh	£400
17		Berehaven	£200
18		Courtmacsherry	£500
19		Clonakilty	£500
20	**Co. Galway**	Barna	£300
21		*Spidéal*	£150
22		New Harbour	£420
23		Ballinacourty	£300
24		St Kitts	£250
25	**Co. Kerry**	Brandon Bay	£400
26		Ballinskelligs	£400
27		Cahersiveen	£400
28	**Co. Mayo**	Killala	£150
29		Bunatrahee	£300
30		Inishturk	£400
31		Tarmon	£400
32	**Co. Sligo**	Roughly Point	£800
33		Pullagheeny	£500
34	**Co. Galway**	*Inis Mór (Killeany)*	£230
35		Duras	£300
36		Clifden	£495
37		Claddagh	£928
38	**Co. Dublin**	Lambay	£900

could be profitably developed around proposed new piers, including Cleggan, Clifden, Roundstone and *Béal an Daingin*, and he was generally sanguine about the prospects for the whole region, which others had virtually written off. He had already designed port facilities for Cork and Sligo and was in the process of constructing the harbour of Dunmore East in Waterford, where he had in his employ a cohort of men skilled in harbour works. During his coastal survey, he had designed other small piers and quays for private clients, so that by the time he started to carry out the works in the West, he was well familiar with what was necessary and possible. In addition to all that, in June of 1822 he was appointed the engineer in charge of implementing the Government's famine relief measures in the Western District (comprising most of Connacht) under the *Employment of the Poor Act* of that year.[21] That gave him considerable authority and autonomy in determining what works were to be undertaken, and the way in which they were to progress. He was flexible in his approach, creatively combining his duties under the Fisheries Commission and his Western District responsibilities, so that he achieved more than might at first have been anticipated.[22] Government guidelines, for instance, required that employment of the poor was to be the main consideration when undertaking relief works, rather than the successful completion of any of the works that were started. Nimmo nevertheless succeeded in erecting supernumerary small piers as part of his Western District duties, and in constructing short access roads to others, making them much more accessible and therefore more useful. Overall, he built a total of twenty piers in Co. Galway and north Clare (Table 2.3), instead of the thirteen that the Fisheries Commissioners had anticipated.

The quality of the works constructed was generally very good, well beyond that of local artisanal works, with some of them technically quite sophisticated. All but the 'Western District only' piers were clad in good-quality ashlars, either limestone or granite, laid in regular, level courses without mortar and generally with limestone coping. In contrast, the piers at Derryinver, Leenaun and Clonisle, financed only with Western District funds, were much less sophisticated, being built of stone laid in rough courses in a more open form. When building embankments for the roadway at these sites, Nimmo had taken the opportunity to form roadside quays to which he added short jetty piers. Their integration with the new roads under construction made them exceptionally useful from the very beginning and, unsophisticated though they were, all three named sites remained valuable for a long time. Derryinver pier was built along a rocky outcrop to form one side of a small harbour, which would be enlarged and strengthened later in the century, and is still in use for fisheries today. Leenaun would be transformed too, and it is described in greater detail in chapter 6. Clonisle, from which Connemara marble was exported in the 1820s, remained unaltered until very recently when it was renovated by Galway County Council.

A quay erected at Furbo was mainly the work of Mr Andrew Blake of Furbo House, a member of the Galway Harbour Commissioners.[23] Situated on the rocky eastern edge of a wide beach, it stood entirely detached from the land. Built of massive granite blocks, it is outlined on the six-inch OS map of 1839, roughly as an E-shaped structure, comprising three jetties joined together by a common quay, thereby forming two small, sandy-

TABLE 2.3

Twenty-three piers and quays erected in Co. Galway and north Clare, 1822 to 1830.

PIER	FUNDED BY	TYPE OF STRUCTURE
Leenaun (East Jetty)*	1	Roadside quay and jetty
Derryinver*	1	Quay, pier and small basin
Cleggan	2, 3, 6	Quay and small dock and harbour
Doaghbeg	6	Longshore wharf
Clifden	1, 5, 6	Quay
Roundstone (South Pier)	1, 2, 3, 6	Quay and pier
Clonisle *	1	Roadside quay and jetty
Maumeen *	1	Pier
Costello (Cashla)	2, 3, 6	Pier
Spidéal (an tseancéibh)	4, 5, 6	Quay, pier and harbour
Furbo *	1, 5	Quay (with Andrew Blake)
Barna	2, 4, 6.	Quay and pier
Recorder's Quay, Salthill*	5	Small pier and quay (O'Hara's)
Slate Pier, Claddagh, Galway*	1	Breakwater
Claddagh *	1, 5	Quay and two piers
Moneen, Lough Atalia*	5	Pier (Lynch's)
Rinville (New Harbour)	2, 4, 6	Wharf and jetty
Ballinacourty	2, 4, 6	Pier and small basin
St Kitts (Killeenaran)	2, 4, 6	Pier and small basin
Duras	5, 6	Pier and quay (with Patk. Lynch)
Killeany (Aran)	2, 3, 6	Wharf and pier
Kilronan (Aran) *(an tseancéibh)*	5	Pier
Burren New Quay, Co. Clare	2, 4, 6	Quay and jetty

This table shows the type of structures built and the source of funding. Excluding those marked with an asterisk, all were erected by the first Fisheries Commission. Nimmo was involved in all except Kilronan *(an tseancéibh)*, Recorder's Quay and Moneen Pier.

1 = Government Relief Funds of the Western District; 2 = Dublin Castle Relief Committee Fund; 3 = Mansion House Relief Committee Fund; 4 = London Tavern Committee Fund; 5 = Local contributions; 6 = Commission for Fisheries grant.

bottomed harbours sheltered by the rocky walls. Nimmo helped Blake out with tools and advice, but did not contribute to the design or the cost, which was met entirely by Blake. Later efforts to have Furbo quay improved by the OPW were unsuccessful because of its advanced state of decay. By 1860 it was completely derelict and abandoned, and it is not shown in the second edition (1898) of the six-inch OS map of the area. Today, its remains are only barely recognisable as part of a manmade structure, only one jetty remaining evident and in a very poor state. Not appearing on modern maps, and not easily identified in satellite images, it is, in a way, a 'ghost' structure, evident only on careful on-site examination. Another such 'ghost' quay, shown in the first edition of the six-inch OS map but not in the second, lies west of Spidéal in Travore (*Trá Mhór*) Bay, below Cashel House (sometimes called Tully Castle), a ruined residence of the Blakes of Tully.[24] This quay, too, was detached from the land and completely submerged at high tide. Only a heap of rocks marks the site today.

In contrast to these two, Nimmo's Slate breakwater in Galway City (called Nimmo's Pier locally) is quite remarkable for its dominance of the entry to Galway port, its substantial structural integrity and the quality of its sea face, clad in beautifully hand-curved [*sic*] Aran limestone. (Nimmo's brother George was supervisor at the quarry on *Inis Mór* where the curved slabs were hand-worked.) It will be dealt with in a later chapter, along with the Claddagh quay and piers, of which it is now an integral part.

While building the piers was very useful in providing work, and their long-term effect was good in other ways, the acceptance of funding from charitable sources was, in hindsight, a travesty; it ensured that the Government could keep inviolate the principle of a local contribution, no matter how overwhelming the distress, destitution and despair of the local inhabitants might be, or might become. Accepting contributions from elsewhere in lieu of a *local* contribution emphasised the fictional nature of its 'local' aspect and appeared to reward the failure of the landowners to meet their responsibilities.

It also established a precedent that would be followed again in later periods of distress: no matter its duty and obligations towards its citizens, or the immensity of destitution and the depth of the people's misery, the Government would henceforth rely on the kindness of strangers to meet its own responsibility for its citizens' very survival, rather than breach its

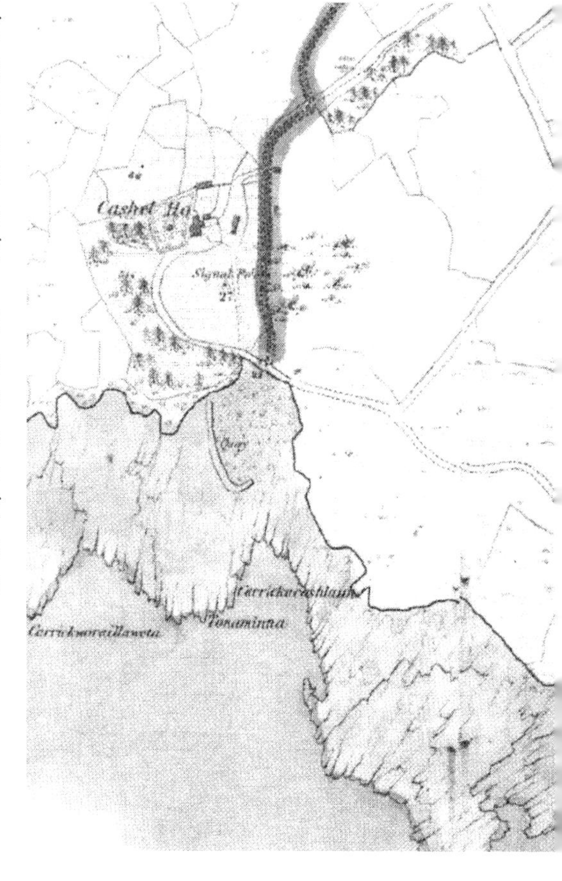

FIGURE 2.1. The quay indicated at Travore (*Trá Mhór*) in the first edition of the six-inch OS map. It does not appear in later editions of the map. Reproduced with permission of the Ordnance Survey of Ireland.

self-imposed fiscal policy. As we shall see in another chapter, it was the people of Canada and their Government, and other benefactors, who would provide all, or a large part of, the local contributions for certain piers erected throughout the country during the distress of 1880, allowing the Government to keep its *laissez-faire* policy sacrosanct.

The End of Relief Measures

Considering what was successfully achieved during the period of famine – a total of twenty-three new piers and quays constructed in Galway and north Clare – it is hard to believe that the Government's and the private charities' funding had lasted only for a very short period. The intensity of the famine was at its height only for the duration of spring and summer of 1822. That is not to say that the ordeal of the poor was short-lived: they started the decade desperately poor at very best, and remained destitute, weakened and distressed for years, so that relief works started under the 1822 Act had to be extended and continued almost to the end of the decade. But for the present, the ongoing relief works and the contributions by the charities largely ceased in August 1822 when the Government decided that the worst was really over and special efforts were no longer needed.

There were to be no new relief works started after that date without specific sanction. Fortunately, construction of the piers funded by the Commissioners for Fisheries, and not entirely dependent on relief funds, could continue for some years more. Up to the end of 1823, the Commissioners had provided their share of the cost of piers out of a special Government allocation, separate from their own voted annual income of £5,000 (the latter provided under section 66 of the 1819 Act), which they had not drawn down since their appointment.

In January 1824, they drew down the arrears, then standing at more than £15,000, which they immediately distributed among their various fishery projects.[25] Over £2,000 was allocated to the completion of certain piers in the West and South that were already partly constructed, and almost £7,800 was granted for additional new piers and harbours (often called 'harbours of refuge' or 'safety harbours') on the east, south and north coasts. In this way, the Commissioners endeavoured to rebalance their expenditure towards counties that had not received financial aid during the 1822 crisis.

They continued to make supplementary funds available during the rest of the decade to complete most of the fishery piers already underway and to start new ones wherever the requisite local contributions were raised. But the Commissioners continued to complain: 'The difficulty of procuring the necessary contributions from those whose private interests must be promoted by their erection, has tended in some measure to retard the progress of similar works, which, if executed, must prove of equal utility to the coast fisheries of Ireland.'[26]

Mitigation of the Local Contribution by an Act of 1824

Clearly, the local contribution of one-half of the cost remained a serious obstacle. By an Act passed in June 1824,[27] the Commissioners were formally required and empowered to spend £4,500 of their annual grant of £5,000 'in the encouraging and assisting the building, making, or repairing of piers or quays at such ports and places on the sea coast as shall appear to be most necessary', but only in those cases where not less than one-fourth of the cost of the work was paid by the person receiving the Commission's grant. Once any grant-aided pier was built, the recipient of the grant was expected to undertake all the necessary maintenance of it on an ongoing basis, in advance of receiving financial recompense for the requisite outlay.

Although the Act reduced the over-ambitious requirement of a one-half local contribution to one-fourth, it did not have any immediate beneficial effect. In their seventh report, for the year to 1826, the Commissioners repeated their complaint that, in the erection of fishery piers:[28]

> [T]hey have been much impeded in their progress by local difficulties, chiefly arising from the impossibility of procuring the necessary contributions in aid of their formation or their future repairs as prescribed by the Act of the 5th Geo. IV C. 64. These impediments, it is much to be lamented, too often occur in those parts of the coast where poverty predominates.

The Act of 1824 was the very first in which Irish fishery piers were expressly mentioned in the legislation and it represented the first legislative sanction by Government to funding them from the public purse. However, no new funds were granted: the Act simply directed that the Commission's existing annual vote be allocated mainly to piers and quays. It also formally established the statutory basis for the local contribution. Overall, the Act was a positive development that endorsed the earlier action of the Commissioners, and ensured funding for the completion of most of those piers already under construction. The reduction of the local contribution to one-fourth, however welcome, was tacit acknowledgement that there were not sufficient resources 'in those parts of the coast where poverty predominates' to meet any greater level of local support.

Although supportive of the general principle of a local contribution, the engineer James Donnell highlighted another of the Act's negative aspects:[29]

> I venture to observe that the clause contained in the section of the Act of the 5th Geo.4 C. 64, which requires local contributions in aid of every grant, works badly. It has the effect of fixing the construction of harbours on inferior instead of superior sites: local contribution is only to be obtained where the individual subscribing will derive personal advantage; he therefore, in general, proposes the site as the condition of his subscription; and though a preferable site may not be very remote, if local contribution cannot be obtained for it, the Board cannot decide in its favour, but must either submit to grant for the inferior site, or give up the measure altogether. The principle of local contribution is good, but its universal and indispensible application practically renders the selection of the best sites impossible.

The Act also applied to Scotland in specifying a minimum one-fourth local contribution, but there seems to have been no great difficulty with the requirement there, and in many Scottish cases the amount contributed locally exceeded the statutory minimum. In Scotland's case, the relevant piers were to be built exclusively for fishery purposes, and not for general trade. That particular restriction did not apply expressly in Ireland and it would have greatly negated the benefit of the piers had it been imposed here.[30] But years later, the three-fourths free grant contributed by Government for fishery piers in Ireland would be reduced to two-thirds in cases where the piers had some commercial use, in addition to the fisheries.[31] This extra restriction would have a further retarding effect on the size of Irish fishery piers as the century progressed.

Exit Alexander Nimmo

Nimmo's engagement by the Commissioners ended in 1825, but his final report was not submitted until 1826.[32] By then he was professionally over-extended, being involved in the design and construction of more than forty piers and harbours throughout the country, along with many other public and private works in Ireland and England. The Commissioners therefore appointed James Donnell, whom they had previously employed 'as occasion may require', to a new position of Harbour Engineer.[33] He had already reported on the state of the piers in 1824 and, as a first official duty, he was now charged to revisit them in 1825–26 with a view to bringing them to completion within newly determined costs.

Donnell was thorough and was less benign than Nimmo in his response to the pitiable human conditions prevailing in some localities. For example, he expressed dismay that during the distress Nimmo had allocated money to certain piers that was not accounted for in the fisheries budget. Donnell's report, dated 1 November 1826, appeared in the Eighth Report of the Commissioners,[34] as did Nimmo's final report of the coastal survey, dated 9 November 1826. The latter's charts of the coast from Mayo to Clare, which were published separately, contained the principal coastal and hydrographical information for Co. Galway and Galway Bay.[35] His written report, drawing heavily on his Bogs Commission report of 1813, extended and enhanced by his experience in his Western District relief works of 1822 to 1824,[36] reflected his landward interests as much as his marine interests, and had a comprehensive rural economic slant.

The End of the First Commission for Irish Fisheries

In May of 1826 the Fisheries Commission came to its scheduled end and was extended to 5 April 1830 by a new Act.[37] In the following year, 1827, the Commissioners, aware that there was little likelihood that their term would be extended again beyond 1830, resolved 'to minutely revise the state of their funds, to disappropriate all unclaimed grants, to ascertain what portion of these funds had accumulated under the 66th section of the Act of the 59th of Geo III, C. 109 ... and what portion of them had accrued since the passing of the

Act of the 5th of Geo IV, C. 64'.[38] Having reviewed this information and received legal advice, they found themselves in a position to commit £10,000 to a general-purpose 'Irish Fisheries Loan Fund'. It would be used for many years to assist fishermen to buy and repair their boats and gear. They applied the remaining balance of their funds to the repair and improvement of the existing fishery piers, for which they had Nimmo's and Donnell's latest recommendations to hand.

These necessary works were carried out in 1828, and Donnell was instructed to make a final examination of the outcome in 1829. When he reported back, the Commissioners expressed great satisfaction with the results, for agriculture as well as fisheries, and went on unexpectedly to claim:

> They rarely experience any difficulty at present in eliciting local contributions in aid of their grants for those purposes; and while, in the early stage of their proceedings, there was an unwillingness to contribute even the minimum rate of aid required by the Act, the applications of proprietors willing to co-operate in a much larger proportion have within the last year far exceeded the funds at the disposal of the Board. The numerous advantages attainable by a more general application of the principle of local contribution have been abundantly proved ... The Commissioners are of opinion, that by some modification of the present local regulations, with some additional powers ... a very excellent system of permanent encouragement [of the coast fisheries] may be produced.[39]

That comment sounded like a belated plea for the Commission to remain in existence, and Donnell's report certainly appeared to bear out their bullishness: since their appointment in 1819, a total of 106 proposals for piers had been dealt with in the whole country. Fifty-five had been completed or were making good progress; ten were still under, or awaiting, consideration; thirty-five proposals had been rejected, and six had been abandoned. The rejections arose largely from lack of sufficient funds on the Commission's part (eighteen cases); rejections and abandonment arising from inadequacy of local contributions numbered ten cases.[40] In Co. Galway and north Clare, twenty-three piers and quays had been successfully erected, including those that were not funded by the Fisheries Commission but by Nimmo's Western District famine relief funds (Table 2.3). Two others (Recorder's quay and Moneen pier) were, for the most part, funded privately.

Since the Commission was scheduled to end in 1830, the Commissioners found themselves in a difficult position in 1829: piers started that year, or already in progress, or those for which contracts were already entered into, were unlikely to be taken fully to completion before the termination date.[41] It was necessary therefore for a new Bill to be introduced, empowering them to enter into new contracts, to arrange the continuation of existing works, and to regulate the operation of the Fishery Loan Fund beyond the impending 1830 deadline. The necessary Bill was duly passed into law in June 1829 as an *Act to amend the several Acts for the Encouragement of the Irish Fisheries*.[42] Early in

1830, Donnell made a final report on all the works, both finished and in progress. The Claddagh fishery piers and Claddagh quay, the only marine works outstanding in Co. Galway and north Clare, were still in progress, at a cost to the Commissioners that year of £1,100.

In the period since his 'final' report of 1829, a further seven piers around the whole country had been completed, fifteen were still in progress and three, fully approved, were yet to start.[43] With this positive information, the Commissioners closed their eleventh and final Report to Parliament and went out of office. It was the end of the first official, entirely autonomous, authority specifically dedicated to the administration of the Irish fisheries. The remainder of the century would see many efforts to re-establish an Irish fishery authority that was independent of other government departments, and that would have its own engineering staff.

Assessment of the Piers of the First Commission for Irish Fisheries

Since its establishment in 1819, the Commission had received 111 applications for piers and quays nationally. It had spent over £23,000 on sixty-five; the rest had been rejected or the applications abandoned. Of the sum expended, one half had come from the contributions of the charitable committees and other private sources.[44] Later, there would be criticism of the small size of the structures that were built, especially in Co. Galway, and the fact that virtually all of them dried out at low water. These features were not entirely inappropriate or unsuitable for the the small fishing boats of the time, the vast majority of which were currachs and rowboats, and they avoided the enormous cost of erecting floating docks or deep-water piers. For whom, or for what other kind of fishing boats, would larger, deeper ports really have been needed in Galway or Galway Bay at that time? Criticism would come, too, of where some were sited: there was little evidence that they were particularly convenient to known fishing banks, and often there were no access roads leading to them. The road to Rinville pier at New Harbour, for example, was not made until many years after the quay and pier were erected. At St Kitts (Killeenaran), the pier stood detached from the adjacent land and could not be accessed by road, even if a road had existed. Finally, there would be criticism that many of them were used more for agriculture, trade and general communication than for the fisheries. This last was certainly true, but not necessarily a negative feature, as Donnell recognised and expressed trenchantly:

> The finished works answer the expectations in extending the fisheries and promoting employment, and some of them far exceed the anticipations of general improvement, by the introduction of agricultural commerce and merchantile enterprize into districts which before were totally unproductive of useful employment, from the want of loading or landing places, as well as the want of roads for the transit of agricultural

> produce, and of the merchandise introduced in return for it ... Numerous small and cheap structures, bringing accommodation for employment and industry to the doors of the population in wild and sequestered districts, where extensive tracts of mountain land lay uncultivated, were better calculated to promote employment, and improve the condition of the labouring class, than a few expensive refuge or asylum harbours would be on a few points of the coast.[45]

That was praise indeed, expressed by someone who had not always approved of Nimmo's designs, funding and execution of the works. It acknowledged the contribution of the first Commission for Fisheries to the development of the maritime and agricultural infrastructure of the west coast during the difficult decade before the formation of the Irish Board of Public Works (the OPW). It was a view that would continue to be voiced for the rest of the century by those who best understood the needs and nature of the indigenous coastal communities. Nimmo's works had laid a foundation for all future pier building in Co. Galway, and places like Claddagh, Cleggan, Clifden, Roundstone and Burrin New Quay in Co. Clare would remain pivotal sites for fishery and other maritime developments right to the present day. Outside of these, few new locations of major importance would develop later in the century. Those piers that were built later – there certainly was a superabundance of them – were generally small and were satellites to Nimmo's piers; few of them would attain, or retain, the importance of his original locations.

The piers and quays of the first Fisheries Commission were generally very well constructed of good ashlars and proved to be remarkably long-lasting; most are still in use to this day. The administrative conditions attaching to their financing with public money – maximum grants of three fourths of cost; a cap on total expenditure per site; the demand for a local contribution; the acceptability of charity as a source of that contribution; and the requirement for a continuing maintenance charge – proved just as long-lasting. The story of the piers for the rest of the century, and their very size and suitability, would be determined largely by these administrative conditions.

A list of all the applications for piers and quays made to the Commissioners for Irish Fisheries between 1821 and 1830 for all counties of Ireland, and their outcomes, is given in Appendix B. Of 111 applications from fifteen counties, fifty-six were completed or in progress in thirteen counties. Fifty-five had been rejected or had failed to be undertaken for various reasons. A further nine would be started later, to make the final total of sixty-five completed or nearly so.

Piers of the 1822 Famine

West Connemara

The three most significant and most resilient piers erected by the Commission were located in west Connemara at Cleggan, Roundstone and Clifden.

Cleggan Harbour

Cleggan was one of Nimmo's chosen locations for a village that would be built around a new pier of his own design. His chosen site was a small bight on the south shore, near the mouth of Cleggan Bay at the place where his new road from Clifden terminated. The place belonged to his friend Thomas Martin of Ballinahinch. There was a boggy hollow there, which Nimmo proposed to cut into and to excavate as a small dock of about 100 feet by 60 feet. On the north side of the entrance to this dock, he designed a sheltering pier, 200 feet long and fourteen feet high at the head, projecting east-west, with a stony, sloping sea-pavement on its seaward, northern face. Opposite this on the south side of the entrance to the hollow, he built a wall projecting south-north, completely closing off the bog hollow from the sea, but leaving the narrow entrance cut into it.[46] His coloured sketch of the site is conserved in the National Library of Ireland.[47] An uncoloured copy was engraved and published in his report to the Fishery Commissioners.[48] Construction was put in the hands of Alexander Hay, a young architect in his employ. Tolerable progress was made initially, but during a storm in October 1823, the entrance to the dock filled up with beach gravel and the pier was severely damaged. By then the money had run out and all further work had to be suspended. New funds were approved in January 1824 and Nimmo sent to Dunmore in Co. Waterford for Alexander McGill, a well-experienced harbour builder, to take over supervision of the project. McGill successfully completed the work, widening the south-north wall in an eastwards direction and sloping it down to the sea. That prevented further encroachment of beach material into the dock entrance which, located between the pier and the newly widened wall, was converted into a short stone-lined entrance channel.

Despite some ongoing problems, Nimmo wrote to the Commissioners in March 1824: 'As a result of work at Cleggan, last week there were constantly above 100 boats there, busily engaged with the herrings. The road into Connemara is quite cut up by the horses of the fish jobbers. The new road I have made admits carts to pass with salt from Roundstone to Cleggan.'[49] The engineer James Donnell visited Cleggan in 1825 and suggested some small alterations to the dock. He also drew up a new plan for the entrance channel, mainly a further widening and raising to full height of the south wall, which he rightly called 'the beach retaining wall'. The plan is shown in his report to the Commissioners.[50] Although his alterations were not implemented immediately (due to a shortfall in the local contribution), Donnell found the works in good order when he examined them again in 1829.[51]

In another engineer's report in 1833, Cleggan was reported to be 'in a ruinous state', although useful when it had been in good order.[52] Repairs were duly carried out in 1835–37. Cleggan did not receive any further public financial support for almost another fifty years. It vested in the OPW in 1848, as required by the second Piers and Harbours Act,[53] and it was listed in the schedule of piers that vested in the county by an Act of 1853.[54] Apart from these, Cleggan did not appear in published records during the famine period or its immediate aftermath, and would not feature again until 1880. Its history from then is given in chapter 11.

Roundstone Harbour

Roundstone pier was one of the most successful of Nimmo's marine works in Co. Galway and probably his own favourite site; as early as 1813, he had identified the place as a suitable location for a village. Because of complaints about damage to property during the initial construction of the pier in 1822, he bought-out the lease of the farmer occupying the land and became tenant himself under Thomas Martin of Ballinahinch. As originally conceived, the harbour works consisted of a wharf 150 feet long, with a short approach road leading to it from the main road at its landward end, and a jetty sixty feet long at its seaward end.

Work commenced in 1822 under Alexander Hay. Granite was brought in by boat and the wharf was completed roughly to high tide level the following year. Alexander McGill took over as supervisor in 1824 and greatly enlarged the works, using funds provided by Nimmo from his Western District allocation, so that the wharf was extended finally to 217 feet and the jetty pier to 150 feet. The place was soon busy with the transport of seaweed and kelp, the landing and curing of herrings in season, the sale and distribution of salt and general trade. Situated near Slyne Head, it was a place where sailing boats could shelter before attempting passage around the Head, or rest up, having completed it. It was in considerable use when Donnell visited in 1824, but the pier still needed a parapet and its deck needed to be raised above high tide level. That necessitated a new coping layer, work that was carried out in 1825.[55] When he visited again in 1826, the sea-pavement of the pier needed some trifling repairs; otherwise it and the quay were satisfactory.

Sites for houses had been laid out nearby and had sold quickly, so that a village was already springing up.[56] Nimmo built a storehouse and workshop there for his government-funded road projects and these buildings formed the nucleus of the new village. He must have been pleased with what was now taking shape. His brothers, John and George, and his other assistants, lived there as work and circumstances permitted, and the Nimmo brothers came to live there permanently after Alexander's death in 1832. Alexander is thought by some to be buried in Roundstone but there is no hard evidence for this. Nimmo's wharf and pier would eventually comprise only one half of the full harbour of Roundstone, the later development of which will be described in chapter 11.

Clifden Harbour

Clifden harbour was to be the last of Nimmo's harbour works to be brought to completion. In 1821, John D'Arcy, the founder and proprietor of Clifden, then called Clifton, applied to the Government for money to erect a quay at the proposed site of the village. He was given a grant of £230 on condition that he contributed a like sum. D'Arcy did nothing that year, and when famine occurred in 1822, Nimmo was instructed by Government to start relief works immediately in the district. About one mile west of the village, there was a small gravelly spit and island at a place called Doaghbeg, where fishermen used to unload their catches. Straight away, Nimmo built a low wall along the outer, western edge of the spit, making a temporary but substantial landing wharf at a cost of £50.[57] Boats could also lie behind the spit for greater security, sheltered from the open sea by the new wharf wall and the island, so that the wall served a double purpose.[58]

Having first set men to work on this, he turned to the main quay proposed by D'Arcy, and drew up a detailed design for the work. It consisted of a very obtuse V-shaped, longshore quay: the arm above the angle was in shallow water and this was to be a quay for small boats that could lie aground there at low tide; the arm below the angle extended into deeper water and this was to be a ship quay, capable of accommodating trading ships of a larger and deeper sort, at most states of the tide. The angle between the two sections was necessary for the structure to follow the natural contour of the shore at the site. Altogether the whole structure, a plan of which is shown in the 8th Report of the Commissioners,[59] was designed to be 600 feet long, built in stone, at an estimated cost of £445.

The plan soon changed when D'Arcy started construction. The boat quay section was excavated deeper than originally planned, so that costs rapidly increased beyond budget. Nimmo had to contribute an extra £342 from Western District relief funds to maintain operations and eventually 490 feet of the quay was reportedly constructed. When Donnell visited in 1825, the boat quay above the angle was substantially complete for 175 feet, with a further 114 feet needing to be raised about two feet to the designed height. The angle and the quay below it – the proposed ship quay – had not yet even been started.

Donnell recommended that the boat quay, already almost fully built, should be completed and the proposed ship quay should be abandoned entirely. In 1827, he drew up a plan and specifications for perfecting the works in accordance with this new scheme.[60] John Killaly surveyed the works in 1830 and drew up another new plan for further developments, which also is still extant.[61] There are indications on this that Killaly proposed to excavate the seabed close to the ship quay wall in a way that would be done many years later. However, nothing further was done in 1830 and Clifden quay remained in this incomplete state for a significant number of years, although Captain Boileau of the Coastguard carried out some relief works on it in 1830–31;[62] these were probably essential repairs after the severe storms of that winter.

Nimmo's original plan for Clifden would not be taken to completion until much later in the century, as we shall see in chapter 11. But even from its earliest days, the quay, as it then stood, proved more useful for general trade, some of which was reportedly with Liverpool and with ports in North America. Its location, far inside Ardbear Bay, was inconveniently distant from the fishing grounds so that it was never of much use for fishermen. It would, however, become a convenient place for the landing and dispatch by rail of lobsters and shellfish towards the end of the century, which will be discussed in a later chapter.

FIGURE 2.2. John Donnell's plan for Clifden, 1827. OPW/8/84, with the permission of the Director, National Archives of Ireland.

The Piers in Inner Galway Bay

Four of the Fishery Commission piers built during 1820s were situated in inner Galway Bay: Rinville (New Harbour); Ballinacourty; St Kitts (Killeenaran); and Burrin New Quay in Co. Clare.

Rinville (New Harbour)

The clean, reasonably well-sheltered anchorage at this place near Ardfry was known as New Harbour from at least the seventeenth century. It is named *Havreneuve* on a French maritime chart of 1690, as a place for the safe careening of ships.[63] Boats moored there, awaiting entry to Galway port, or awaiting suitable outbound sailing conditions, having cleared customs from Galway. The naval vessel HMS *Spy*, lay at anchor here for periods in the 1730s.[64] There were no landing facilities or other manmade structures, although the site is recorded as a quay in Larkin's 1819 map.[65] Here, Nimmo designed and built a wharf, 150 feet long, extending 60 feet out from the shore. It had a short, twenty-three-foot long jetty at

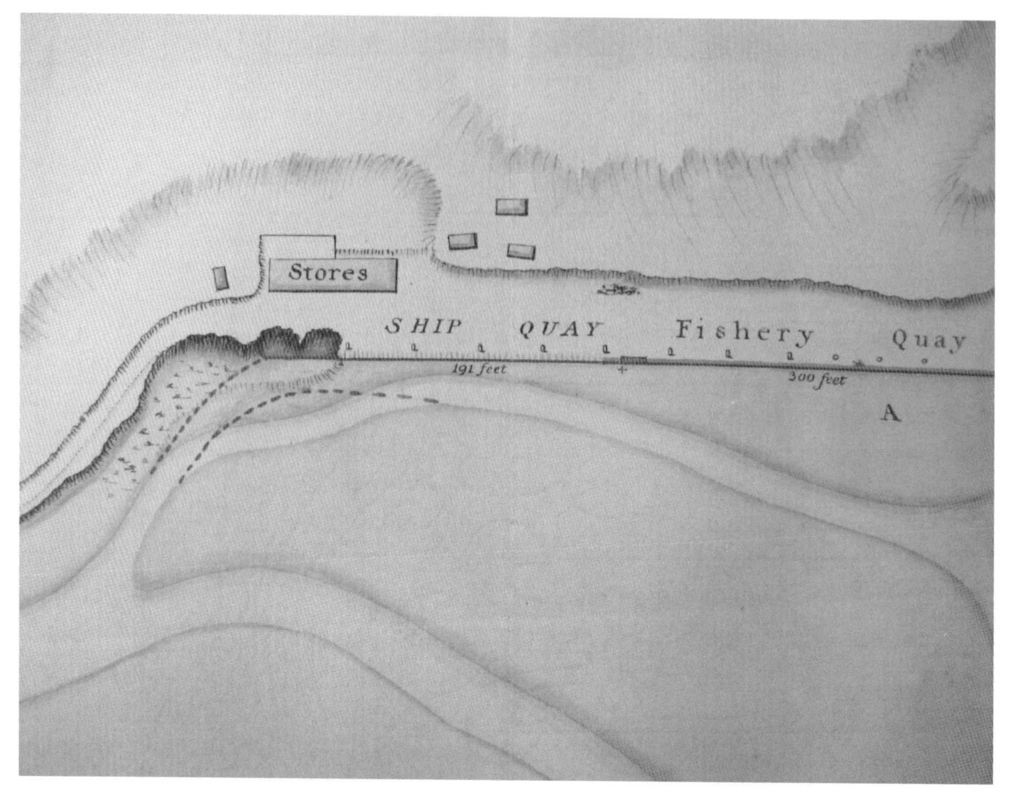

FIGURE 2.3. John Killaly's plan for Clifden, 1830. OPW/8/84 with permission of the Director, National Archives of Ireland.

the seaward end, and a shingle slipway at the landward end. The structural work was faced in cut limestone, laid in regular courses and was generally well made. It was a simple enough structure, a plan of which is shown in the 8th Report of the Fisheries Commissioners,[66] and a coloured manuscript plan drawn by Nimmo is conserved in the National Archives.[67]

The new quay was an immediate success, especially for the turf trade, despite the fact that there was no roadway to it for many years. The engineer, James Donnell, who had succeeded Nimmo by then, recommended some small repairs in 1827, but otherwise the works were structurally sound. Later, the deck, normally submerged at high Spring tides, was raised some feet, and the entire structure has been further capped with concrete in recent years. Otherwise it remains essentially as Nimmo designed it, and today it is in constant use for leisure and yachting purposes, the site being the location of the Galway Bay Sailing Club.

The Short life of Ballinacourty Pier and Harbour

The pier at Ballinacourty was much less successful and is one of the few structures by Nimmo that is no longer extant. The exact site was located at Carrowmore, on a small peninsula projecting into inner Galway Bay, sheltered by Mweenish Island and Island Eddy, part of the estate of Walter Blake. The precise location of the pier is named '*Poulnalui*' ('*Poulnaloinge*'? = 'ship pool') and elsewhere 'Ship Pool' on Nimmo's outline plan.[68] It, too, was recommended as a careening site in the French maritime chart cited earlier; indeed that chart indicates a remarkably good knowledge by its authors of the coast of inner Galway Bay. Nimmo considered this place to be potentially one of the finest sites for a boat harbour in the inner bay, having, as he wrote, 'fourteen feet at low water, with mud and grassy banks'.[69]

Today, *Poulnaloinge* is very shallow and unsuitable for any but the smallest craft; the whole shore area dries out at low tide and forms an extensive strand used for intertidal oyster cultivation. On the west side of the little peninsula, there was a small lagoon, separated from the sea by a gravel spit. (It is still shown like this in the second edition of the Discovery Series of OS maps, dated 2003.) Nimmo proposed to make a cut, west to east, through the spit into the lagoon and erect a stone quay wall 154 feet in length along the south side of the cut. The lagoon would then become a useful lying-up dock for small boats, the quay would provide a wharf, and an 80-foot extension of the quay, decreasing in height down to low water, would act as a convenient half-tide landing wharf. If desired, this sloping extension could be built up later, he wrote, to form a full-height quay right

FIGURE 2.4. The remains of Nimmo's Ballinacourty pier (RH 151X) in 2014. Photo NPW.

down to low water. It was an ambitious but straight forward plan; a coloured copy with specifications signed by Nimmo is conserved in the National Archives[70] and an outline plan is shown in the 8th Report of the Fisheries Commissioners.[71]

Funds were provided for the work in 1822 by the Fisheries Commission, the Dublin Castle Relief Committee and the London Tavern Relief Fund. Messrs Plunkett and McDonagh, contractors, also signed the coloured plan, suggesting that they were the ones initially chosen to carry out the construction. But the work never advanced very far, perhaps understandably, as only £276 in total had been made available for it.

In March 1824, extra funds were allocated, Nimmo drew up new specifications, and the work was put out to tender once again. Plunkett and McDonagh now offered to build it for £500, and different contractors, White and Coen, for £389. It appears from the extant records that the latter were then the successful tenderers. However, progress continued to be very slow; Alexander Hay, Nimmo's supervisor, reported little advance by September 1824. By then all the extra money had allegedly been fully spent.

When Donnell visited towards the end of that year, he confirmed that work was at a standstill and that the money was indeed gone.[72] His report shook things up: by June 1825, Hay was able to report that the work was substantially complete to Nimmo's new specifications. Donnell confirmed this when he revisited in January 1827 and found only minor repairs were needed.[73] He included a plan of the works as they then stood.[74] Nimmo, too, had confirmed in November 1826, that the work was finished to his satisfaction and he proposed that the northern side of the cut could be given a groyne or pavement, should the harbour begin to be well utilised. It never was. As early as mid-1827, it needed more money for repairs. These were done, and bollards were added, but there is very little evidence that the wharf and dock ever served any useful function after that. Donnell, who was not given to positive overstatement, reported that it was a place for landing and curing fish, but that is the sole reference to its use in the fisheries that has been found in the records.[75]

The documentary confirmation that the construction was fully completed to both Nimmo's and Donnell's satisfaction is important in light of the subsequent fate of the structure, which was one of the harbours that was virtually demolished by the extreme weather of 1830–31.[76] The next we hear of it was in an engineer's report of 1833, where it is recorded as 'never properly finished, owing to which the works have been completely destroyed'.[77]

A structure, labelled 'new quay', is shown at the site in the first, 1839, edition of the six-inch OS map of the area, but in a way that suggests it had almost disappeared by then. A 'ghost' quay today, the final remnants of Nimmo's Ballinacourty works are evident as a straight line of cut stone blocks lying on the shore leading down to the sea from the lagoon. The cut into the lagoon has been closed off completely with shingle, and the lagoon itself has dried out, now evident only as a large, boggy depression in the land. The line of stones on the shore marks the last vestiges of this quay.

In an earlier work,[78] this writer incorrectly implied that Lynch's (Carrowmore) pier, situated on the eastern side of the little peninsula, was the modern representation of

Nimmo's structure. The precise site of Lynch's pier was shown as a quay in Larkin's 1819 map, but no manmade quay or pier existed there until the 1880s. The present Lynch's pier certainly superseded Nimmo's from a functional, but not from a physical, point of view; the documentary and physical evidence confirm the precise location, full completion and ultimate dilapidation of Nimmo's original pier. Lynch's pier, also called Carrowmore or Ballinacourty pier, is an entirely different structure, was built many years later, and will be treated in a later chapter (chapter 6). When accessing the records, Ballinacourty pier in Co. Galway is not to be confused with Ballinacourty pier in Co. Waterford, which was transferred from the OPW to Waterford County in 1854.[79]

St Kitts Pier and Unique Harbour at Killeenaran

St Kitts is the name given in the official records to the pier built in 1824 at the village of Killeenaran on the south shore of inner Galway Bay. At that time the village, on the estate of Arthur St George, was said to have 300 inhabitants. Nimmo designed an L-shaped pier, consisting of a jetty 120 feet long with a short return head. Its most distinctive feature is a stone-arched bridge incorporated in it near its landward end, under which small rowboats could pass into a natural sea-pond alongside, sheltered by a raised storm beach.[80] Nimmo intended this pond as a dock or 'harbour of refuge' for small craft. Work started in 1822, but it was soon abandoned, only partly completed, when funding for relief works was terminated.

At the request of the Fishery Commissioners, Nimmo made new plans in 1824 for the completion of both St Kitts and Ballinacourty, just across the bay from it. New contractors, White and Coen, took on the work at both sites.[81] When Donnell visited in late 1824, St Kitts was almost finished, but was still not attached to the land, and had no road access to it. Soon after that, White and Coen repudiated the contract, probably because the money had run out again and they saw little prospect of the works ever being restarted. Donnell did not approve any further work there, on the grounds that the pier was useful only for turf and seaweed landings, and he felt that only the persons engaged in those trades, not the Commissioners for Fisheries, should pay for its completion.[82] The connection to the land – in effect the making of a causeway 50 feet in length – was eventually made, but it did not improve the pier's usefulness for fisheries.

As it stands now, the pier is straight for most of its length, with a right-angled, upstream turn at the head which has a set of cut limestone steps. The quay face is made of good limestone ashlars, and the soffit of the arch is well executed in the same cut stone, well worthy of note; the sea face of the pier, steeply battered at the angle, is made of local rough stones set diagonally on edge to form a sloping sea-pavement. The sea-pond that was to form the 'harbour of refuge' is located at the exposed, seaward, side of the pier, still protected by a stony storm beach; no boat larger than a small rowboat or canvas currach could ever possibly have entered it, due to the low elevation of the entrance arch. The problem is that the clearance under the arch diminishes as the tide rises; by the time the water level rises sufficiently to allow boats of even shallow draft to pass through, the overhead clearance is insufficient for any but the very smallest craft.

The pier itself, and the sea-pond, dry out at low water and can only be approached from about mid-tide onwards. The clearance problem may have been overcome originally by frequent dredging underneath and around the arch, but all in all, it is difficult to see how the entrance to the pond was ever fit for purpose. The general design was one that Nimmo proposed also for the Claddagh in Galway and for Old Head Pier in Clew Bay, but St Kitts was the only location where a design of this type was actually implemented. As the only example of its kind, with the original design and specifications dating from 1824 still extant, the structure is unique and merits care and conservation. It is still in use by a small number of boats.

Duras Quay and Pier

This place is noted as a 'quay' in Larkin's 1819 map of Galway, when it was a natural landing place with no evidence of any built structures. Patrick Lynch of Duras Park commenced building the original, 150 feet long, quay wall in 1823, using locally available, roughly shaped stones laid in level courses without mortar. To build the jetty pier, a much more demanding task, he sought the help of the Fisheries Commission. Nimmo designed and supervised the construction of an angled pier of fine limestone ashlars without any pointing, laid nine courses deep. It is about 100 feet long from the root to the elbow, and a further forty feet from the elbow to the head. The ashlars are restricted to the wharf face and head; the sea face is made of roughly hewn stone blocks set horizontally in the lower courses and on edge in the upper two. The sea face is steeply battered at the elbow, less steeply elsewhere. Viewed from close up, the pier head is quite commanding. A storm wall about 3 feet high surmounts the sea edge of the entire length of the pier.

Lynch was sufficiently happy with the structure to insert a commemorative plaque (to himself!) in the inner face of the storm wall, which reads: 'This pier was erected by Pat^{k} M Lynch Esq. of Duras Park. A D 1823'. It failed to acknowledge that the Commissioners for Irish Fisheries had contributed £157 to its erection and that Nimmo, too, had helped. Because of this omission, Duras was never mentioned subsequently in the reports of the Commissioners, and therefore the place has little documented history of use in the fisheries. It was used mainly in shipping goods and agricultural produce between the south coast of Galway Bay and Galway city and it gave its name to the steamer SS *Duras* that plied between Galway city, north Clare and the Aran Islands later in the century. Although in very little use today, the structure is in good, original condition and is a fine example of masonry work from the 1820s. The ashlar wharf and head, the turning of the quay wall at its junction with the pier masonry, and the battered sea face are all features of note and well worthy of conservation.

The small harbour is fully silted up today, making it appear almost incongruously land-locked at low tide. Lynch's plaque is still in place, halfway along the storm wall, still legible

FIGURE 2.5. Duras Pier and Quay. Top left: the quay; bottom left: sea face of the pier; top right: where the pier abuts on the quay; bottom right: the fine limestone ashlars of the wharf face of the pier. Photos NPW.

but almost totally obscured by weeds during summer. Lynch's family was associated with four separate piers – at Moneen (Lough Atalia), Ballyloughaun, Carrowmore and here – and this is the only one bearing a plaque commemorating the family's contribution to local piers and harbours.

The Piers at Burrin, Co. Clare

Nimmo's only quay on the north Clare coast of Galway Bay, one of his most commercially successful, was at Burrin near the mouth of Aughinish Bay. It consisted of a longshore quay, 105 feet in length and extending 72 feet out from the original shoreline, with a 70-foot jetty pier at its seaward end.[83] Built out from the land as much as it was, there was an appreciable depth of water at the quay and jetty for most of the tidal cycle, adding greatly to their usefulness. It soon became known as Burrin New Quay, to distinguish it from an older quay it replaced, located some few hundred yards further west, close to Vernon Mount Lodge. The latter quay subsequently became known as Burrin Old Quay, and it is shown in the first edition of the six-inch OS map.[84] Today, nothing remains of Old Quay; when the Flaggy Shore road was built through the site, it obliterated entirely the quay and its surrounds. In 1986, a stone-lined rectangular enclosure, measuring 48 feet by 24 feet, the remnants of an old oyster pond dating from about 1868, still existed at the bottom of the field fronting Mount Vernon Lodge. The pond was cut off from the seashore when the road was built, but it still filled with sea water at high Spring tides.[85] One wonders whether the oyster pond might have been the innermost part of the original Old Quay? It has now been filled-in and fully grassed over since it was first recorded in 1986, but some simple excavation might shed further light on the possibility that it is, indeed, part of Burrin Old Quay.

In October 1883, Rev. H.W. Skerritt of Finavara and J.L. Baggott of Vernon Mount Lodge applied for a quay to be built at a site 'as nearly as possible opposite the police barracks at Old Quay'.[86] They believed such a quay would be good for trade and for visitors on excursion, as Nimmo's quay at New Quay was said to be dilapidated. The directors of a company running a steamboat service in Galway Bay were refusing to let their ship enter New Quay 'owing to her having met an accident there because of deficient accommodation on a former occasion', they alleged. The proposed quay would also be useful, they claimed, for bringing young oysters to the Burrin Oyster Fisheries beds and for landing turf from Connemara. The application eventually proved unsuccessful and Nimmo's New Quay remains to this day the only functioning pier and quay extant at Burrin.

Trouble in the Bay

Burrin New Quay was a pronounced success from the very beginning. Agricultural produce was loaded there for transport to Galway and a packet service sailed to the city throughout the summer in its early years. Donnell said of it that 'I place it in the second class, as a mere fishery harbour, but for general utility it stands in the first class', observing that 'it is a substantial and useful work, but sometimes too small to contain the numerous craft seeking shelter in it'.[87] The fishermen of Co. Clare, who used it regularly, complained that Galway fishermen (from Claddagh) imposed their own rules on fishing times and seasons in the Bay, so that the Clare men could not fish or, if they did, they could not sell their fish in Galway 'except at the risk of their lives'.[88] They asked the Fishery Commissioners to place a fishing vessel, 'about the size of a Torbay trawler' in Galway Bay, manned by officers and crew

> who are fishermen, some from Scotland and some from England and some from the east coast of Ireland and provided with every implement of the most approved kind

for taking and curing fish to investigate the Bay and its fish, in order to expose the stupidity of the Galway fishermen, who will not fish for herring before the 12th of August, despite their abundance in the Bay.

The memorial was signed by eminent landowners and merchants, *inter alia*: Thomas Browne, Francis Gore, Bindon Blood, Joseph Skerritt, John Ormsby Vandeleur and Robert Persse snr. Complaints like this, against the customs and traditions of the Claddagh fishermen, were common during the early nineteenth century. Claddagh village, poor as its facilities were before Nimmo came, was the centre of the fishing trade in Galway at the time, and its people were notoriously superstitious regarding when and where to fish in the Bay. However, they were hardly stupid, as alleged. It is far too often overlooked that there was no market of any serious size for fresh fish anywhere in the West, and very poor transport for perishable products made it impracticable for fresh fish to be taken elsewhere for sale. Galway city was the only real market outlet, and it is understandable that the Claddagh fishing community might have tried to monopolise that small market for itself in any way it could. Attempting to retain a monopoly for themselves by controlling fishing activity in the Bay is not far removed from dubious competitive commercial practices in other ventures – although two wrongs never made a right.

Bindon Blood, one of the memorialists, was also engaged at the time in building a road from north Clare through the Burren to Ennis and Limerick that might have given the Clare fishermen a different market outlet; but the Limerick fishermen would hardly have been any more welcoming of competition from outsiders than the Claddagh fishermen. Nimmo had great plans for Claddagh, which he first advanced in 1822, but which were not completed until the time of the Great Famine, long after he had died. For that reason, and to keep the narrative of the Claddagh intact, the full account of the Claddagh fishery piers and quay, home port of the alleged offenders, is deferred to chapter 4.

Some Piers of the Aran Islands

Killeany Pier and Quay

Neither Claddagh nor Killeany, the latter located on *Inis Mór*, had been on the initial list of thirty-three piers drawn up by Nimmo for the Fishery Commissioners. It was only after a deputation of Commissioners had visited the West that these, and three other sites, were added to the list. Nimmo had already been to the island and, on seeing the distress there, had immediately set men to work, under the supervision of his brother George, quarrying and dressing ashlars for use in public works on the mainland.[89] George's wages sheets for this enterprise are extant. They show that labourers were paid eight to thirteen pence per day, and stone masons one shilling and sixpence to two shillings and sixpence per day. That was very good money then, and no doubt it greatly helped the recipients who were wretchedly poor.

The quality of the stonework, as evidenced by the hand-curved slabs on the sea slope of the Slate pier in Galway, has already been remarked upon. When the Commissioners added Killeany to the list of sites to be funded, Nimmo drew up appropriate plans for a pier and quay. The village was mainly a fishing community, with a population that Nimmo estimated to be about 1,000 persons. Cod and ling were landed there and dry-cured on the limestone pavement, an activity that continued into the late twentieth century.[90] The total sum allocated for the pier and quay was only £230, well below the average of £414 per site allocated over all the thirty-eight sites. Approximately £116 came from the Commissioners for Fisheries and £114 from the Dublin (£54) and the London (£60) charities.[91]

Welcome as it was, the money would obviously not last long, especially for the costly structures that Nimmo usually designed. A pier of 245 feet with a canted head was built and joined to the village by a roadway quay, 326 feet long, all clad in hewn limestone. The money did indeed run out after only a few months due, Nimmo claimed, to his plans being misunderstood or not being followed properly. In January 1824, a further £462 was granted and the work started up again in October under the direction of James Donnell. According to him, he modified Nimmo's plans, put the work out to contract, and monitored its progress; it was finally completed by the end of 1826.

The new pier had an immediate effect in increasing the number and the class of boats that gathered at Killeany for the herring fishing; at one period, there were reputedly over 100 boats moored along the roadside quay.[92] Donnell even reported that in 1829 the place was not big enough for all the boats that moored there.[93] By 1833, there were reportedly thirty-nine hookers frequenting the place, and about 350 persons employed in the fish industry.[94] Although well sheltered from the open sea, the site had a number of features that created difficulties from the very beginning: the approach to it was through a very narrow channel, with rocky ledges on both sides; it was situated inconveniently far inside Killeany Bay; the tide ebbed to a considerable extent, so that boats could only move in and out around high tide; and it was some distance from the village of Kilronan and therefore was of little value to fishermen of that place. Nevertheless, it achieved its purpose reasonably well and seemed to stimulate fishing activity in its early years, as well as having an immediate effect in alleviating distress. We will return to Killeany again, and to the other piers of the Aran Islands, in chapter 12.

Maumeen Pier, Tír an Fhia, Greatman's Bay

All the funding sources, Government and charitable, were of one mind that the funds granted for piers were designed to give only instant and temporary assistance 'without encouraging a spirit of indolent reliance on extraneous relief.'[95] It was inevitable therefore, that funding would come to an end as soon as the conditions of distress passed away, which they started to do by the autumn of 1822. The Government indicated that no new grants would be made after 16 August of that year, and the London Tavern Committee resolved to wind down its charitable support at the same time.[96] Construction was, in consequence, peremptorily terminated and eventually abandoned at a number of sites at the end of the summer, with hugely detrimental physical and human consequences. The pier at Maumeen in Greatman's Bay, funded with charitable donations and with Government relief funds administered by Nimmo, is a good example. Francis O'Flaherty, agent of the landowner Arthur St George, had drawn up a memorial for a pier to be erected there, which he submitted to the authorities in June 1822. A 'memorial' was the term used for a formal petition to have a pier built; it was usually signed by many supportive residents in a locality, especially the landowners and their agents. O'Flaherty's memorial is not extant, but his letter of 12 June 1822, informing St George about it, is conserved in the archives. 'Seven persons have died of actual starvation, four of them on your estate ...' he wrote, before advising St George to solicit relief from the London Tavern Committee. He then went on:[97]

> I have received £50 from the Bishop of Tuam with which sum I have commenced to make a commodious harbour in Greatman's Bay. £300 pounds would finish it. There are upwards of fifty poor famished creatures at it now, but in about a week I think they will be stronger and therefore work better. We have been obliged to feed some of them gradually they were in such a reduced state.

Two days later, St George wrote asking the Fisheries Commissioners to grant £100 towards the work, but there is no evidence he offered anything from his own resources.[98] An undated minute in the same file signed 'William Burke, Chairman and J. Smyth, Honorary Secretary' (probably of a local relief committee) records: 'We recommend that £200 be paid by Government to A. St George, Francis O'Flaherty and Fr O'Flaherty, to be used by them, under the directions of Mr Nimmo, in building a quay at Crompaun Edward, otherwise St George Harbour, Gorumna Island.'[99] Nimmo designed a pier almost 500 feet in length with a return head of 100 feet, the plan of which is shown in his report, published by the Fisheries Commissioners.[100] The requisite rock mole had been put fully in place (a major task employing many labourers) and the pier had been walled for about half its projected length when the official funding ceased abruptly, work stopped, and the unfinished structure began quickly to fall into decay. Later, Lieutenant White of the Coastguard tidied up the site.[101]

In 1837, the OPW, assisted by William Pierce of Clifden as site engineer,[102] finished off the section already completed, leaving the pier at about half the originally designed length, but without a return head. The first, 1839, edition of the six-inch OS map shows clearly the completed and uncompleted sections, and they are also clearly recorded in the second, 1898, edition of that map. Even today, the mole for the unfinished section is clearly visible at low tide, extending out from the present pier head, and turning to the left, marking Nimmo's intended, but never completed, return head.

As it stands, the whole structure is an evocative relic of a project undertaken by starving men, supported by the human decency of some others, but doomed to non-completion by

FIGURE 2.6. View, from the head of Maumeen pier, of the original mole, lying uncompleted since 1822. Note that it turns to the left, marking Nimmo's planned return head.. Photo NPW.

the sudden withdrawal of funding in a forgotten famine almost 200 years ago. It is one of the most visually attractive, socially evocative and culturally important of all the piers in Connemara. It is still in use today, a tribute indeed to those famished men who built it; nothing else commemorates them.

The Piers and Quays near Galway City

Barna (Lynch's) Pier

The Fishery Commission's contribution to Barna pier involved restoration rather than the initiation of something entirely new. The site was originally a landing place in the townland of Freeport, a name that suggests it may have been a long-established toll-free outport adjacent to Galway city. Marcus Lynch built the first pier recorded there in 1799,[103] at a time when illicit trade and smuggling were commonplace on the Galway coast. One record suggests that the original pier was a timber structure, little more than a groyne sheltering a natural landing place. James Donnell, who first visited the location in 1824, gave a different account of the site:

> It was of considertable extent and utility. The pier extended about 470 feet and had a lighthouse on the head; it had besides quays and walling surrounding an interior dock, which quays measured about 620 feet in length; the pier was demolished in one night by a storm in consequence of the neglect of some trifling injury it had previously sustained; and the other walls are greatly dilapidated.[104]

There is no evidence that Donnell had ever seen the place before Nimmo's work had commenced and his comments made in 1824 may be a description of the unfinished state of Nimmo's works as Donnell found them. Otherwise, his comments were based on accounts by others, possibly inaccurate. A pier, lighthouse and dock as substantial as he described could hardly have been so completely demolished in one night.

Whatever the case, the original pier of Marcus Lynch was in ruins long before Nimmo's arrival in 1822. Nimmo gave £150 of Western District funds to a General Elly to start work on it, largely as a relief measure for the poor of the area. Elly was a visitor residing there 'for the benefit of the bathing' who 'very kindly' agreed to administer the money on Nimmo's behalf.[104] The Fishery Commission and the charitable bodies added about £277, enabling substantial reconstruction to be carried out. This amounted to a completely new pier being built in stone to a length of 360 feet. That was about three quarters of Nimmo's designed length, but the pier head was left unmade. Nimmo drew a sketch plan for the completion of the work; it is shown in the 8th Report of the Fishery Commissioners.[105] When work was substantially complete, the pier afforded reasonable shelter, especially in westerly and southwesterly winds, but the quay walls were imperfect in places and the pier head was still entirely unmade. That was the state of affairs that faced Donnell when he visited in October 1824.

Donnell drew up a new plan for the completion of the pier head and the repair of the walls, shown in the 10th Report of the Fishery Commissioners.[106] As well as showing Nimmo's work, the plan shows a small, walled lay-by at the mouth of the stream that enters the sea beside the root of the pier. Some local opinion holds that this was the location of a mediaeval landing place or dock, which gave the ancient name 'Freeport' to the townland. If so, it does not appear that any vessel bigger than a currach or a small hooker could ever have loaded or offloaded there. Nimmo's plan in the Commissioners' report shows a small, unlined pond in the stream near its mouth, whereas Donnell's plan shows a stone lining there. This discrepancy suggests the stonework was a later construction and not a mediaeval artefact. Today, the stream mouth is arched over by a cut-stone arched bridge of unknown but not ancient date, and the site of the alleged dock upstream of it has been obliterated by modern development.

Donnell's plan is useful in showing the state of the pier head as it existed in 1824: there was an accumulation of large stones at the sea end of the uncompleted pier, which had probably been placed there as material for the intended pier head and lighthouse.

The Commissioners for Fisheries considered bringing the head to completion in 1828 but, there being no prospect of any local contribution, nothing was done.[107] A full account of the later development of Barna pier will be given in chapter 3.

Recorder's Quay, Salthill

Some years before Nimmo's arrival, James O'Hara of Lenaboy erected a small pier at Salthill, a little west of Galway city. Shown in the 1839 and the 1898 editions of the six-inch OS map,

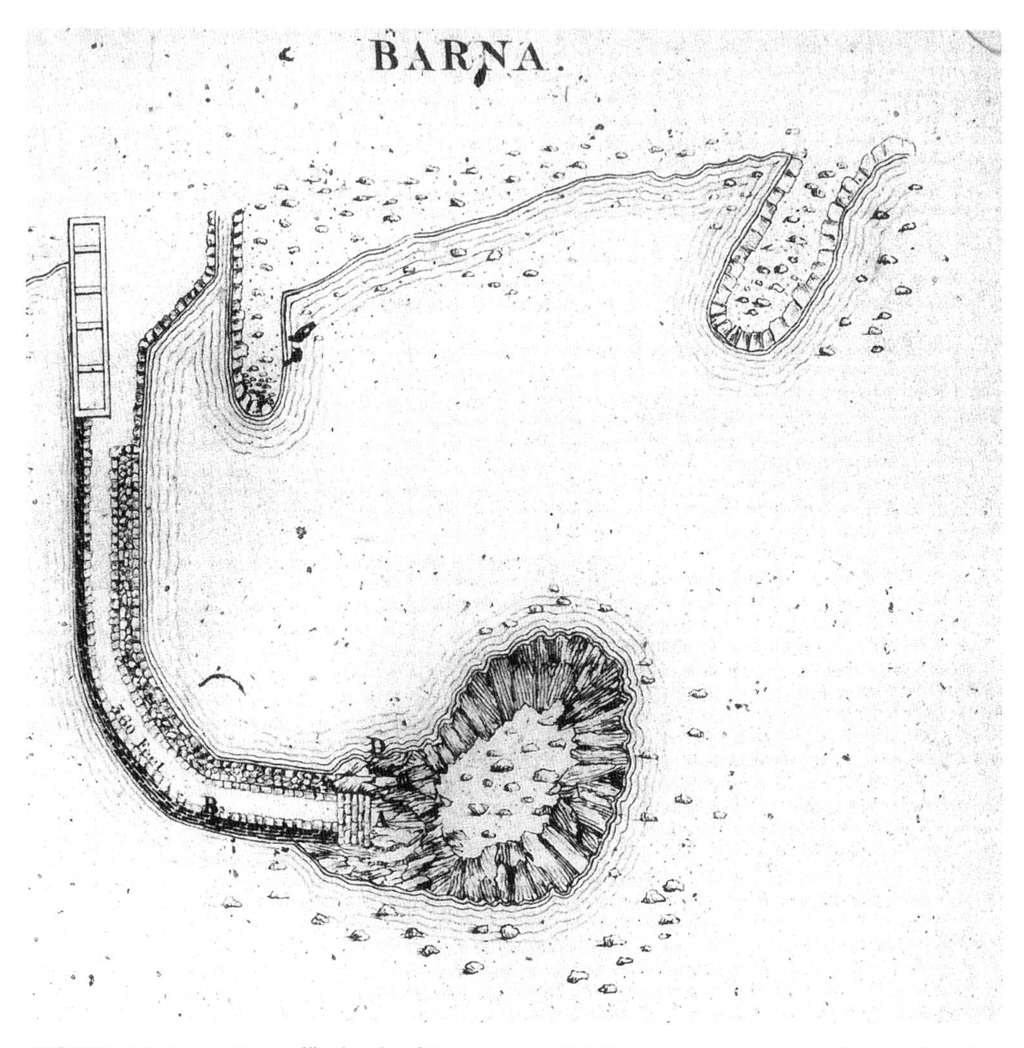

FIGURE 2.7. James Donnell's sketch of Barna pier, 1829. The pier was substantially complete, but the head was not finished. Note the stone-lined entrance to the small stream and the absence of any bridge over it. Reproduced from the 8th Report of the Commissioners for Fisheries, BPP 1827 (487).

it was known as Recorder's Quay, since O'Hara held the official position of Recorder of Galway, a public legal office then of some importance in the governance and legal system of the city. According to Nimmo,[108] the pier was too small to be of much use to hookers, but no doubt it served the needs of smaller boats reasonably well. O'Hara held lands in Connemara and the pier may have been built to facilitate transport and communication between the city and those lands. It lost its importance and usefulness once the city docks and Claddagh piers were developed. Its exact location has been entirely filled in and the shore reclaimed in the modernisation of the suburb of Salthill, so there is no longer any trace of it. It was located close to present-day Beach Road, broadly where the Galway Business School is today.

Moneen (Lynch's) Quay, Lough Atalia, Galway City

One mile to the east of Galway city, a pier was erected during the 1820s, principally for landing turf and seaweed.[109] In the only documentary record found of it, the pier is named 'Oranmore', which is the name of a village more than five miles southeast of the city. 'Rinmore' (Renmore today) is the name of a townland that is only one mile east of the city, beside Lough Atalia, and the pier there, known as Moneen pier, situated at the head of the sea lough, is almost certainly the pier in the record. The locality nearby is named Moneenageisha (*Móinín na gciseach*, 'the grassy patch of the wicker baskets'), probably from the wicker baskets used to carry away the considerable amounts of turf and seaweed reportedly landed there. Because it was not exclusively, nor even primarily, for fisheries, it received no funds from the Fishery Commission, but was financed in part as a relief work by the Dublin and London charities; for this reason it is never mentioned in any official fishery reports, then or later. It is located beside Renmore Lodge (now the Holy Family School), the residence of the Lynch family of Renmore who were large landowners and important merchants of the city. That family provided most of the cost of construction, so the pier was known until recent times as Lynch's Pier.[110] It is still extant and, having been partially restored in recent years, is an attractive feature at the head of the lough, although few persons are really aware of its existence or its origin. It is built of regular, generally large but uncut stone. Coped with large, rough-cut limestone blocks and grassed on the deck, it extends outwards about 70 feet and then angles to the left for a further 45 feet. There is one set of stone steps placed at the angle.

Now within the boundary of the present city, and associated with the locally noted Lynch family (one of the tribes of Galway), it merits greater local attention and appreciation as an an artifact of the severe local famine of 1822, well described in the letters of H.S. Persse, a member of another well-known Galway merchant family of the time.[111]

A Pier Proposed for Salthill

Power le Poer Trench, Anglican Archbishop of Tuam, leased a summer house at Seamount in Salthill. In June 1827 he wrote to the Fishery Commissioners requesting that a pier and landing place be erected at Blackrock, Salthill, for the benefit of the local fishermen.[112] The exact site was 'just opposite this house' but in order to avoid any suggestion of personal advantage he was careful to point out that he had no permanent interest in the house, his tenure being due to expire in October 1828. His declared motive was that the proposed pier would 'leave behind a permanent service to the many poor fishermen' of the area. He wrote again to the Commissioners in July, this time generously offering to pay the full local contribution, and to go security for the cost of the pier's maintenance for three years. Henry Cashel, the local Inspector of Fisheries in Galway, supported the application, as did William King, the regional Inspector of Fisheries for the western region. Cashel surveyed the site and estimated that a pier and landing place would cost £300.

James Donnell, by then the Fishery Commission's harbour engineer and a man not much impressed by speculative memorials, was sent to examine the place. He immediately pointed out that the site was completely unprotected and was located only two miles from Claddagh to the east, and two miles also from Barna to the west, both locations with good sheltered harbour facilities for fishermen. Accordingly he did not favour the application, expressing the opinion that the money could be better spent at either of the alternative piers mentioned. The Commissioners took his advice and the application was rejected in April 1828. The site is now a bathing and diving place, popular with Galway residents and visitors.

Le Poer Trench was a concerned humanitarian and this was only one small example of his efforts on behalf of the poor during his tenure of the Anglican See of Tuam, from 1817 to 1839. Some other landed and influential persons also made applications for piers during this time and were similarly unsuccessful. For example, Bindon Blood, a landowner in North Clare, applied in July 1822 for a pier at Glenina in that county but was told that all the funds had been used up by then.[113] That was just one month before all relief funds were officially terminated.

Piers in Oranmore Bay

Roscam Quay

Some four miles southeast of Galway city is the simple, grassy pier at Roscam, called Roscom in the record.[114] This pier was also funded by a charity, not by the Fishery Commissioners, although about forty-five small fishing boats were said to use it. Its main inward traffic was in turf and seaweed and its special importance was in the outward traffic of marble (black limestone) from the nearby Doughiska quarry from which a road leads directly downhill (a very useful topographical detail at the time) to the pier. The site is named *Pollachusláin* (meaning 'the pool by the grassy bank' – *cuisleán*, or possibly 'by the castle' – *caisleán*) in the first six-inch OS map of the area. This is one of the most rustic and picturesque of the old landing places in Inner Galway Bay, comprising a broad, grassy surface, a loose stone quay wall and very primitive stone bollards. There is no boat traffic

FIGURE 2.8. Moneen pier at the head of Lough Atalia in Galway city. The steps are located in the angle of the pier where the vertical line is seen. Photo NPW.

at the place today and it is little known. A locked gate blocks access, possibly indicating it is now privately owned.

St Mary's Port

There is another quay and landing place located east of and reasonably close to Roscam, at the very head of Oranmore Bay, north and west of the river called the Frenchfort river. This is a primitive vernacular structure, consisting of a low, short, grassy, much-dilapidated quay made of rough blocks, with a small, cleared landing slip alongside. The deck is grassed and has one primitive stone bollard. Shown on the first edition of the six-inch OS map as St Mary's Port, and in the second edition as St Mary's Quay, it is still evident today, both in Google Earth images and on ground survey, but it seems to be largely unknown to persons living in the locality. No information on its use has been found. Located close to the junction of the Galway to Oranmore road leading south and west, and to the Oranmore to Claregalway road leading north and east, the quay was very convenient for the distribution of goods to and from places in east Galway.

Water-borne transport was the most convenient method of distributing heavy goods in Co. Galway at the start of the nineteenth century; what could not be carried by boat had to be carried by people, along roads that were little more than mere tracks. For this reason, landing places that were as close as possible to the origin or to the destination of goods to be transported were much to be favoured over those further away. Oranmore is situated at almost the furthest inland reach of Galway Bay, which may account for the existence and possible original importance of St Mary's Quay, and also the pier at Oranmore Castle, although there is little documented record of their use.

In the late eighteenth century, east Galway was a major sheep run, and one or both of these piers (and others in the locality like Tyrone and Roscam) were places where wool was surreptitiously loaded on to small boats to be taken to rendezvous with the larger boats of smugglers waiting in Casheen Bay and other bays of Connemara.

St Mary's Port is particularly vulnerable to being demolished by eventual development of the area behind it that borders the important road to Galway city. Roscam, bordering the same road but further off it, appears to be less vulnerable.

Oranmore Castle Pier

An application for a pier at Oranmore Castle was submitted to the Fisheries Commissioners in 1827 and rejected by Barry Gibbons, then a young engineer, on the grounds that the site was located too far from the fishing grounds.[115] To sail, or row, from the proposed pier site to the fishery, through Oranmore Bay, which dried out at low tide, would take

FIGURE 2.9. St Mary's Port in Oranmore Bay. The slip is in the foreground. Behind it is the rough wharf wall with the single upright bollard. Photo NPW

far too long and be much too unreliable to make such a pier worthwhile. Gibbons did not even recommend a preliminary survey, although he observed that large amounts of turf and seaweed were landed at Oranmore – probably at St Mary's Port. If the proposed castle site could be used to load limestone as a back cargo to Connemara, it might possibly be developed as a commercial quay, he thought. However, its suitability for commercial rather than for fishery purposes doomed the application for Fishery Commission funds to certain failure. The exact site beside Oranmore Castle was one where a pier had existed from mediaeval times and which is labelled as 'pier' in the first edition of the six-inch OS map, although no manmade structure is shown on that map. The old pier was privately owned by the Blakes of Oranmore Castle, who charged tolls for its use. The large, cut-stone pier present there today was built much later in the nineteenth century and its history will be described later (chapter 6).

Endnotes

1 A. Nimmo to Commissioners of Irish Fisheries 30/1/1821 in 4th Report of the Commissioners of Fisheries, BPP 1823 (283).
2 46 Geo III C. 153.
3 2nd Report from the Select Committee on the State of Disease, and Condition of the Labouring Poor, in Ireland, BPP 1821 (646).
4 59 Geo III C. 109.
5 5th Report of the Commissioners of Fisheries, BPP 1824 (283) p. 5.
6 1st Report of the Commissioners of Fisheries, BPP 1821 (32).
7 8th Report of the Commissioners of Fisheries, BPP 1827 (487) appendix 9, p. 18.
8 2nd Report of the Commissioners of Fisheries, BPP 1821 (646).
9 3rd Report of the Commissioners of Fisheries, BPP 1822 (428).
10 4th Report of the Commissioners of Fisheries, BPP 1823 (283), p. 5.
11 H. Townsend to the Tavern Committee, 4/6/1822, in Smith, *Report on Distress.*
12 1st Report of the Commission of Inquiry into the Irish Fisheries, 1836 (Dublin: Alex Thom for HMSO, 1836), appendix 12, p. 38.
13 Smith, *Report on Distress..*
14 H Stafford Persse's letters from Ireland 1821–32. In J.L. Pethica and J. C. Roy, *Irish Narratives*, (Cork: University Press, 1998).
15 C. McNamara, 'This Wretched People' in King, C., and C. McNamara (eds), *The West of Ireland. New Perspectives on the Nineteenth Century* (Dublin: The History Press, 2011), pp. 13–34.
16 4th Report of the Commissioners of Fisheries, BPP 1823 (283), p. 5.
17 H. Townsend to the Tavern Committee, 4/6/1822, in Smith, *Report on Distress.*
18 A. Nimmo to Commissioners of Irish Fisheries 30/1/1821 in 4th Report of the Commissioners of Fisheries, BPP 1823 (283).
19 Smith, *Report on Distress.*
20 A. Nimmo, in 4th Report of the Commissioners on the Bogs of Ireland, BPP 1813/14 (131), appendix 12.
21 *An Act for the Employment of the Poor in Certain Districts of Ireland,* 3 Geo IV C. 34. (1822).
22 See Wilkins, *Alexander Nimmo.*
23 P. Melvin, *Estates and Landed Society in Galway* (Dublin: Edmund Burke, 2012).
24 See Melvin, *Estates*, p. 278.

25 5th Report of the Commissioners of Fisheries, BPP 1824 (283), appendix 14 and p. 5.
26 6th Report of the Commissioners of Fisheries, BPP 1825 (385), p. 3.
27 *An Act to amend the several acts for the encouragement and improvement of the British and Irish Fisheries,* 5 Geo IV C. 64.
28 7th Report of the Commissioners of Fisheries, BPP 1826 (395), p. 3.
29 Donnell J, in 8th Report of the Commissioners of Fisheries, BPP 1827 (487), appendix 11, pp. 21–22.
30 1st Report of the Commission of Inquiry into the Irish Fisheries 1836, Appendix 12, p. 38.
31 Minutes of the Crichton Committee of Enquiry, BPP 1878 [C. 2060], queries 1707 and 1757.
32 Nimmo's report is dated 9 November 1826. It is published as Appendix No. 10, in the 8th Report of the Commissioners of Fisheries, BPP 1827 (487), pp. 18–47.
33 8th Report of the Commissioners of Fisheries, BPP 1827 (487).
34 Donnell's report, dated 1 November 1826, appeared as Appendix No. 11, in the 8th Report of the Commissioners of Fisheries, BPP 1827 (487), pp. 47–75.
35 A Nimmo. Maps of Galway Bay, Costello and Greatman's Bays and the Harbours of Roundstone and Birterbuy with Part of the Coast of Galway from Slyne Head to Mynish Island. National Library of Ireland. Maps, M011, M143, M144.
36 4th Report of the Commissioners on the Bogs of Ireland, BPP 1813/14 (131), appendix 12; Wilkins, *Alexander Nimmo*, chapters 5 and 10; Villiers-Tuthill, K., *Alexander Nimmo and the Western District* (Clifden, Co. Galway: Connemara Girl Publications, 2006).
37 *An Act to amend an Act of the fifth year of His present Majesty, for amending the several Acts for the Encouragement and Improvement of the British and Irish Fisheries,* 7 Geo IV C. 34, section 4.
38 8th Report of the Commissioners of Fisheries, BPP 1827 (487), p. 3.
39 10th Report of the Commissioners of Fisheries, BPP 1829 (329), p. 5.
40 J. Donnell in 10th Report of the Commissioners of Fisheries, BPP 1829 (329), appendix 10, p. 16.
41 11th Report of the Commissioners of Fisheries, BPP 1830 (491), p. 3.
42 *Act to amend the several Acts for the Encouragement of the Irish Fisheries,* 10 Geo IV C. 33.
43 J. Donnell, Report in 11th Report of the Commissioners of Fisheries, BPP 1830 (491), appendix 9.
44 10th Report of the Commissioners of Fisheries, BPP 1829 (329), appendix 10, pp. 16–29.
45 J. Donnell in 11th Report of the Commissioners of Fisheries, BPP 1830 (491), appendix 9, p. 15.
46 A. Nimmo, Report in 8th Report of the Commissioners of Fisheries, BPP 1827 (487).
47 NLI, maps, 16 H 5 (5)
48 8th Report of the Commissioners of Fisheries, BPP 1827 (487), Plate 6 .
49 A. Nimmo to Commissioners of Irish Fisheries, 2/3/1824. NAI OPW/8/83.
50 10th Report of the Commissioners of Fisheries, BPP 1829 (329), Plate 10.
51 10th Report of the Commissioners of Fisheries, BPP 1829 (329).
52 1st Report of the Commission of Inquiry into the Irish Fisheries 1836. Appendix 19, p. 124.
53 10&11 Vict C. 75, sect. 6.
54 16&17 Vict C. 136.
55 J. Donnell, Report in 8th Report of the Commissioners of Fisheries, BPP 1827 (487).
56 J. Donnell, in 10th Report of the Commissioners of Fisheries, BPP BPP 1829 (329), appendix 10, p. 22.
57 Costs are given by Nimmo in his Estimate for Works at Clifden. NAI OPW8/ 84.
58 A. Nimmo, Report in 8th Report of the Commissioners of Fisheries, BPP 1827 (487), p. 37.
59 8th Report of the Commissioners of Fisheries, BPP 1827 (487), Plate 7 and pp. 54–5.
60 10th Report of the Commissioners of Fisheries, BPP 1829 (329), Plate 13 and p. 40. The original of the plan and estimate by James Donnell is extant. NAI OPW8/84.
61 Plan of Clifden Quay by John Killaly, dated 1830. NAI OPW8/84
62 1st Report of the Commission of Inquiry into the Irish Fisheries 1836, appendix 19, p. 124.

63 J. Conroy, Galway Bay, Louis XIV's Navy and the 'Little Bougard'. *Journal of the Galway Archaeological and Historical Society,* 49 (1997), pp. 36–48.
64 L.M. Cullen, 'HMS *Spy* off the Galway Coast in the 1730s: The Politics and Economics of Wool Smuggling.' *Journal of the Galway Archaeological and Historical Society,* 65 (2013), pp. 27–47.
65 W. Larkin, *Map of the county of Galway.*
66 8th Report of the Commissioners of Fisheries, BPP 1827 (487).
67 NAI OPW/8/267.
68 8th Report of the Commissioners of Fisheries, BPP 1827 (487), Plate 12, figure 2.
69 8th Report of the Commissioners of Fisheries, BPP 1827 (487), p. 44.
70 NAI OPW/8/17.
71 8th Report of the Commissioners of Fisheries, BPP 1827 (487).
72 J. Donnell, in 8th Report of the Commissioners of Fisheries, BPP 1827 (487), appendix 11, p. 58.
73 10th Report of the Commissioners of Fisheries, BPP 1829 (329), appendix 10.
74 10th Report of the Commissioners of Fisheries, BPP 1829 (329), appendix 10, p. 32.
75 J. Donnell, in 8th Report of the Commissioners of Fisheries, BPP 1827 (487), appendix 11, p. 58.
76 Paul Duffy, personal communication, 2016.
77 Return of the Small Piers of Ireland, dated 17 February 1833, BPP 1835 [573].
78 Wilkins, *Alexander Nimmo*, p. 370.
79 23rd Annual Report of the OPW, BPP 1854–55 [1929], p. 25.
80 8th Report of the Commissioners of Fisheries, BPP 1827 (487), appendix 10, p. 45, Appendix 11, p. 59 and Plate 13, figure 1.
81 A. Nimmo to Commissioners of Fisheries, 2/4/1824. NAI OPW8/17.
82 J. Donnell, in 8th Report of the Commissioners of Fisheries, BPP 1827 (487), appendix 11, p. 59.
83 8th Report of the Commissioners of Fisheries, BPP 1827 (487), Plate 14, figure 1.
84 Six-inch OS map of Clare, sheet 3, first edition.
85 N. P. Wilkins, *Ponds, Passes and Parcs* (Dublin: Glendale Press 1989).
86 NAI OPW/8/273.
87 J. Donnell, in 8th Report of the Commissioners of Fisheries, BPP 1827 (487), appendix 11, p. 59.
88 NAI CSORP 1827/1591.
89 NAI OPW/8/11.
90 P. O'Toole, *From Aran to Africa* (N.P.: Nuascéalta Teó 2013).
91 J. Donnell, 8th Report of the Commissioners of Fisheries, BPP 1827 (487), appendix 11.
92 8th Report of the Commissioners of Fisheries, BPP 1827 (487), appendix 10, p. 47.
93 J. Donnell, 10th Report of the Commissioners of Fisheries, BPP 1829 (329), appendix 10, p. 22.
94 Return of the Small Piers of Ireland, dated 17 February 1833, BPP 1835 [573].
95 Smith, *Relief of Distress*, p. 5
96 Smith, *Relief of Distress*, p. 19.
97 O'Flaherty, F., to A. St. George, 12/6/22. NAI OPW8/ 153.
98 A. St. George to Fishery Commissioners, 14/6/22. NAI OPW8/153.
99 NAI OPW8/153.
100 8th Report of the Commissioners of Fisheries, BPP 1827 (487), plate 9, figure 1.
101 First Report of the Commissioners of Inquiry into the State of the Irish Fisheries. appendix 19, p. 124.
102 William Pierce to OPW, 29/7/1837. NAI OPW8/166.
103 Return of Small Piers erected on the Coast of Ireland in Report of the Select Committee on Public Works in Ireland, BPP 1835 [573], appendix 4, pp. 306–08.
104 A. Nimmo, Report in 8th Report of the Commissioners of Fisheries, BPP 1827 (487), p. 41.
105 8th Report of the Commissioners of Fisheries, BPP 1827 (487), plate 11, figure 1.

106 10th Report of the Commissioners of Fisheries, BPP 1829 (329), plate 5.
107 10th Report of the Commissioners of Fisheries, BPP 1829 (329), Appendix 10.
108 A. Nimmo, 8th Report of the Commissioners of Fisheries, BPP 1827 (487), Appendix 10, p. 41.
109 1st Report of the Commission of Inquiry into the Irish Fisheries 1836, appendix 19, p. 124.
110 Personal communication, Paul Duffy, 2016.
111 Pethica. and Roy, *Irish Narratives.*
112 NAI OPW/8/44.
113 NAI CSORP 1822/2237.
114 1st Report of the Commission of Inquiry into the Irish Fisheries 1836, appendix 19, p. 124.
115 Report of Barry Gibbons on Pier for Oranmore, 5 October 1826. NAI OPW/8/11. NOTE: This reference number, which relates to Oranmore, appears under Killeany in the archive.

CHAPTER 3

ENTER THE OPW: 'WE REMAIN, AS A FISHERY BOARD, TOTALLY INOPERATIVE'

[The small piers] appear to have been productive of great advantages to the neighbouring districts, not only as affording encouragement to the fisheries but also accommodation of the landing and shipment of produce, goods, manure etc. on which considerations they seem to merit the attention of Government.

Col. John Fox Burgoyne, 1832[1]

Transfer of Fisheries and the Piers and Quays to the Directors of Inland Navigation

In January 1830, as they entered their last three months in office, the Commissioners wrote to the Lord Lieutenant advising him of the iniquities they believed would result from the cessation of certain aspects of the fishery legislation. Continuance of the actual laws and regulations concerning trawling, seining and the herring fisheries was essential, and compliance could be assured only by vesting them in some authority, rather than allowing them to lapse entirely, which would lead to legal chaos.

As for the fishery piers, the Commissioners, or their potential successors, would have adequate funds to complete those already in course of construction, but there were ' sundry appplications made to us for aid to erect other piers, in other places, where it would be most desirable to have them, but upon the erection of which we cannot enter, without some annual renewal, or other augmentation, of the fund available to these purposes.'[2] Their bullish optimism that they might be continued in office was still clearly undiminished.

The Fishery Loan Fund, which provided loans for the purchase of boats and gear by poor fishermen, also needed to be maintained, since many had yet to complete their loan repayments. Finally, since the Scottish fishery authorities, unlike the Irish, were to remain in office and not be dissolved, the Commissioners expressed the fervent hope '... that the same means of assistance and protection that have been, or may be, afforded to the British fisheries, will be extended to those of our own country ...' The Government, however, had no intention of continuing an independent Irish fishery authority. It did agree, on the other hand, that the fisheries could hardly continue without regulation, and the fishermen's debts certainly needed to be repaid.

A temporary expedient was hit upon: reform was in the air and, pending the establishment of a planned new Commission for Public Works, responsibility for the Irish fisheries would be passed *pro tem* to the Directors-General of Inland Navigation. The latters' existing duties were overseeing and regulating the construction of canals and other inland waterways that were used for navigation and transport. They had no particular knowledge or experience of fisheries, but they were skilled in the management of public funds and could administer the fishery funds adequately for the time being.

The necessary legislation to effect the transfer was enacted on 16 July 1830, three months after the Fisheries Commission went out of office.[3] The fishery laws, but not the Fisheries Commission, were revived and the powers of the late Commissioners were vested in the Directors-General of Inland Navigation by section 6 of the Act. The balances of all the funds in the late Commissioners' hands were transferred to the Directors-General, including the balance, £16,422, in the piers fund.[4] This may seem like a large sum, but the Commissioners left commitments outstanding of £15,627 for piers sanctioned or in progress, so that the unappropriated balance in the transferred fund was only about £800.

Section 7 of the Act empowered the Lord Lieutenant to appoint suitable persons to complete the construction of the piers begun before 5 April 1830 and, under section 8, he could direct the payment of an extra £13,000 for their completion. This sum was to be provided out of the Consolidated Fund in Ireland over five years on a reducing scale: £4,500 in the year commencing April 1830; £3,500 in 1831; £2,500 in 1932; £1,500 in 1833; and £1,000 in 1834, the final year. In handing the fisheries over therefore, the Act facilitated and funded the carrying to completion of all the piers already in train, but it provided no new funds for any other aspect of the fishery laws and their administration. It was also clear that there would be no further money for piers beyond the reducing fund of £13,000, or any for the commencement of new works.

It was altogether a neat solution: the Act had cleanly terminated the Fisheries Commission, provided adequately for the completion of all piers in progress, ensured the proper management of the Fishery Loan Fund, and closed off any further expenditure on fisheries or fishery piers once the reducing fund was exhausted.

The Winter of 1830–1831

In the scheduled wider reforms, the Directors-General were themselves soon superseded and their Board extinguished, but not before they had spent £9,164 of the piers fund on essential works during 1830–31.[5] A return made to Parliament many years later[6] indicated that they had spent £14,372 during those two years, but that seems to be excessive; there was still £11,395 remaining in the piers fund when it was transferred to the OPW in

October 1831, the first £4,500 instalment of the reducing fund having been drawn down by then.[7] Whatever the case, 1830–31 was a year of very extreme weather, causing great damage in the West.[8] Storms were particularly severe, causing numerous bridges and piers to be washed away along the whole Co. Galway coast, and leaving immense destruction in their wake.[9]

Barna pier and Nimmo's pier at Ballinacourty were among the works most seriously damaged. Agriculture was badly affected too, and famine prostrated the people once again. Consequently, the Viceroy approved a scheme of immediate relief works, to be carried out under the direction of the Coastguard. The construction of a new north pier at Roundstone was also attempted to a limited extent in 1830, reportedly with unspecified charity funds.[10] This involved little more than making a short, crude groyne that did nothing to improve the harbour. *An tSeancéibh* at Kilronan, the earliest pier at that place on the Aran Islands, was built in 1831 by men paid with meal provided from charitable sources. *Spidéal* (old pier) was repaired in 1831 'in return for food to relieve the starving poor'[11] and Barna pier was repaired in 1832 'to a considerable extent by an outlay of £70 given from the subscription of the Coastguard for the relief of the poor in that year'.[12]

Perhaps it was support like this from charitable and emergency sources that accounts for the sum of £14,372 reportedly expended in the 1865 Return mentioned above. Reports certainly confirm that officers of the Coastguard were active in the works and generous too: in some cases they contributed their own money towards building small piers and quays, very much to their credit. The reason for undertaking all these works was the relief of distress, rather than any serious plan or effort to improve or enhance the general fisheries infrastructure.

The OPW Takes Over

The Irish Board of Public Works was established by the *Act for the Extension and Promotion of Public Works in Ireland,* passed on 15 October 1831,[13] and all the powers and duties of the Board of Inland Navigation, including their recently acquired fisheries responsibilities, were transferred to the new Commissioners of Public Works, along with many other public enterprises then in progress. The principal purpose of the Act was to consolidate various existing public boards, many of which comprised bodies of amateur, unpaid commissioners, into one new public authority.

The new, salaried Board of Public Works would administer the existing Consolidated Loan Fund; inland navigation; fisheries; roads and bridges; public buildings in Dublin; the Phoenix Park; Dunmore and Kingstown Harbours; and other responsibilities committed to it. The new Board – hereafter referred to as the Board, or Office, or Commissioners of Public Works, or simply the OPW – consisted of three Commissioners. The first Chairman, Colonel (later General Sir) John Fox Burgoyne (1782–1871), a military engineer, received a salary of £1,000 *per annum*. The other two commissioners, Brook T. Ottley, a military accountant and John Radcliff, RE, an engineer/architect who was previously a Director of Inland Navigation, each received £600. They were allocated an initial fund of £550,000 for their operation. All advances of money they proposed to make under the Act, whether as loans or as free grants, required the express prior approval of the Lords of the Treasury (sections 17 and 64 of the Act), so that in practice the OPW functioned as an arm of the UK Treasury in Ireland, effectively superseding the Irish Executive under the Viceroy and the Chief Secretary in Dublin Castle.

The new Commissioners took a minimalist view of their fisheries responsibilities: 'The principal duties which have devolved upon us under this head were, the completion of several Fishery Piers, and the collection of the balances due by poor fishermen on advances heretofore made to them out of the Fishery Loan Fund.'[14]

The balance in the Piers Fund (hereafter referred to as the legacy pier fund) was then £11,395; to this was added the second (£3,500) instalment of the reducing fund, making a total initial sum of almost £15,000 for the business of finishing the fishery piers. This would increase to £20,000, once the remaining annual instalments of £2,500, £1,500 and £1,000 were drawn down, but the fund could only be used exclusively to complete or repair piers that were started prior to April 1830. The balance in the Fishery Loan Fund for boat and net repairs, augmented by some small receipts, was £7,156. The new OPW therefore had a grand total of over £27,000 for all fishery purposes.[15] They spent £3,737 completing piers during 1832, the first full year of their operation. That was to be the largest annual outlay they would make for fishery piers for a whole decade.

One of their earliest actions was to commission yet another engineer's survey of the state of the piers and quays, despite the fact that Donnell had made three such surveys, and Nimmo two, in the previous five years. On receipt of the requested survey report, the Board commented favourably on the general utility of the small piers that the late Fisheries Commissioners had built. They forwarded the report to Treasury on 27 February 1833. The stated condition of the eighteen piers in County Galway and north Clare was hardly encouraging. Five of them – St Kitts, Duras, Ballinacourty, Costello, and *Spidéal* – were reported to be of little or no use to anyone. Ballinacourty was 'completely destroyed' – most likely by the storms of 1830 – and Costello was '... in a state of dilapidation, not worth the cost of repairing it'.[16]

The piers that were proving useful were generally more valuable for the transport of turf and seaweed than for fisheries, as the OPW comment at the head of this chapter confirms. The Claddagh quays and Slate pier in Galway city were among the most useful and successful, although the engineer remarked about the Claddagh fishermen that '... with all the advantages of good boats and accommodation, there are few places where the fisheries are carried on with so little enterprise'.

The Board, nevertheless, sanctioned a loan of £300 to Galway city on surety of a presentment, to extend the Claddagh quay northwards; the Board had already approved a loan of £17,000 to the city for the development of the commercial docks under the engineer Hamilton Hartley Killaly, and the works proposed at Claddagh were seen as complementary to that development. The other useful piers were Barna, Rinville (New Quay), Clifden,

FIGURE 3.1. The Lower Pass of *Béal an Daingin* as it is today. The T-shaped structure towards the top is the 1836 wharf with its causeway joining it to the shore at Annaghvaan. The long road on the right is the modern road, causeway and bridge over the Pass, first erected in 1890–91. The old intertidal 'paved road' is evident winding towards the T wharf to the left of the modern road. Photo by Michael Gibbons, reproduced with permission.

Cleggan and Roundstone. A summary of the engineer's report was published as an appendix to the Report of the Select Committee on Irish Public Works in 1835.[17]

Keeping the Existing Piers in Order

Small allocations to existing piers, made out of the legacy pier fund, would continue for a decade. (The fund was constantly being replenished with repayments of interest and principal on loans given earlier; no other new money was voted to it.) The amounts spent for various purposes on the different Galway piers are shown in Table 3.1, together with the total spent (£15,345) on all fishery piers throughout Ireland. County Galway accounted for about 9 per cent of the total expenditure. Except for Barna, all the sums were relatively small, hardly more than the most basic maintenance would demand. New piers still could not be built, due to the restriction limiting the fund to structures started prior to April 1830.

TABLE 3.1

Annual sums (in pounds, rounded) spent on the repair and maintenance of fishery piers in Co. Galway from 1832 to 1842, together with the total amounts spent in each year on all fishery piers in Ireland

	Pier	1832	1833	1834	1835	1836	1837	1838	1839	1840	1841	1842	Total
1	Claddagh		40										40
2	Rinville			20									20
3	Barna			34		1,500						200	1,734
4	Clonisle			50									50
5	Derryinver			80									80
6	Slate Pier			64	181	7							252
7	Rosroe			10	13								23
8	Leenaun (old)			40									40
9	Roundstone			16	193								209
10	Cleggan				50	17	24						91
11	Maumeen				20	40	170	80	50				360
12	Clarinbridge											400	400
Total, Co. Galway			40	314	457	1564	194	80	50			600	3,299
Total for all Ireland		**3,737**	**2,619**	**1,833**	**155***	**2,903***	**1,648***	**187**	**391**	**992**	**280**	**600**	**15,345**

*These sums include the costs of the Fishery Enquiry paid from the piers fund (£64 in 1835, £1,386 in 1836 and £1,238 in 1837).

Data collated from OPW reports, 1833 to 1843, including the sums spent on Barna and Clarinbridge piers from Section 85 funds (see text).

Completely separate from the fishery duties transferred to them by sections 105 and 106, the 1831 Act empowered the Board of Public Works to make grants for harbours, piers and quays that they considered useful or necessary for the fisheries and for other purposes (section 85). Grants and loans under that section were subject to the same conditions as those for roads and bridges, the proposers being obliged to pay one half of the estimated cost, with no grant to exceed a total of £1,000. In addition, the recipients had to satisfy the OPW that funds existed, or could be made available, for the continuing upkeep of the works. The Board could spend up to £50,000 of the £550,000 allocated to it for 'section 85' purposes, such as road, bridge and pier grants.

It was from this source that work on Barna pier was funded in 1836, one of only a very small number of grants issued for fishery piers under that section. The OPW had agreed, for example, to fund *Béal an Daingin* quay under the section, but as it transpired, the Lord Lieutenant stepped in with relief funds. The most significant effect of sections 62, 85, 105 and 106 of the Act was to place responsibility for the erection of all harbours, piers and quays on the coast of Ireland, whether for fisheries or for trade and commerce or for both, into the hands of the OPW exclusively, where it would remain for the next fifty years at least.

In their second Annual Report, the Commissioners of Public Works began to show enhanced interest in the fishery piers, which they increasingly realised had greater social importance than simply facilitating fisheries. It was very desirable, they stated, that:[18]

> These very useful works should be placed on a permanent footing, by the establishment of a legal authority for their future maintenance, and for the purpose of enforcing a system of order and regularity by the parties making use of them, and which is now too often infringed. There are also some instances in which additions and improvements might be made at a comparatively trifling expense, calculated greatly to encourage an export by the small occupiers of land, for which the situations afford many facilities.

By the following year, they were beginning to find certain aspects of the fisheries burdensome, especially disputes about the modes of fishing and suchlike, and they were 'entirely without the means or power of acting on such occasions; for although the Act transferred the powers of the Commissioners [of Fisheries], no funds were provided for carrying them into effect.'[19] Therefore they recommended to Treasury that an investigation of the whole fisheries brief should be carried out. This should be done, they proposed, by qualified and experienced persons who might make suggestions for legislative measures that would ensure a revival of the fisheries as a means of providing work for the labouring classes.

The Select Committee on the Board of Works, Ireland, 1835

The whole operation of the OPW came under review in 1835, when a Select Committee of Parliament was established to enquire into the amount of advances made by it, the regulations and purposes that were served by them, and other relevant issues.[20] The Committee, comprising fifty Members of Parliament, sat between March and August 1835, interviewing forty witnesses, including eminent engineers like David Aher, William Bald, William Cubitt, Richard Griffith and Charles Vignolles, all experienced men who knew and had worked in Ireland over the previous twenty-five years. When asked to name places in Co. Galway that would benefit from new piers, William Bald, for example, showed no hesitation in making proposals: [21]

> In Connemara, at Killeen on the west side of Costelo [*sic*] Bay, at Ternee [*sic*] west side of Great Man's Bay, at Rustreen [*sic*] in Kilkieran Bay, at Ard Castle in Ard Bay; to clear the Pass of Stradle; this would only cost about £40, and would then be fitted for ships of 200 tons to pass through, and do away with the necessity of going around Macdara Island, which is attended with much difficulty and delay; to complete the north pier at Roundstone, one at Bunowen Bay, one at Ballinakill; to finish the quay at Clifden; one at Tully, near the entrance of the Killary Harbour ... and to widen the channel to the new pier at Killaney [*sic*] in the island of Arran.

His proposals, which included many sites previously identified by Nimmo, elicited no immediate response. Later in the century, all his suggestions would be implemented, with the exception of the clearance of the Pass of Straddle (*Bealach na Srathrach*). This is a sea passage between Mason Island and Mweenish Island that greatly shortened and made less hazardous the inshore sailing route along the Connemara coast. Clearing it fully would be an ambitious project, even today.

In general, the Select Committee said little enough in its report regarding fishery piers. Early in its proceedings, Colonel Burgoyne had indicated that his Board was trying to have established a separate special enquiry to consider the whole fisheries business, including the piers and that information seems to have taken the urgency from the topic in the Select Committee's extensive brief. Nevertheless, one of the Committee's recommendations regarding piers was that the requirement for local contributors to be responsible for their maintenance should be dispensed with, in favour of tolls which, it said, should be leased out by the Board.

Another recommendation was that Grand Juries should have the power to authorise expenditure on small piers, quays and harbours, which would be recouped by levies on the districts or baronies benefiting from them. While ignoring entirely the need to build new piers to aid or encourage the fisheries, the Committee made some desultory reference to the need for safety harbours to facilitate commercial coastwise trade, and it made no reference at all to the urgent issue of the regulation of fishing activities. One small, but useful outcome of the proceedings was the publication, as an appendix, of the engineer's report of 27 February 1833, summarised earlier. That report had alluded to the unsatisfactory state of some of the piers, and in his evidence Richard Griffith gave a jaundiced opinion of why the piers were in this state: [22]

> We have had melancholy examples of small piers erected at the Government expense which have been almost all carried away, which shows the impropriety of building *very small piers* [Griffith's emphasis] ... The money expended rarely exceeded £500 or £600, which was quite inadequate for the erection of piers capable of withstanding the Atlantic waves.

This complaint about the inadequacy of the piers would remain a bone of contention for the rest of the century.

The Committee's main overall conclusion was that the setting up of the OPW had been 'of much and essential service to Ireland', although the prescribed terms of its loans and the interest charged on them had been excessive. It recommended an increase in the £550,000 general allocation to the Board; the amelioration of the terms for individuals availing of loans; reduction of the interest rate charged; and enlargement of the repayment period, all to be subject to the agreement of Treasury.

The OPW, it must be said, came out of the investigation well. Discussion had shifted between those who championed a non-interventionist, *laissez-faire* approach and more liberal others, who believed that measured state assistance, along with a sustainable local contribution, was the better strategy, particularly in the circumstances prevailing in Ireland. Setting up the OPW with an allocation of State funds for use as loans and grants had been, in the words of the Committee, an 'experiment', and one that had succeeded so far.[23]

The Select Committee issued its first report in June 1835. Within days, a Bill was brought in at Westminster by William Smith O'Brien MP, Thomas Wyse MP and Andrew Lynch MP of Barna (all of them members of the Select Committee) 'to empower Grand Juries in Ireland to raise Money by Presentment for the Construction, Enlargement or Repair of Piers and Quays'.[24]

The actual aim of the Bill was to permit the public funding of piers and quays in *inland waterways* (Smith O'Brien was MP for Limerick, and Wyse was MP for Waterford, both districts with important inland navigations, the River Shannon in Smith O'Brien's case and the River Suir in Wyse's), and to empower the Grand Juries to present for a moiety of the cost. As it stood, the legislation did not allow the Grand Juries to present for any piers or quays at all, not even for ones located on the sea-coast. The proposers brought in an amended version of the Bill[25] two months later, within days of the publication of the Select Committee's second report.

Neither this Bill nor the earlier version progressed successfully through Parliament. Instead, after many Members had spoken in the House of Commons during 1834 and 1835 in support of a special investigation of the Irish fisheries, the Lord Lieutenant moved on 2 November 1835 to set up a full, formal Fisheries Enquiry.

The Fishery Enquiry of 1835–6

The persons appointed as Commissioners of the *Enquiry into the State of the Irish Fisheries* were the Secretary (Edward Hornsby) and the three Commissioners of the OPW, together with six other persons, among them J. Redmond Barry, previously the Inspector General of Fisheries for the southern region under the Fisheries Commission of the 1820s. They were tasked to report on the facilities available for fishing and the existing state of the fisheries along the coast; to investigate the causes that affected them injuriously; to indicate measures to promote their extension; and to suggest laws for their encouragement. It was a remit that was broad and comprehensive, and the fishery piers and harbours were included within it. Oral evidence was taken between December 1835 and February 1836 at different locations around Ireland, in the Isle of Man and in Liverpool.

Further evidence was elicited by way of written replies to specific queries sent out to interested parties. Other documents consulted, and published with the report, included Nimmo's coastal survey reports, Donnell's report on the piers and quays (chapter 2) and the 1833 survey of the piers summarised earlier. The Enquiry reported in October 1836.[26] The published evidence recorded in the minutes has been a comprehensive source of information (not all of it fully reliable) on fisheries and other aspects of life in pre-famine Ireland, and it is still consulted and quoted widely today.

The first of its seventeen recommendations proposed that the entire body of fisheries legislation should be repealed and replaced by a single new Act, incorporating all the legislative provisions deemed necessary to be retained. The superintendence and regulation of the Irish fisheries, it continued, should be vested in a public department that reported annually to Parliament. It seems the OPW was the department implicitly favoured for this role.[27]

Six of the remaining recommendations related to fishery piers: increased financial aid should be given for their construction and maintenance; repairs should be carried out on existing structures using the balance remaining in the old legacy piers fund; fishery piers and quays receiving public money should be made public property, with rules and regulations regarding their use; funds for piers should be applicable to the provision of moorings and lights; moderate tolls should be established to defray the necessary costs of maintenance; and, finally, the cess-payers (rate payers, essentially the Grand Juries), should be allowed to levy the county, barony or district, as appropriate, with such sums as were necessary to make up the local moiety for any work that was to be done on the piers and quays, boat slips, capstans, lights, and so on. These recommendations were largely what the Select Committee on the OPW had already outlined very briefly a year earlier.

In 1836, the OPW was openly admitting that it was 'nearly inoperative' as a Fisheries Board because of, it claimed, 'the existing Acts of Parliament being inapplicable to their being exercised with much degree of utility'; this was not the Board's fault.[28] This, and the uncertainty of the laws themselves, was acknowledged in the report of the Enquiry, but a detailed presentation of the discussion of the actual laws need not detain us here.

One year after the Enquiry, the OPW's report on their fisheries activity was just a short, laconic statement: 'We remain, as a Fishery Board, totally inoperative; a Bill, however, for regulating matters connected with the Sea Fisheries of Ireland, founded on the Report presented to Parliament last session, is in a course of preparation'.[29] That may well have been so, but a Bill could take quite some time to pass through the legislative process.

The Lord Lieutenant, George Howard Viscount Morpeth, did eventually bring in a Bill in 1838, the text of which was reprinted in 1867 when reference was made to it at another Select Committee at that later date.[30] One of the proposals notable for our purpose in Morpeth's Bill was that regulation of the fisheries should be given to a new body but responsibility for the fishery piers and harbours should stay firmly with the OPW. That was the first ever official proposal to sever fishery piers from the general fishery administration. After first and second readings, the Bill was postponed a number of times and eventually it was not proceeded with any further. J. Redmond Barry expressed his suspicion that it had been scuttled by pressure exerted on the Government by a delegation of Scottish fishery interests headed by the Duke of Sutherland.

One day, Barry alleged, the Duke and his delegation had waited upon the Government concerning the Bill, and on the very next day, the Bill was postponed indefinitely.[31] Nor was that the first time Scottish interests had objected to the development of the Irish fisheries: on 8 April 1820, the 'Magistrates, Town Council, Principal Inhabitants and Fish Curers of the town of Fraserburgh in Scotland' had written to Treasury objecting to the bounties for Irish fishermen introduced by the 1819 Act. They were worried that the bounty would give an advantage to the Irish fishermen – the Scots themselves had enjoyed bounties of almost £270,000 from 1809 to 1819,[32] while the Irish had received nothing! – and they claimed their own prospects of carrying on the fisheries with advantage 'will be entirely blasted, and thousands of the labouring class dependent on their enterprise and perseverance, rendered miserable.'[33] Their avowed concern for the Scottish labouring class clearly did not embrace their brothers of the Irish labouring class whose condition was even more miserable than their own.

The Beginning of the Modern Administration of the Irish Fisheries

Having contributed £2,688 to the costs of the Fishery Enquiry,[34] the OPW still had a balance of £9,371 in the legacy fishery account, of which £6,634 was in the Piers Fund and £2,737 in the Fishery Loans Fund. So it was hardly a lack of finance that had kept the Board 'inoperative' while awaiting new legislation. In the meantime, the conduct of the fisheries and of the fishermen went into sharp decline. The Board reported conflicts arising in situations where 'differing modes of prosecuting the fishery have thwarted the interests or aroused the passions and prejudices of persons who have considered themselves injured thereby.'[35] There was little the OPW could do to prevent such occurrences, beyond distributing notices through the Coastguard, explaining the existing law so as to minimise conflict as best it could.[36]

The only pier the Board addressed at this time was that at Glandore, Co. Cork, home of J. Redmond Barry. Barry had erected the first Glandore pier in 1828 before the village was fully established, and the new works in 1840–1 extended that original pier and quay wall. Since Barry had been advising the OPW on fisheries matters in an informal, unpaid capacity from 1831 onwards, this was a small enough reward for his dedicated service up to then. Barry was a significant landowner in the Glandore district and the founder of the village and its fishing tradition.[37]

Passage of a new, much hoped-for Irish Fisheries Bill was interrupted by the fall of the Whig Government in 1841 and its replacement by the Tory administration of Robert Peel. Despite its distaste for 'interfering' in the business of fisheries, the new Government could hardly leave the fishery laws in anarchy and chaos, especially when the recommendations of the Fishery Enquiry were available to address the situation. Therefore, without any further delay, Parliament approved a new Act in 1842, entitled *An Act to Regulate the Irish Fisheries,*[38] initiating the modern phase of Irish fisheries legislation and regulation. It came not before time: by 1841, the OPW had removed all mention of fisheries from its Annual Reports, except to record a balance of £8,179 in the legacy fisheries account at the end of that year.[39]

The Irish Fisheries Act, 1842

The new Act was meant to bring to Irish fisheries the fruits of the Fishery Enquiry's deliberations. It commenced by repealing all the earlier fishery laws, comprising twenty-six separate statutes or parts of statutes going back to the fifth year of the reign of Edward IV (1465), in so far as they related to the Irish fisheries. There was, however, an extremely important exclusion clause relating to fishery piers and quays: [40]

> Nothing herein contained shall be construed to repeal any enactments or provisions of the said Acts or any of them, which relate to piers and quays or assisting poor fishermen or any powers in respect thereof now vested in the Commissioners of Public Works in Ireland ... but that all such enactments and provisions relating to piers and quays and the assistance of poor fishermen ... shall remain in effect.

In a single stroke, the Act completely severed the fishery piers and quays from all other aspects of fisheries and their management, exactly as Viscount Morpeth had recommended in his aborted Bill of 1838. This left the OPW in a somewhat peculiar position, as we shall see. The Act went on to set down a new fisheries dispensation in 115 sections. These were mainly prescriptive, technical regulations as to the place, manner and time of fishing, the registration of boats, the types of gear to be used, the erection of dams and weirs in freshwater and so on, all topics of relevance principally to close inshore fishing, to fishing for salmon with fixed nets and to the oyster fisheries.

The Commissioners for Public Works were made Commissioners for the Fisheries under the Act, as presaged in the Fishery Enquiry recommendations. They performed their fishery duties through the offices of two salaried Inspectors of Fisheries whom they were permitted by the Act to appoint. One of these was William Thomas Mulvany (1806–85), already a seasoned member of the OPW engineering staff and the other, J. Redmond Barry (1789–1879).

Having full statutory responsibility for the fisheries placed on them was not, it would appear, something the Commissioners entirely welcomed. They started work slowly, claiming they did not quite understand the general tenor of the Act, which was based on the principle of non-interference, and they were confident that their role in fisheries would come to an end within one or two years, once the new arrangements became accepted. That was the reason, they believed, why the OPW, an already very busy department, was given the fisheries 'instead of organising a new Board for the purpose, as would have been required had a principle of more active and minute interference been adopted'.[41] Their position was understandable: the thrust of the new legislation was mainly a matter of inspection, monitoring and regulation, with nothing concrete involving engineering or structural works as these are usually understood. The Act encompassed many aspects of the fisheries that the Commissioners were not particularly familiar with or experienced in, and, most importantly, it did not grant any funds at all for carrying out the fishery duties and functions.

The opening balance in the new vote for fisheries was a mere £157.[42] All in all, the Board was given responsibility for fisheries without any financial resources other than the salaries of the Inspectors, and it is hardly surprising that the Commissioners felt decidedly uneasy about their new role. The Act required that they submit an Annual Report to Parliament on their fishery duties, completely separate to their Annual Reports covering their other responsibilities.

From 1842 on, therefore, the OPW produced two distinct Annual Reports each year. One covered their multiple public works functions and responsibilities, the *Annual Reports of the Board* (olim *Commissioners*) *of Public Works,* which continued each year until the foundation of Saorstát Éireann. The other, entirely separate, annual report covered their fisheries responsibilities, variously titled the *Annual Reports of the Commissioners of Fisheries Ireland* (in 1843 and 1844); later the *Annual Reports of the Commissioners for Public Works in re the Fisheries of Ireland* (in 1845 and 1846); later still, the *Annual Reports of the Commissioners of Fisheries, Ireland* (1847 to 1861); finally, the *Reports of the Deep Sea and Coast Fishery Commissioners, Ireland* (1862–65). All the annual reports, both general and fishery, appeared under the signatures of two or more of the Commissioners of Public Works. Nothing emphasises more clearly the separation between the Board's new fisheries responsibilities under the 1842 Act and all their other functions under the 1831 Act, than this double approach to reporting.

Fishery Piers under the New Fisheries Administration

The exclusion clause regarding fishery piers quoted above caused a further unusual anomaly. Because of it, the Commissioners, *acting in their capacity as Commissioners for the*

Fisheries under the 1842 Act, were statutorily precluded from having anything to do with fishery piers and harbours. But *acting in their capacity as Commissioners of Public Works under the 1831 Act*, they continued to have full and exclusive responsibility for them. In seeking OPW support for any proposed fishery pier, the public therefore could not expect, or even seek, official assistance, encouragement or advice from Fishery Inspectors Mulvany and Barry. The Board, for its part, when considering the suitability of any proposal for a fishery pier, was not obliged to consult, and, in fact, rarely if ever consulted its own Fishery Inspectors as to the pier's suitability or desirability; it was simply not a legitimate concern of the Inspectors. This was hardly a satisfactory arrangement for either fisheries or fishery piers, nor, indeed, for the OPW or the Inspectors.

As we shall see, this division would remain a contentious issue for many years and would flare up eventually into a full-blown interdepartmental discord when the two bodies, the OPW and the Inspectors, were fully separate entities. In fact, it would be a further twenty-seven years before the Inspectors of Fisheries (an independent body by then) would be given, by statute, a small role in decisions concerning fishery piers and quays.

It needs to be emphasised that neither the 1831 nor the 1842 Acts imposed any obligation or positive duty on the OPW to initiate the construction, repair or maintenance of any fishery piers or quays. The Board, however, still had the old legacy fishery funds at its disposal, which it could, and did, use to make small allocations for emergency ongoing repairs where these were absolutely essential. At the end of 1839 there had been a legacy balance of about £8,790 available.[43]

The Board spent nothing in 1842, but it gave £100 for repairs to Barna pier in 1843 among other small disbursements. The balance then stood at £7,403. Having been omitted since 1842, a brief mention of the legacy funds appeared in the 13th Annual OPW Report, published in 1845. The Board indicated that Treasury had sanctioned the expenditure of further small sums towards six small fishery piers, none of which were in Co. Galway or north Clare.[44] Although they spent £744 on these, other sums had been paid into the fund, resulting in a healthy new balance of £7,459. In the Annual Report for 1845,[45] this had grown to £8,768 through the accumulation of interest and repayments of principal of old loans. The Board therefore spent £1,968 repairing thirteen more piers (none of these was in Galway or north Clare), leaving a new balance of £6,800.[46]

The following year, 1846, the Board reported that the fund had been applied once again to repairing small piers and harbours and that 'now all funds are totally absorbed'.[47] But Appendix 12 of the Report shows that the total expenditure that year on eleven named piers (none in Co. Galway or north Clare) was £3,561 and the balance carried forward was a tidy £3,985. The old legacy fund made another appearance in an Appendix to the 16th Annual Report: in the year to the end of 1847, a total of £3,774 was expended from it on fourteen piers, one of which was Ballyvaughan in north Clare.[48] With further accruals, a balance of £496 still remained at the end of 1848.[49] For a fund that was supposedly 'totally absorbed' by 1846, it had proven remarkably resilient, but it was finally merged with a new piers and harbours account in 1849.[50]

Little could the original Commissioners for Irish Fisheries have imagined that the fund they set up for piers over twenty years earlier would last so long! It had been the mainstay of the various repairs and renovations carried out on piers in the otherwise lean period from 1832 to 1846; without it, nothing at all would have been done, as there were no new funds voted for fishery piers during that entire period. Indeed, as we shall see in the next chapter, the paucity of new piers might well have continued indefinitely into the future, were it not for the catastrophe of the Great Famine and the Government's response to it.

The OPW Warms to Fishery Piers and Quays

From 1842 to 1846 the legacy fund had provided about £12,000 without any declared participation of, or input from, the Fishery Inspectors. All the other fisheries' functions of the OPW, however, were carried out and reported on by the two Inspectors, Mulvany and Barry, although the published Fishery Reports were actually signed by the Commissioners alone.

In 1846, Mulvany was elevated to a full Commissionership for Public Works and his post as an Inspector was filled by William Joshua Ffennell (1799–1867). This change did not alter the Board's responsibility for the piers and quays in any way, but Mulvany's new elevated status as a Commissioner seems to have improved the Board's perception of the value and importance of them. On the prompting of the Inspectors, the Board began to make important suggestions regarding piers as suitable investments in the fishing industry, such as it then was: [51]

> Two accessories might still be added of very great value: one of them, the establishment or improvement of small piers and quays, and places of shelter; and there are many that could be made available and most valuable for a comparatively small outlay ... For any general extension of assistance to the fisheries, the promotion of these two objects [the second was the improvement of access roads to landing places] would be far preferable to any attempt at direct pecuniary aid to the trade itself or interference by expensive local, personal establishments.

In their next Fishery Report, dated 1845, they were even more forceful:

> The progress of improvement in the fisheries of Ireland is, doubtless, materially checked by want of small piers and harbours on the coast; and we are of opinion that every reasonable encouragement and assistance should continue to be afforded, with a view to remedy this palpable evil, in conjunction with an improved organisation of system in the collection of local funds applicable to the same purpose. [52]

They went on:

> The only proper and effectual remedies to these drawbacks on the prosperity of the Irish fisheries will be found to consist in the construction of piers and harbours for the protection of the fishermen and their boats and the throwing open [of] good markets for the disposal of their produce ... The most desirable object of combining capital with labour would be ... promoted by proprietors of lands and villages bordering on the coast contributing liberally towards the providing of suitable piers and shelter harbours within reasonable distance of the reputed fishing grounds.

These were cogent pleas for the support of fishery piers and the encouragement of a fishing industry coming from the Commissioners, although they retained political correctness in emphasising the need for local proprietors to make liberal contributions to the works. It is ironic that the case was finally being made for funding of new piers and quays solely as an infrastructural aid to the fisheries, and not as relief measures, just as the catastrophe of the Great Famine was about to explode and convulse the country.

The Early Piers of the OPW

The first works of the OPW in the West, leaving aside the construction of the commercial docks at Galway City and a new quay at *Garraí Glas* near the Claddagh, were the construction of a wharf and the clearing of the sea passage at the Pass of *Béal an Daingin* (Bealandangan) and the completion of Barna pier.

The Pass of Béal an Daingin

When Nimmo first surveyed Connemara in 1813, he recommended the establishment of small villages and towns at locations that were easily accessible by land and water. One such place he favoured was Bealandangan, the place on the mainland not far from Carraroe from which people proceeded on foot along intertidal trackways into the islands of Annaghvaan, Gorumna, Lettermore and others that make up *Ceanntar na nOileán* ('the island district'):

'I should, on the whole,' Nimmo wrote,

> prefer Bealandangan, over against the south point of Rossmuck, which has depth enough of water for a small port and lies in the centre of the communications by land or water ... [T]he formation of a quay there, and beaconing and clearing part of the channel, are necessary steps in the agricultural improvement; a good boat-creek on the spot may be enlarged into a tolerable wharf.[53]

Strangely enough, he did not attempt to initiate any structural work there during his time building piers and quays in the 1820s. The so-named Pass of Bealandangan was a narrow, rock-strewn channel, joining the upper end of Greatman's Bay to Kilkieran Bay and it was indeed an important nexus in the communications network of the area. Small boats could navigate through it at high tide from the upper reaches of one bay to the other, avoiding the long, circuitous and dangerous passage around Golam Head that was otherwise the only sea route to Galway and elsewhere. But the pass was dry at low water and navigable only around high tide due to the masses of rock that obstructed it and the conflicting currents they created. Tidal flow was swift in the channel and the varying currents made navigation particularly difficult when sailing or rowing. If the rocks were cleared away, boat passage would be easy from about half flood to nearly half ebb.

However, the pass was essential to foot traffic as well: people and livestock used to cross the channel on foot around low tide, following a well-used intertidal track that connected the island of *Eanach Mheáin* (Annaghvaan) with *Béal an Daingin* village on the mainland. Dual-use passes like this were common in Connemara early in the nineteenth century; many would later be replaced by elevated causeways that closed off the passages for boats, or that incorporated bridges to facilitate the passage of small boats.

Nimmo's original plan for clearing the pass and erecting a wharf there was one of the first entirely new works that the OPW undertook in the West. In their second annual report the Commissioners made a strong plea to Treasury, calling attention to the extreme hardship suffered by boatmen using the pass and the frequency with which they lost cargo during their passage through the pass.[54] So bad was the situation that the boatmen offered to give their own labour free to help with implementing any improvements the OPW might undertake. Their poverty precluded them from contributing money to the enterprise. Hamilton Hartley Killaly made a detailed survey of the place on behalf of the OPW, and in June 1833, he drew up a preliminary, indicative plan and estimate for suitable works. The plans, signed by Killaly, are reproduced in the second annual report of the OPW.[55] He proposed to clear the channel of rocks and to erect two piers in the lower section of the pass, one on either side of the cleared channel.

The western pier, on the *Eanach Mheáin* side, was to be 150 feet long and that on the eastern, mainland side, 48 feet. Both would have 'warping posts' on them to assist in moving boats through. The upper section of the pass (above the modern road bridge) required only the removal of the impeding rocks, and no pier or quay was necessary there. The estimated cost, including a paved intertidal roadway leading to, and crossing, the channel near the piers, was £580.

The Commissioners agreed to grant half this sum, the maximum they were permitted to allot under section 85 of the Act, on condition that the other half was made up locally. Local landowners, Mr St George and Mr John O'Shaughnessy, offered £50 each towards the work and the Coastguard generously agreed to subscribe a few pounds. Unfortunately no other contributors came forward, so the total local sum did not make up the full one half required and in consequence no action was taken immediately.

The matter of Bealandangan was raised again in 1835 at the Select Committee on the Board of Works, Ireland. Alexander Nimmo Junior – nephew of the late Alexander Nimmo – once again explained the benefit that clearing the pass would have, claiming that great loss of life and cargo occurred there regularly. He stated that some improvements had already been made by a Mr O'Flaherty, but they needed to be continued and extended; further rock clearance could be done for £200.[56] In his evidence, Martin O'Malley claimed that fifty laden boats, at least, passed through the place every day of the year, carrying turf or seaweed. He also said that 500 boats, on average, passed through each day in summer, a hardly believable volume of traffic.[57] Both witnesses emphasised the importance of the pass for the locality and reputedly for places as far away as Galway and the Aran Islands. Almost as an afterthought, they claimed its development would be helpful to the fisheries as well. Towards the end of that year (1835), the OPW brought the matter before the Lord Lieutenant who, they were aware, had some unappropriated relief funds still at his disposal. He readily offered £700 to the Commissioners to take what action they thought necessary. A detailed new map was drawn up, largely following Killaly's scheme, although differing from it in some features. It was illustrated by sectional drawings of the upper and lower passes, showing the depths of water that could be expected once everything was finished. These were probably drawn by Henry Buck, the supervising engineer for the OPW, since Killaly had already emigrated to Canada.[58]

Starting in early 1836, using day labour, rock was extensively cut away at the lower pass by blasting with gunpowder.[59] A long (approx 350 feet) landing pier/wharf was erected on the western, Annaghvaan side of the channel, at a place now called *Tithe na Cora* ('weir houses'). It was joined to the dry land there by a narrow causeway about 150 feet in length, built out across the shore, the whole forming a T-shaped structure. Killaly's proposed eastern pier on the opposite side of the channel was not built, but a substantial stone mound was erected to mark the mouth of the channel. At the upper pass, all the impeding rocks were cut away to below the level of low neap tide.

All of these improvements, shown in a large-scale map and diagrams published in the fifth Report of the OPW,[60] were of invaluable help to boats navigating the pass. Pedestrian needs were also addressed: a so-called 'paved roadway' was laid down from the mainland at *Béal an Daingin* village (starting near the present-day short slip dating from 1920, near *Tí Darby*), through the intertidal zone to the stone marker mound. It then crossed the channel at low neap tide level and passed through the western pier/wharf by means of a short, low underpass tunnel. In this way, the needs of the boats were met from half flood to half ebb and those of pedestrians from three-quarters ebb to one-quarter of the flood.

Today, the original intertidal 'paved road' is still evident and its 'paved' surface, consisting of small, even, broken granite stones, is easily exposed by cutting, or simply pushing aside, the seaweed attaching to it. The underpass tunnel is still *in situ*; the lintel stone on its eastern face bears the incised inscription, still legible: 'B.P.W. A.D. 1836'.[61]

The pier/wharf and linking causeway vested in the OPW in 1846[62] and transferred to the county in 1854.[63] Bealandangan does not feature in any Returns to Parliament regarding fishery piers for many years after that. In fact, it was never formally a fishery structure at all. When the OPW first proposed to fund it in 1832, it was to facilitate the coastal boat trade under section 85 of the 1831 Act; and when the Lord Lieutenant actually funded it in 1836, the money came from relief funds rather than fishery funds. When the

Fishery Inspectors made their first mention of Bealandangan in 1879, it was only to state that it was 'in very bad order, but useless for fisheries'. In 1884, however, an application for repairs and improvements was submitted to the OPW and the Board's engineer visited the site. He estimated that £500 was necessary for renovations involving repairs to the pier and approach, fixing mooring rings, repairing beacons and erecting new ones.[64]

After a public enquiry, a free grant out of the Sea Fisheries Fund (see chapter 7) for the whole amount needed was recommended.[65] Work started the following year and was completed in May 1887 at a total cost of £416.[66] A further small amount (£19) was needed the following year, bringing the final cost to £435, a saving of £65 on the estimate.[67] An elevated causeway, road and bridge were erected just north of the original wharf, later in the century, between and separating the upper and the lower passes and closing off access to all but very small rowing boats. The final marine work at Bealandangan was the construction, at a cost of £91, given by the Congested Districts Board, of a small slip beside *Tí Darby* in the village in 1920.[68]

The existing structures at Bealandangan Pass are notable artefacts from a time when movement throughout Connemara and its islands was far more difficult and hazardous, whether on foot or by sea, than is the case today. That fact, the Canadian connection through Killaly (see chapter 4), and the fact that the works were the very first OPW venture in Connemara make them significant monuments in the maritime heritage of the county.

Barna Pier

Barna pier had been left uncompleted when the Fisheries Commission went out of office. Damage inflicted by the storms of 1830–31 was made good by relief works carried out by the Coastguard in 1832. Their funds were too small to undertake anything beyond immediate essential repairs, as was the sum of £34 expended on the pier by the OPW in 1834.[69] The year 1836 proved to be much better for Barna: the OPW gave a grant (No. 26) of £1,000 for the repair of the pier, with the landowner, Andrew Lynch MP, providing another £500.[70]

Lynch had chaired the Select Committee on the Board of Public Works in 1835, which had given the OPW a clean bill of health; that obviously did him no harm.[71] The grant sanctioned by Treasury came from the OPW's main fund, under section 85 and not from fishery funds; £500 of it was drawn down immediately. There was some difficulty finding a contractor competent enough to do the necessary work, which for that reason did not commence until 1837, when Treasury advanced the second instalment of £500.[72] During 1837 and 1838, a total of £1463 was spent on the work, including, at long last, the completion of the pier head. All was fully completed by 1839,[73] by which time Barna pier was a busy place.

Lynch had also applied to the OPW in 1836 for a separate loan of £3,000 to improve his lands by drainage and to build new roads. A loan of £1,950 was sanctioned for these

FIGURE 3.2. The pedestrian tunnel through the wharf wall at *Béal an Daingin*. The inscription is incised on the lintel stone but is not obvious in this picture. Photo NPW.

purposes in 1837 and the work got underway immediately. Together with the reconstruction of the pier, it must have given much-needed employment to people in the Barna locality. By 1839, the OPW was contemplating a new road from Barna to Cloniff near Galway city, and started by spending a small sum surveying the intended line. Funding for the road was sanctioned by Treasury in 1840: a free grant (No. 70) of £663 was accompanied by a loan (No. 196) of the same amount repayable by the Grand Jury of Galway city; Andrew Lynch gave £232 from his own resources. Later, a further grant (No. 81) of £337 and loan (No. 208) of the same amount was made to the Galway County Grand Jury to complete the road, which was successfully finished in 1842. It is now the L1321 road joining Barna to the Galway/Moycullin road.

Lynch applied again to the OPW in July 1842, seeking yet another loan of £200 for further repairs to the pier'[74] an engineer named Edward Russell having confirmed to him that extra backing was needed for the sea face. Russell drew a coloured plan of the pier as it then stood.[75] There is no sign in it of a bridge over the stream beside the pier root, nor, indeed, of the supposed mediaeval walled dock. The Board was not permitted at the time to commit money from the Piers Fund for upkeep and maintenance purposes, the law expressly stipulating that this duty was to fall on the original local contributors. This meant the OPW could not normally undertake repairs, even where an engineer certified them as essential to keep old piers in a useful state. However, for the OPW to do nothing meant that further dilapidation could be expected. When approached, therefore, Treasury would usually sanction the expenditure of small sums for essential repairs to those piers deemed by the Board to be most worth preserving.

A free grant of £100 was sanctioned for Barna pier in 1842, with Lynch providing another £100. This time the grant was not without special conditions. A manuscript schedule of tolls to be charged, signed on 5 December 1842 by Chief Secretary for Ireland, Henry Goulburn, set out a scale of charges for the use of Barna pier.[76] Vessels carrying sand were charged two pence; every boat landing seaweed paid sixpence, with a further two pence per day if it stayed beyond the first day; boats with vegetables (most likely potatoes) paid one penny; with slates, three pence; and with turf, two pence. The tolls were meant to pay for the upkeep of the pier, quay and harbour. (Piers constructed in open sites exposed to the swell, as Barna was, suffered repeatedly from storm damage and needed constant refurbishment.) That Barna was regarded mainly as a 'commercial' pier, rather than one for fisheries only, is apparent from Treasury's insistence on these tolls. They also give some indication of the kind of goods that were likely to pass through the place in the years before the Great Famine.

The pier's usefulness for fisheries was not, however, entirely overlooked: a new local bye-law specified that fishing vessels were to get preference in berthing there. Another bye-law prohibited the use of spikes or grapples as mooring devices driven into the stone work of the structures, and admonished users to avail of the mooring rings provided. With finance for maintenance assured in this way, Russell provided detailed specifications for the works in January 1843[77] and they were all done soon after.

Clarinbridge Old and New Piers

Under section 85 of the 1831 Act the Board could, and did, give grants and loans from their main funds in support of proposed commercial piers and quays that did not patently support fisheries. One such was a pier at Clarinbridge for which they granted £400 in 1842.[78] The Galway Grand Jury contributed £200 and T.N. Redington of Kilcornan gave £400.[79] Almost £900 was spent immediately, and the pier was completed in 1843 by a contractor, Richard A. Gray.[80]

Today there are two pier structures in Clarinbridge. One is an old, now abandoned, vernacular quay (named Tobarnagloragh pier) situated beside an old ruined mill on the right bank of the River Clarin as it exits the village. Much used in early times for landing seaweed and turf, it was built by county presentment before 1833[81] and is indicated in the first OS six-inch map of 1839.

Named 'Cloranbridge' in the record, thirty fishing yawls of about three tons each were said to frequent it (not all at the same time, one assumes), resulting in the employment of about seventy persons. The other is a very fine, cut-stone jetty pier located slightly further down the estuary. This is the one built in 1842; it was not in existence when the survey for the 1839 six-inch OS map was carried out, but it is shown, with its approach road (today just a grassy track), in Mulvany's map of the Oranhill Drainage District in 1846.[82] Tellingly, there is no reference to this good pier in later fishery documents. Clearly it was funded, built and treated as a pier of exclusively commercial usefulness rather than for fisheries, and possibly also as predominantly a private venture of the Redington family. For similar reasons, Duras pier, on the south shore of Galway Bay, built in 1823 by Patrick Lynch of Duras Park with practical help from Nimmo and financial assistance from the then Fishery Commission, was also largely ignored by the fisheries authorities in later years.

Ballyvaughan Pier

Nimmo investigated Ballyvaughan as a possible site for a pier in 1824 and was not much impressed.[83] He regarded Ballyvaughan Bay as too shallow for any but the smallest boats, and the quay already in existence, made by the villagers on land owned by Mr MacNamara, was too dilapidated for him to bother restoring. The most significant relevant feature of the site was the presence there of a Coastguard station, located beside the old quay. Nothing was done then, but attitudes were more positive ten years later at the Fishery Enquiry of 1835–36.[84]

The local Coastguard Officer, Lieut. V.P. Hunter, stated in evidence that a new quay at Ballyvaughan would be a decided benefit to the Clare fisheries and could be provided for £500, in his opinion.[85] By then the village had thirty-seven houses with 235 inhabitants, some employed in the seasonal herring fishery, and the old quay no longer fulfilled the needs of the place.[86] A plan and sketch of a new pier were drawn up in 1836 by Richard Grantham (1805–91) the outgoing County Surveyor of Clare. (Grantham resigned the surveyorship in September 1836 and travelled to England to work under Isambard Kingdom Brunel, going on to make a fine reputation for himself as a consultant engineer.) James Boyd (*c.* 1781–1853) replaced Grantham as Clare County Surveyor and he carried forward Grantham's plan for the pier. Boyd remained Clare County Surveyor from September 1836 until December 1845.[87] Grantham's plan was referred to the OPW in 1837, supported by Boyd, with Major MacNamara indicating that he would provide £250 towards it.

The OPW responded in October, recommending a grant of £750 for the work, but for some unknown reason the proposal was not taken any further.[88] In 1841, the Grantham plan and sketch were again given to MacNamara to bring before the Grand Jury in the hope that it would make a presentment for the work; MacNamara indicated that there would be a local contribution (from himself) of £250. For a second time the plan was not to be progressed with, on this occasion 'for want of funds' by the county.

Other requests to the Clare Grand Jury around that time – for county courthouses – were also refused due to lack of funds.[89] However, Ballyvaughan pier did not fall permanently by the wayside and in 1844 the engineer Edward Russell (whom we encountered at Barna) was sent by the OPW to investigate the site and the advisability of proceeding with the work. Russell drew a new map, still extant, showing Grantham's proposed pier, the site of the old quay in front of the Coastguard Station, and his own proposal for a long, straight pier sited a little to the west of Grantham's proposed site.[90] Russell's site had the advantage of being nearer to deep water than either of the others.

The proposal now faced its third and final obstacle, an objection of a kind that would recur on occasions with some Irish piers. William Fishbourne, agent of the Duke of Buckingham, wrote from Kelvingrove, Co. Carlow, to W.N. MacNamara MP, complaining that Russell had been instructed to report on one particular spot only, without reference to any other part of the coast. 'This I certainly cannot think was right', he complained, 'and if the Duke of Buckingham is to be expected to subscribe, I think the site on His Grace's property is, if I am not much mistaken, the largest in that locality'.[91]

The Duke owned about 6,000 acres in the Barony of Burrin, including Clareville House near Ballyvaughan village. Fishbourne had already intimated to the OPW that His Grace would offer £10 towards the cost of a new survey to include the Clareville site.[92] Russell's original survey had indeed been confined to the location beside the Coastguard Station, because the application in hand concerned that site. MacNamara was owner of the land at that precise place and was the only person who had offered to contribute towards the cost of building the pier. The person making the local contribution was normally the one who chose the precise site; that was the rule then, and it would be clearly and expressly enshrined in legislation just two years later.

Nevertheless, the OPW directed Russell to return to Ballyvaughan and carry out another survey 'to ascertain if a more suitable site could be discovered for a pier and quay than that close to the Coast Guard Station'. Russell did as instructed in March 1845 and he reported that he 'examined particularly Clareville, an attempt having been made to construct a rude pier there'; he made no mention of Fishbourne's correspondence. He found both the MacNamara site and Clareville to be the only feasible sites for a pier in the whole area. The Clareville site was, he wrote, well sheltered, once the boats were inside Gall and Green islands. The MacNamara site was more windward, but slightly deeper and less obstructed. In a carefully worded report,[93] he seemed personally to favour the MacNamara site, but did not express his preference outright, choosing only to say that it was the opinion of the fishermen and the local inhabitants that the MacNamara site was, for them, the clear place of choice.

Fishbourne was greatly peeved by the new report. He wrote again to the OPW, this time complaining that Russell had visited and done his survey without telling him in advance.[94] He, Fishbourne, had intended to be on site to put Russell straight on certain practicalities. Russell had spent only one day there, Fishbourne claimed, and had not even spoken to the Duke's bailiff; he had not visited Clareville at high tide, and there were other alleged deficiencies in his survey. In a minute recorded by the Board one week later, Russell made a spirited response to these criticisms: he had examined Clareville at high and low tide, spending over twelve hours there on his first day, and returning to the site again early the next morning; he had taken soundings all over the bay using the Coastguard boat; Captain Beaufort had done a tracing of the soundings that had confirmed Russell's good general approach; he had spoken with the fishermen and the inhabitants in great detail and it was clear what they considered the best place and also the one closest to the village.

The OPW accepted this explanatory minute and, on 18 January 1846, Russell conveyed the OPW's decision: if a county presentment raised £300 and a local contributor gave the same sum, the Board would recommend a grant of £600 for a pier at the site selected by him. On 17 March 1846, Treasury sanctioned the grant of £600 to be taken from the legacy piers fund. Because the money came from this source rather than from the Piers and Harbours Fund (see chapter 4), and although the pier was built during the years 1846 to 1848, Ballyvaughan pier was not, strictly speaking, a Great Famine pier: only the piers funded under the Piers and Harbours Acts of 1846 and 1847 were regarded and recorded as piers of the Great Famine.

One month later, on 24 April 1846, Barry Gibbons drew a new site map (still extant)[95] and drew up specifications for the new structure. Tenders were invited and the OPW accepted the offer of William Brady to do the work for £1,127. Everything was completed to satisfaction by 1848 under the local supervision of J.R. King and the overall direction of Barry Gibbons.[96] The pier vested in the OPW in 1849.[97] Following repairs costing £200, funded by presentment and carried out under the supervision of Robert Cassidy of Skull [Schull, Co. Cork] in April 1852, it was transferred to Clare County by the Grand Juries Act in 1853.[98] The Duke of Buckingham sold his Clareville Estate in 1848 to Col. Henry White who would later become Baron Annaly.[99] The latter's agent for the estate would be William Lane Joynt, who came to reside at Clareville House and became a noted supporter of small fishery piers and quays (see chapter 6).

FIGURE 3.3. Loughrask quay and pier at Callahavreedia near Ballyvaughan in north Clare (CS 023) as it is today. This structure pre-dates the main pier at the village. Photo NPW.

Today there are two piers at Clareville, just outside Ballyvaughan village in the townland of Loughrask. One (CS 024) is a long (approx 400 feet) narrow, linear structure, about three courses high, extending straight out into the sea. It is not shown in the first edition of the six-inch OS map of the area. The second comprises a short quay with an angled pier (CS 023), located on the shore at a place named Callahavreedia, close to the main road at Clareville. This structure, comprising a quay and a curved pier, is shown in the first edition of the six-inch OS map (1842) of the district. Both this pier and the narrow linear structure have gates at their respective roots opening in to the grounds of Clareville House.

The pier shown in the early OS map and numbered CS 023 in the Clare Heritage Survey is most probably the original structure mentioned by Russell as 'a rude pier' constructed at the place.

Endnotes

1 1st Annual Report of the OPW, BPP 133 (75), p. 3.
2 Commissioners of Irish Fisheries to Chief Secretary, 14/1/1830, in 1st Report of the Commission of Inquiry into the Irish Fisheries 1836, Appendix 5, pp. 27–28.
3 *An Act to revive, continue and amend several Acts relating to the Fisheries,* 1 Wm IV C. 54.
4 11th Report of the Commissioners of Fisheries, BPP 1830 (491).
5 Return to Parliament, BPP 1842 (394).
6 Further Return to Parliament, BPP 1865 Sess. I [207].
7 1st Annual Report of the OPW, BPP 133 (75), p. 18.
8 1st Annual Report of the OPW, BPP 133 (75), p. 3. ; NAI CSORP 1830/5039.
9 Paul Duffy, personal communication, 2016.
10 J.S. King, in 1st Report of the Commission of Inquiry into the Irish Fisheries 1836, p. 223.
11 Lieut J. Kemp, CGO, in 1st Report of the Commission of Inquiry into the Irish Fisheries 1836, p. 223.
12 H. Buck, CE, in 1st Report of the Commission of Inquiry into the Irish Fisheries 1836, p. 102.
13 1&2 Wm IV C. 33.
14 1st Annual Report of the OPW, BPP 1833 (75), p. 3.
15 1st Annual Report of the OPW, BPP 1833 (75), p. 18.
16 1st Report of the Commission of Inquiry into the Irish Fisheries 1836, Appendix 19.
17 Select Committee on Public Works in Ireland, BPP 1835 [573], appendix 4, pp. 306–08.
18 2nd Annual Report of the OPW, BPP 1834 (240), p. 5.
19 3rd Annual Report of the OPW, BPP 1835 (76), p. 4.
20 Select Committee on Public Works in Ireland, BPP 1835 [573].
21 Select Committee on Public Works in Ireland, BPP 1835 [573], query 3076.
22 Select Committee on Public Works in Ireland, BPP 1835 [573], queries 2620–21.
23 Select Committee on Public Works in Ireland, BPP 1835 [573], report, p. 8.
24 A Bill to empower Grand Juries in Ireland to raise Money by Presentment for the construction, enlargement or repair of Piers and Quays, BPP 1835 (331).
25 A Bill (as amended by the committee) to empower Grand Juries in Ireland to raise Money by Presentment for the construction, enlargement or repair of Piers and Quays, BPP 1835 (548).
26 1st Report of the Commission of Inquiry into the Irish Fisheries 1836.
27 A.R.G. Griffiths, *The Irish Board of Works, 1831–1878* (New York: Garland Publishing Inc., 1987).
28 5th Annual Report of the OPW, BPP 1837 (483), p. 4.
29 6th Annual Report of the OPW, BPP 1837–38 (462) p. 10.
30 Select Committee on the Sea Coast Fisheries (Ireland) Bill, BPP 1867 (443), appendix, pp. 270–71.
31 J.R. Barry, in Select Committee on the Sea Coast Fisheries (Ireland) Bill, BPP 1867 (443), queries 3276–3289.
32 Return to Parliament, BPP 1842 (394).
33 1st Report of the Commission of Inquiry into the Irish Fisheries 1836, appendix 4, p. 26.
34 See 4th, 5th and 6th Annual Reports of the OPW.
35 9th Annual Report of the OPW, BPP 1841 Session I (252).
36 10th Annual Report of the OPW, BPP 1842 [384].
37 J. Coombes, *Utopia in Glandore* (Butlerstown: Muintir na Tíre, 1970)
38 *An Act to regulate the Irish Fisheries,* 5 & 6 Vict. C. 106.
39 11th Annual Report of the OPW, BPP 1843 [467], p. 30.
40 *An Act to regulate the Irish Fisheries,* 5 &6 Vict. C. 106, Section 1.
41 1st Annual Report of the Commissioners of Fisheries, Ireland, BPP 1843 (224), p. 1.
42 1st Annual Report of the Commissioners of Fisheries, Ireland, BPP 1843 (224), p. 11.

43 8th Annual Report of the OPW, BPP 1840 (327).
44 13th Annual Report of the OPW, BPP 1845 [640], p. 4 and p. 33.
45 14th Annual Report of the OPW, BPP 1847 [762], p. 33.
46 14th Annual Report of the OPW. BPP 1847 [762], appendix 8.
47 15th Annual Report of the OPW, BPP 1847 [847], p. 4.
48 16th Annual Report of the OPW, BPP 1847–48 [983], p. 98.
49 17th Annual Report of the OPW, BPP 1849 [1098], appendix 10, p. 159.
50 18th Annual Report of the OPW, BPP 1850 [1235], p. 96.
51 2nd Annual Report of the Commissioners of Fisheries, Ireland, BPP 1844 (502), p. 4.
52 3rd Annual Report of the Commissioners of Public Works, in re the Fisheries of Ireland BPP 1845 (320), p. 2.
53 A. Nimmo, in 4th Report of the Commissioners for the Bogs of Ireland, BPP 1813/14 (131), appendix 12, pp. 200–01.
54 2nd Annual Report of the OPW, BPP 1834 (240), pp. 5–6.
55 2nd Annual Report of the OPW, BPP 1834 (240), plate 1.
56 A.Nimmo, Jnr., in 1st & 2nd Reports of the Select Committee on Public Works in Ireland. BPP 1835 [573], queries 2205–2211.
57 M. O'Malley, in 1st & 2nd Reports of the Select Committee on Public Works in Ireland, BPP 1835 [573], queries 3299–3308.
58 NAI OPW/8/245.
59 NAI OPW/8/245.
60 5th Annual Report of the OPW, BPP 1837 (483), plates 1 & 2, p. 30.
61 I thank Roger Derham for bringing this to my attention.
62 17th Annual Report of the OPW, BPP 1849 [1098], appendix E2, p. 281.
63 23rd Annual Report of the OPW, BPP 1854–55 [1929], p. 61.
64 Royal Commission on Irish Public Works (the Allport Commission), Second Report, BPP 1888 [5264-1], list No. 5 handed in by General Sankey, p. 710.
65 Annual Report of the Commissioners of Fisheries for 1885, BPP 1886 [C. 4809], p. 7.
66 Annual Report of the Commissioners of Fisheries for 1887, BPP 1888 C. 5388, p. 16.
67 Annual Report of the Commissioners of Fisheries for 1889, BPP 1890 C. 6058, pp. 22–23.
68 28th Annual Report of the Congested Districts Board, BPP 1921 [Cmd. 1409], appendix 24, p. 46.
69 3rd Annual Report of the OPW, BPP 1835 (76), p. 34.
70 5th Annual Report of the OPW, BPP 1837 (483), p. 30.
71 Select Committee on Public Works in Ireland, BPP 1835 [573].
72 5th Annual Report of the OPW, BPP 1837 (483), p. 30.
73 8th Annual Report of the OPW, BPP 1840 (327).
74 Lynch to Commissioners for Public Works, 6/7/1842. In NAI OPW/8/38.
75 E. Russell to H R Payne, 5/9/1824. In NAI OPW/8/38.
76 Schedule of tolls at Barna, 5/12/1842. In NAI OPW/8/38.
77 Specifications for Barna pier by Edward Russell, 12/1/1843. In NAI OPW/8/38.
78 10th Annual Report of the OPW, BPP 1842 [384], p. 6.
79 11th Annual Report of the OPW, BPP 1843 [467], grant no. 93, p. 26.
80 NAI OPW8/81.
81 1st Report of the Commission of Inquiry into the Irish Fisheries 1836, appendix 19, p. 123.
82 17th Annual Report of the OPW, BPP 1849 [1098], facing p. 260.
83 8th Report of the Commissioners of Fisheries, BPP 1827 (487), appendix 10, p. 46.
84 First Report of the Commissioners of Inquiry into the State of the Irish Fisheries 1836.

85 First Report of the Commissioners of Inquiry into the State of the Irish Fisheries 1836, evidence of Burton Bindon and V. P. Hunter, p. 224.

86 Lewis, *Topographical Dictionary of Ireland.*

87 Information on Grantham and Boyd is from B. O'Donoghue, *The Irish County Surveyors, 1834–1944. A Biographical Dictionary* (Dublin: Four Courts Press, 2007).

88 Minute of Edward Russell dated 28/01/ 1846. NAI, OPW5 10225/77.

89 O'Donoghue, *Irish County Surveyors*, p. 108.

90 Plan of a pier at Ballyvaughan, dated 1844. NAI OPW5/10225/77.

91 W. Fishbourne to W.N. McNamara MP, 7/7/1845. NAI OPW5/10225/77.

92 W. Fishbourne to OPW, 18/3/1845. NAI OPW5/10225/77.

93 Report of Edward Russell on the proposed pier at Ballyvaughan, 18/3/1845. NAI OPW5/10225/77.

94 W. Fishbourne to OPW, 21/3/1845. NAI OPW5/10225/77.

95 Plan, dated 1846, of proposed pier at Ballyvaughan. NAI OPW5/10225/77.

96 NAI OPW8/ 029.

97 17th Annual Report of the OPW, BPP 1849 [1098], appendix E2.

98 16&17 Vict. C. 136.

99 See Grenville family in the Landed Estates Database, Moore Institute, National University of Ireland, Galway. www.landedestates.ie.

CHAPTER 4

A New Beginning in a Never-to-be-Forgotten Famine

1845–1856

[T]he recent Acts for erection of piers on the coast is not the least important in the results which may be expected from it, whether we regard them with reference to their immediate influence on the fisheries, or their more slow, though more certain effect upon the commerce, trade, and agriculture of the localities in which the works are to be erected ... Nor is it necessary we should repeat that it is not on the ground of positive utility these works are to be judged. They should be considered solely as an effort to obtain a certain amount of labour in return for subsistence, through the medium of money wages ...

Commissioners for Public Works, Ireland, 1848[1]

The Great Famine and the Fishery Piers and Harbours Act 1846

When published in May 1845, neither the Annual Report of the OPW[2] nor their Annual Report as Commissioners for Fisheries[3] indicated any premonition of the disaster that was about to befall the country a few short months ahead; it was late in 1845 when the Board first reported signs of potato blight in Kilkee. The living conditions of the poor, always precarious, were about to become seriously imperilled. To its credit, the Government of Robert Peel in Westminster was not dilatory in initiating legislation, and three Bills to address the impending famine were introduced in Parliament and passed into law in March 1846. These were laws to facilitate an increase in public works so that the destitute could earn money to buy food, rather than be reliant on direct food handouts. It was a formula that Peel had found useful when faced with a famine that occurred while he was Chief Secretary for Ireland in 1817, when he succeeded in getting the Loan Act[4] and later the Moiety Act[5] passed by Parliament.

One of the three new Acts of 1846 was concerned with piers and harbours: *An Act to encourage the Sea Fisheries of Ireland, by promoting and aiding with Grants of Public Money, the construction of Piers, Harbours and other Works.*[6] This became known colloquially as the Piers and Harbours Act 1846, and it set down a template for the funding, construction and management of fishery piers and harbours for the next forty years.

Some Details of the Piers and Harbours Act 1846

Section 1 appointed the Commissioners for Public Works as Commissioners for the execution of this Act. They were given £50,000 in order to make grants and loans for erecting, repairing or improving small piers, harbours, quays, landing slips and places, approach roads, cuts, beacons and lights and other works on the sea coast of Ireland that they deemed necessary for the encouragement and promotion of the sea fisheries, subject to a strict set of conditions:

> No individual grant for any work was to exceed a total of £5,000;
>
> The amount of any free grant was not to exceed three fourths of the total actual cost of the work;
>
> Grants were to be made only on condition that the residual one-fourth of the cost was secured by the county, the district or the owners of land or others in the area;
>
> The amount of the residual one-fourth could be advanced by the Commissioners to the county, the district or individual as a loan, to be repaid at such interest and in such manner as Treasury approved;
>
> The works were to be executed, constructed, maintained and kept in repair by and under the direction of the Commissioners for Public Works; the cost of maintenance and repair was to be recouped out of the tolls, rates and rents accruing from the works;
>
> The works done and lands purchased or necessarily acquired for them, were deemed to be, and were to be taken into, public ownership and for that purpose, absolutely vested in the Commissioners for Public Works and their successors.

Any resident, proprietor or occupier of land in any place adjacent to the sea coast could apply by memorial to the Commissioners requesting that a new pier or quay be erected, or an old one renovated, with a grant under the Act. On receipt of such memorial, the OPW would undertake a preliminary survey and examination in order to prepare a report on the site, the works envisaged, the structural plan, the estimated cost and all other circumstances of the case. Once that report was prepared, the Commissioners could decide to recommend a grant, indicating the appropriate amount to be advanced. The report and the recommendation were then to be forwarded to Treasury for sanction. When Treasury sanction was notified, the Commissioners would prepare a declaration as to how much of the total cost should be given as a free grant, and how much as a loan, identifying which entity – county, district or individual – should be charged with the surety and repayment of the loan portion. All the plans, loan arrangements and so forth were then to be advertised locally for public inspection. If necessary, objections were to be heard and adjudicated in public and, eventually, when all the requirements of the Act were completed, a final notice was to be printed in the *Dublin Gazette*. After publication, and in the absence of legitimate objections, grants and loans could be issued and the Commissioners could commence and proceed with the agreed works, either by direct labour or by contract.

The Act contained many sections – eighty-seven in all – but the level of detail, the technical complexity, the extent of bureaucracy and degree of supervision can be gauged reasonably well from the above résumé of its salient, practical requirements. The process was lengthy, cumbersome, complicated and excessively bureaucratic; it would not be easy for the public to avail of, or for the Commissioners to implement quickly. Legislation that had been conceived with commendable speed and decisiveness was quickly smothered in a morass of bureaucracy.

Section 64 had particular, very long-term import of a unique character: under it, all existing piers, harbours, quays and landing places that had been erected wholly or partly with public money in support of the fisheries in the previous forty-five years, and which were not now private property, but were still useful for fisheries, were declared to be public property and, after a twelve-month period, were to vest in the OPW and be treated in the same manner as the new works done under this Act.

The OPW Starts Work with the Famine Piers

Before 1846 was out, the Commissioners had received 125 memorials for piers – so many that they found it necessary to employ an extra engineer, Barry Duncan Gibbons (*c.* 1798–1862), to assist in implementing the scheme.[7] Gibbons, who was then resident engineer at Kingstown (Dun Laoghaire) Harbour, had gained his experience as an engineer while completing Nimmo's pier at Kilrush in Co. Clare, and at Dunmore Harbour in Co. Waterford in the years immediately after Nimmo's death. He now made the necessary preliminary surveys and reports on eighty-nine of the sites put forward for piers. The Commissioners selected thirty-four of these, which they recommended to Treasury for sanction. The estimated total cost of these was almost £80,000, of which about 60 per cent was to be given as free grants, about 32 per cent as loans to the applicants, and 8 per cent was to be contributed in other ways from the localities. Twelve sites received free grants of 50 per cent, nine were given 67 per cent, and nine more received 75 per cent. Even at the height of the Famine, the law was obviously being implemented with greater strictness than absolutely necessary (the Act allowed free grants of 75 per cent). A sum of £1,321

TABLE 4.1

Works carried out in Co. Galway during the Great Famine, under the Piers and Harbours Acts of 1846 and 1847.

Place	RH No.	Works	Applicant	Estimate	Free Grant	Other funds*	Actual amount spent
Kilronan (Aran Islands)	181	Pier and slip	J. Digby	£2,400	£1,600	£800	£1,607
Bunowen	030	Pier	J.A. O'Neill	£6,400	£4,800	£1,600	£4,256
Ballinakill (Keelkyle)	013	Wharf and slip	F.J. Graham	£900	£450	£450	£410
Claddagh	145A	Quay	Rev. J.R. Rush and others	£4,000	£2,667	£1,333	£2,658
Curhowna (Errislannan)	022	Dock	J.S. Lambert	£800	£535	£265	£572
Kilkieran	072	Pier	Waste Land Society	£3,000	£1,500	£1,500	£1,004
Rosroe	004	Quay and slip	Gen. Thompson	£800	£600	£200	£565
Roundstone (north pier)	040X	Pier 'repairs'	T. Martin MP	£1,200	£600	£600	£35 (work abandoned)
Tarrea	159	Pier and harbour	M. Blake	£2,100	£1,050	£1,050	£1,481

*Other funds include the local contribution and loans under the Acts, along with other contributions.

Source: data from 15th Report of the OPW. Sundry amounts of a relatively trivial nature were also spent at Ardfry (New Harbour) (£15), Barna (£30) and Clifden (£18). RH No. is the number in the Ryan Hanley survey.

was spent immediately on thirty-two of the approved proposals; another £482 was spent on Gibbons's preliminary surveys of the remaining sites that had been surveyed but not yet forwarded to Treasury for sanction. These latter sites, none of which was in Galway or north Clare, are listed in the 15th Report of the OPW.[8]

The nine applications that were sanctioned in Co. Galway are listed in Table 4.1. Four were for completely new, relatively large piers, at Bunowen, Kilkieran, Kilronan and Tarrea. At all of these places, impressive structures were erected, which remain today the most obvious monuments of the Great Famine in the maritime heritage of the county. One proposal was for repairs at Roundstone; another for renovation of the existing piers at the Claddagh; and three were for smaller, new structures at Keelkyle (Barnaderg in Ballinakill Harbour), Rosroe in Killary Harbour and Curhowna (Errislannan). Two more Galway memorials were still awaiting approval, a pier at Oranmore and the extension of the quay at Clifden.[9]

More Legislation, More Money

The 1846 Piers and Harbours Act was the first statute to grant public funds specifically for the erection of fishery piers and harbours in Ireland since the demise of the old Fisheries Commission in 1830. Its principal purpose was to provide for relief works, rather than to initiate a rational, planned response to infrastructural deficiencies in the fishing industry. When addressing section 64 of the Act, which required the older existing piers to be vested in the OPW, the Board noted that this duty would take up much of their time and add to their workload, since the old piers were located all over the country, many in the most remote places, and many were derelict. They took the precaution therefore of flagging the need for new legislation to fund repairs and maintenance once the older piers would vest. Many were in places where they could never sustain tolls and levies on the boats frequenting them, due, not least, to the poverty of the boatmen. At places where tolls might be successfully levied, the amounts collected were unlikely ever to be sufficient to cover the cost of collection, leaving the matter of repairs entirely out of the question.[10]

As to the new piers, it had not taken long, nor taken many memorials, for the whole £50,000 granted by the Act to be committed, and there were still many memorials in hand that needed to be addressed. New legislation alone was not, therefore, all that was required; extra funds also needed to be voted urgently. In response, a new enactment, *An Act for the further improvement of the Fishery Piers and Harbours of Ireland*,[11] was passed in July 1847, providing a further £40,000 to be spent for the purposes of, and under the same conditions as, the 1846 Piers and Harbours Act. The aggregate amount made available for grants and loans was therefore £90,000, and the Commissioners had received 134 memorials.

Within a year, they had fully dealt with ninety-five, had reported on thirty-three others without doing a full survey, and six had been withdrawn. (Memorials could be withdrawn, or sanction refused, for a variety of reasons, such as excess cost, commercial rather than fishery usefulness, unsuitability of the site, failure of the local contribution and so on.) Treasury sanctioned a total of forty-five (about one-third), and work started on thirty-seven of them, using contractors in twelve cases. Thirteen works were fully completed by 1849.[12] Nationally, by the end of 1850,[13] thirty-one works were completed, three were almost complete, six were still progressing, three were suspended, one was not yet started (Brandon in Co. Kerry) and one had been withdrawn (Glenarm in Co. Antrim). Eventually, a total of sixty-six structures were sanctioned and sixty-four successfully completed in all counties during the Great Famine; they are all listed in Appendix C. New memorials were constantly coming in, and others were being withdrawn from the list, but the nine new sites sanctioned in Co. Galway remained unchanged, with work proceeding steadily on all but one. The exception was Roundstone, where the scheduled 'repairs' were listed as suspended.[14]

This achievement was not overwhelmingly impressive, given the scale of the destitution and distress, but nevertheless it was a creditable enough outcome for the peak period of the Great Famine. Most importantly, the number of men employed daily according to the

OPW Reports had been significant. Over the four years, 1847 to 1850 inclusive, a total of 147,954 men were reportedly 'employed daily'. The number varied from twenty-eight to 1,207 in 1847, 3,300 to 6,296 in 1848, 2,766 to 5,287 in 1849 and 1804 to 4210 in 1850 (4.2). Generally speaking, fewer men were engaged in the more inclement months of November to February and greater numbers in the more amenable months, April to October.[15]

Caution is necessary, however, in interpreting the published figures, a caveat that is explained and discussed in greater detail in chapter 6. The Commissioners were quietly satisfied that things had gone reasonably well, and in their 20th Annual Report for 1851[16] they published a map of the various locations where new piers and quays had been built, with a brief explanatory commentary on a number of them. Sketch plans of forty-three of them were also published in the Report; these help greatly in confirming the precise locations and the kinds of structures erected. Substantial new stone piers and wharves, in some cases said to extend to 10 feet below low water of Spring tide, were erected throughout thirteen counties.

All those in Galway were substantial works and are still important to this day in the inshore fisheries and general maritime activity in the county. Although not new structures, repairs had also been carried out at the old 1820s piers at Clifden, Rinville (New Harbour, also called Ardfry) and Barna, improving their usefulness; these are included in Appendix C.

TABLE 4.2

Number of artificers and labourers reported as 'employed daily' in Ireland on piers and harbours work during the Great Famine years of 1847 to 1850.

Month	1847	1848	1849	1850	Total
January	28	3,358	2,999	2,731	9,116
February	265	4,091	3,395	2,800	10,551
March	372	5,281	4,423	3,864	13,940
April	492	4,119	3,493	3,148	11,252
May	536	4,990	3,721	4,210	13,457
June	705	6,296	5,287	3,202	15,490
July	1,106	5,083	4,093	3,230	13,512
August	1,057	5,771	3,196	3,544	13,568
September	1,065	4,917	4,420	2,613	13,015
October	1,207	4,644	3,438	2,751	12,040
November	940	5,308	4,162	2,811	13,221
December	749	3,473	2,766	1,804	8,792
TOTAL	**8,522**	**57,331**	**45,393**	**36,708**	**147,954**

Source: data collated from OPW reports 16, 17, 18 and 19.

Further work under the Piers and Harbours Acts continued at a less intense level in 1851 and 1852, by which time all the sanctioned projects were finished, except the pier at Red Bay (Cushendall, Co. Antrim).[17] When the final account of expenditure of the £90,000 allocated by the 1846 and 1847 Acts was given in a Return to Parliament in 1859, it confirmed that a total of sixty-six old and new piers in thirteen counties had benefitted, a total of £74,340 had been spent (the additional funds were the local contributions), and a balance of £15,654 remained in hand.[18] The parsimony exhibited by saving so large a portion of the voted money during a period so full of death, despair and emigration needs little comment.

Gathering in the Older Piers

The other duty laid on the Commissioners for Public Works by the Act of 1846 (Sect. 64) and further elaborated in sections 3 and 4 of the Act of 1847 – the vesting in them of the piers built with public money in the preceding forty-five years – progressed alongside the programme of building new piers. In 1847, the Board initiated steps to vest these older piers, listed in their 17th Annual Report.[19] They included sixteen of the older Galway piers and two (Burrin New Quay and Ballyvaughan) in north Clare. Ballyvaughan had been funded mainly from the old legacy funds, with only £119 added from the Piers and Harbours funds; for this reason, it never appeared in the lists of piers built during the Great Famine, although it was constructed in 1846–7 (see chapter 3).

The Commissioners awaited precise instructions from Treasury as to how best they should proceed with the delicate task of vesting, which could seem to encroach seriously on private property rights. The general manner of proceeding was prescribed in Section 3 of the 1847 Act: they were to identify and describe the structures that were still useful for fisheries and which, in their opinion, ought to be vested. That information was to be published in the *Dublin Gazette*, and in newspapers circulating in the relevant districts, so that persons interested could voice any objections they might have within two weeks of publication. The Commissioners were then to consider the objections and make a final decision regarding which structures would vest, publishing their Order in the *Dublin Gazette*. Only then would the old piers vest in the Commissioners, who would repair and maintain them thereafter with money paid by the tolls, rates and other charges for their use.

In cases where the tolls were insufficient to cover these costs, the Commissioners were to estimate the likely amounts needed and specify the county which was to be charged with making good the deficiency. They could proceed with the necessary repairs using their normal funds, which would be recouped later by an appropriate levy on the relevant county or district. However, they were empowered to pay for the very first repairs of any of the vesting structures, using the balance remaining from the £90,000. This provision ensured that the counties would not be charged with the initial costs of bringing the old piers up to useable standard, although they would be liable for their ongoing maintenance thereafter in the event that the tolls collected were insufficient. The Act was clear that for the future, while the OPW was charged with actually doing the maintenance work, the burden of the costs would fall ultimately on the counties.

Transferring the Piers to the Counties

By 1849 due process had been completed as required, and the list of fifty-seven older piers vesting was published in the *Dublin Gazette* on 16 February.[20] Taken together with the sixty-four new piers erected under the 1846 and 1847 Acts, which had vested automatically under section 63 of the 1846 Act, a total of 121 piers and quays in the whole country (including eighteen of the older ones in Galway and north Clare, plus the eight new ones named in Table 4.1) vested in the Commissioners by 1849. As a result of this, all the piers that had been built with public money were now in public ownership and under the control of a single authority, the OPW.

However, this vesting, on its own, did not answer the question of what was to be done with them in the long term. They needed to be kept in repair and properly maintained into the future, a costly responsibility which it was never intended should rest permanently with the Board of Works, which had no designated funds for the task. As early as 1851, the Board was complaining that 'no efficient provision has yet been made for the care and maintenance of the works undertaken under the Acts'.[21] The problem was that, whereas the Board could determine what counties were to be levied with the maintenance costs, the

TABLE 4.3

Piers in Co. Galway and north Clare transferred to the counties in 1854 under the Act of 16 & 17 Victoria, C. 136.

Ardfry	Bunowen
Duras	Barnadearg (Keelkyle)
Kilcolgan	Errislannan (Curhowna)
Claddagh piers	Rosroe
Barna	Kilronan
Spidéal (old)	Costelo (Cashla)
Maumeen	Leenaun (east pier)
Killeany	
Bealandangan	**North Clare**
Roundstone (south)	Ballyvaughan
Clifden	Burren New Quay
Cleggan	

County Grand Juries were not permitted at the time to 'present' for maintenance works on piers and harbours. By law, Grand Juries could 'present' for certain public works like roads and bridges that would result in the costs being borne by the county, or by a barony, or by a district; the maintenance of piers and harbours was not, at that time, among the activities for which they were legally permitted to make a presentment ('to present'). Legislation was necessary to overcome this deficiency, and, in August 1853, an *Act for enabling Grand Juries in Ireland to Borrow Money from Private Sources on the security of Presentment, and for Transferring to Counties certain Works constructed wholly or in part with Public Money*[22] was enacted to rectify the situation. It directed that the older piers, together with those recently erected, 'shall be deemed and become the public property of the county and be maintained by the Grand Jury'.[23]

Ninety-three fishery piers vesting in the counties were named and listed in a schedule to the Act. The Grand Juries were also empowered to present for money for repairs and maintenance, and they could appoint harbour constables to enforce local bye-laws regarding the use of the piers. Should the Grand Juries fail subsequently in their maintenance duty, the OPW could undertake the necessary repairs on the order of the Lord Lieutenant, with money provided from the Consolidated Fund, and the county would be held liable to repay these advances, with interest, by later presentment. The OPW could also fix the tolls and levies to be charged from time to time for the use of the various structures.

Repair works were commenced in 1853, small sums being expended on thirteen piers in Co. Galway.[24] By the end of 1854, eighty-three of the ninety-three piers named in the schedule had been handed over,[25] including nineteen in Co. Galway and two in north Clare (Table 4.3). The time limit for the final transfer to the counties had to be extended by another Act in June 1856,[26] which allowed five more piers (none in Galway or north Clare) to be transferred later that year.[27]

Some structures would continue to need repairs for a few years after that and most, naturally enough, would need renovation, improvement and extension at later dates. After the transfers were completed, the OPW dropped all mention of fishery piers from the body of its Annual Reports, confining all information to appended financial accounts of expenditure on the completion of six piers, and repairs at ten more. All these involved only small sums.[28]

The new piers of the Great Famine period, added to the older ones erected by the first Commission for Irish Fisheries, constitute the bulk of the piers existing in County Galway and north Clare at the start of the second half of the nineteenth century. After that, the locations of new works would be determined by reference to the distribution of those already existing: new piers and quays, in general, would be located some way from existing ones, filling in the gaps along the coast, as it were, or would be located nearer to known or suspected fishing grounds which, it was hoped, would come to be worked by larger boats as the fishing industry developed. This latter would ultimately prove to be a vain hope.

A list of all the piers in Ireland that transferred to the counties before 1860 is given in Appendix D.

The Piers and Quays Built in Co. Galway during the Great Famine

Bunowen Harbour

Both Nimmo and Bald had proposed Bunowen, many years earlier, as a suitable place for a pier. Located east of Slyne Head, the site was in a bight in which boats could wait out adverse weather before proceeding on the difficult sailing passage around the Head. It was also convenient to known or reputed fishing grounds both within and outside Galway Bay. The proposal had not been acted upon at the time due to lack of funds. In 1839, the local landowner, John Augustus O'Neill of Bunowen Castle, applied to have a pier erected there. Public funds for piers were, if anything, scarcer then, and the application, even if it could be funded, was doomed by O'Neill's inability to provide the requisite one-fourth contribution; his application accordingly lapsed.[29] He applied again by memorial on 26 March 1846, within days of the 1846 Act being passed.

The OPW dispatched Barry Gibbons to examine the site and, following his inspection in June, he strongly favoured the memorial while noting that the poverty of the locality

FIGURE 4.1. Bunowen Harbour. The pier and inclined slip built during the Great Famine are on the right. The breakwater on the left was constructed in 1889–90 under the Sea Fisheries (Ireland) Act of 1883. The return head of the old pier was also made in 1889–90. Photo NPW.

precluded any possibility of an award above £4,000. Nevertheless, on 12 October 1846, he drew up a plan for a pier and slip, costing an estimated £6,371, and provided a sketch of the proposed works in December.[30] It was a positive and successful tactic: having seen the plan and appreciating its potential, a meeting of the local ratepayers unanimously agreed to take a loan in order to make up the extra that was needed for the local contribution. When proof of this agreement was confirmed by Fishery Inspector William Joshua Ffennell, the OPW recommended a free grant of £4,800 and loans of £1,400 (to the district ratepayers) and £200 (to O'Neill).[31] Work, however, did not get underway until 1848, and even then progress was very slow. George O'Neill, a clerk at the site, was replaced in October by John O'Brien from Tarrea; the work speeded up and was nearing completion by 1851.[32] By then, O'Neill was facing bankruptcy and eventually his whole estate was sold through the Encumbered Estates Court in 1853.[33] The pier was not finally completed until early in 1859, the final account showing it had received a free grant of £4,256 and a loan of £1,241;[34] there is no evidence that O'Neill personally received any of the loan funds.

The structure erected consisted of a straight pier with a stone-built slip alongside. The separate wave wall (breakwater) now existing north of the Great Famine pier, which converts the site into an enclosed small harbour, and the return head of the pier itself, are later additions made in 1889–90 at a cost of £2,465 under the Sea Fisheries (Ireland) Act of 1883.[35]

Working conditions at Bunowen had their own peculiar problems, which resulted in a complaint to the Board in 1849. In a letter, the overseer at the time, Michael Cahill, was accused of improperly exacting a levy of one shilling from each boat used to draw stones to the site. That was bad enough, but worse was to follow, as the text of the letter addressed to Commissioner Jones of the OPW describes:

> Firstly, [with the shilling] he buys whiskey to treat the men employed at the crane in order to help his neighbours and make the men so drunk that some are not able to attend the work the following morning and takes substitutes for them until they get sober. These are favourites. Secondly, he employs the servant boys of those men with whom he lodges and allows them occasionally the same trick for their masters and charges their time to the Board of Works. Thirdly, he has at least the fifth tombstone either for sale or for some deceased friends quarried by the labourers and charged to the Board and most generally comes not to his work until 7 o'clock a.m. and instead of forwarding the work he is rather retarding it by not employing the regular poolers but in their stead taylors and weavers who had no employment at their trades and giving them higher wages too, because they understand each other. He also allows his hostess to bring whiskey to the work and divides it, but I do not say he drinks any. His name is Michael Cahill and at [indecipherable] a week plainly shewing his inability as a steward. The above charges can be proved by John Conneely of Mainistir, large island of Aran and by William Joyce.[36]

Since time immemorial, public works were subject to shady practices and unwarranted patronage, and Connemara was not immune to such behaviour. But what, one wonders, were labourers doing with 'servant boys' in the aftermath of the Great Famine? Was life that good for them in Connemara that they had personal servants? There are two letters extant in the archive with substantially the same allegations: the first dated 23 May 1849 sent to 'Mr Walker, Secretary of the OPW' (it was Edward Hornsby who was the Secretary) and the second, quoted above, sent, undated, to Commissioner Jones and received on 18 September 1849. Both are signed by John Conneely of *Mainister, Inis Mór.*

Another worker at Bunowen, a labourer named James McCann, came under criticism of a different sort when he failed to pay his cess bill of four shillings and five pence and 'refused to respect the police in this matter'.[37] His dismissal from the works was sought. We do not know if that happened, but since the cess payers were funding the works in part, it seems unlikely that he was retained. Rivalry among the workers employed at the Slate pier in Galway in the early 1820s had also occasioned letters of complaint about favouritism,[38] understandable in light of the desperate struggle for paid work in distressful times.

Kilkieran Pier and Slip

The memorial for a pier at Kilkieran, at a site called Rusheen on land owned by Nicholas Lynch of Barna, was made in 1846 by the Earl of Devon, Chairman of the Irish Waste Land Society. Rusheen was the site that Bald had recommended to the Select Committee on Public Works in 1835.[39] The Society held a ninety-nine-year lease, from 1841, on 9,562 acres in the locality and was endeavouring to make improvements on parts of its huge holding.[40] Colonel Daniel Robinson, who was Managing Director and general agent of the Society, visited the district regularly. (In the course of his duties he raised a small, voluntary charitable fund for the relief of distress among widows and orphans in the district; it was administered by a committee of Lieutenant Croker of the Revenue Department, Rev. John Lally of Carna and Rev. Anthony Thomas of Ballinakill.)[41]

On receipt of the memorial, the OPW sent George Tarrant, engineer of the Tuam district of Co. Galway, to examine the location. In his report, dated 27 June 1846, Tarrant presented a plan for the site. It comprised a straight pier on the south shore of the inlet, with a breakwater projecting towards it from the northwest shore, so that between them they formed a small, enclosed harbour. The likely cost was estimated at £3,000. Barry Gibbons did not agree with this plan when he saw it: in his opinion, the northwestern breakwater would give no significant shelter, whereas an angled extension of the proposed pier on the south side would achieve a much better result. Gibbons's opinion prevailed, and in October Treasury sanctioned a free grant of £1,500, a loan to the district of £1,000, and a loan to the Irish Waste Land Society of £500, making up the total of the estimate.[42]

The extended, angled pier, and a slip, in lieu of Tarrant's proposed breakwater, were finished by December 1851 at a total cost of £2,347. Colonel Robinson died that year. His

FIGURE 4.2. The pier built at Kilkieran (RH 072) during the Great Famine, as it is today. Photo NPW.

death was a final, fatal blow to the 'improving' ambitions of the Waste Land Society, which was already struggling under the severity of famine and distress in the area. It applied to be wound up, precipitating an urgent letter from Edward Hornsby enquiring as to how the Society's loan would be repaid. His concern proved premature and the loan was repaid fully in due course. Today, Kilkieran pier is more or less as it was originally designed and built, although it was renovated many times later in the century when further slips were added to the harbour.

Tarrea Pier and Harbour

Tarrea, located in Kinvara Bay on the south coast of Galway Bay, had also been favoured by Nimmo as a site for a pier. He had instructed Samuel Jones (1797–1859), one of his assistants at the time, to make a survey of the place in the 1820s. Queried many years later, Jones claimed to have drawn up a plan for an L-shaped pier at the site, which he had given to the landowner, Maurice Blake of nearby Cloghballymore. Nothing was done, however, until Blake resurrected the idea in 1845, writing to Jones for a copy of the original plan. Jones, by then County Surveyor of Tipperary South Riding, acknowledged he had made a plan, but had no copy of it.[43] Blake nevertheless submitted a memorial for a pier to the OPW in January 1846, supported by the local fishermen and boat owners. That was some months before the 1846 Act passed, showing that some landowners were ahead of the authorities in proposing relief works before the Great Famine really took hold. When the Act passed in March, the local parish priest, Fr Thomas Kelly, wrote to the Board supporting Blake's memorial and pleading for the work to be commenced.[44] Barry Gibbons dispatched George Tarrant to examine the site. Tarrant gave estimates for two different structures, the cheaper of which (£2,036) Gibbons approved in June.[45]

The plans of the work are extant in the National Archives.[46] Blake offered to give £525 towards the work, if the Board would give a grant of £1,050 and the district £525, an arrangement that Treasury approved in July.[47] The fishermen and boat owners using Kinvara Bay had nothing to offer in cash, but, in a letter written on their behalf by Fr Kelly, each offered to subscribe 'the freight of their boat' in lieu.[48] By this one presumes they meant to carry construction materials to and from the site for free. Treasury gave final sanction for a free grant of £1,156, with Blake getting a loan of £525 and the local ratepayers a similar amount to make up the local contribution.

Work commenced in July 1848 under the overall direction of Gibbons and the local supervision of J.R. King. Serious storm damage disrupted work in January 1849 and Treasury had to sanction an extra £135 to cover essential repairs in April.[49] By its completion, the pier had absorbed a free grant of £1,118 and loans of the same amount repayable half and half by the district and by Blake.[50]

FIGURE 4.3. Tarrea Harbour today. Note the steep batter of the breakwater wall on the left. Photo NPW.

The harbour today is essentially as it was originally designed and built. It consists of a large, angled, cut-stone pier and wharf on the north side and a substantial breakwater on the south, enclosing a large, excavated basin or harbour. An inclined, stone-paved slip was constructed along the east side, from the root of the pier towards the root of the breakwater. The inner wall of the breakwater is steeply battered. This and the slip were so designed in order to still the swell of the sea inside the basin. As it turned out, internal stilling was not to be the main problem: the harbour entrance was open to the west and northwest, so that it was difficult for sailing boats to enter or leave in heavy weather from those quarters. If the design was possibly inadequate for the particular site, the masonry work compensated by being particularly fine. The ashlars were dressed by stonemasons on site, under the direction of King, and there is an integral inscribed stone plaque bearing the date 1849 embedded in the wharf face, alongside the steps. The stone steps themselves are particularly worthy of note for the form of their construction.

Gibbons was asked by the OPW to explain why the workers at Tarrea were paid one shilling and two pence per day during construction, which was regarded as excessive at the time.[51] While Gibbons's written reply is not extant in the archives, the high quality of the stonework *in situ* amply justifies the response '*res ipse loquitur*' – 'the work speaks for itself'. Tarrea is a fine example of a structure built by local stonemasons and labourers suffering under the yoke of Ireland's worst ever famine; it deserves to be better known and listed as a significant monument in our maritime heritage.

Keelkyle (Ballinakill) Pier and Slip

Ballinakill Harbour was one of a number of locations proposed by Rev. John Griffin when he wrote to the OPW on 9 February 1846 requesting that quays be erected locally for the benefit of his parishioners.[52] Francis J. Graham, son of Robert Graham, who had recently purchased 10,000 acres of land around Keelkyle,[53] supported the idea and he, too, wrote to the Board in April, asking that George Tarrant be sent to inspect a site he had in mind. Tarrant and Gibbons visited together in May. Gibbons considered Ballinakill Harbour to be eminently suitable and convenient for the fisheries, but he was not happy with the precise site at Keelkyle favoured by Graham. It was located at the inner end of the Bay, beside the Clifden–Westport road and adjacent to Graham's own residence at Ballinakill Lodge, but relatively far away from the fishing grounds.

The local inhabitants preferred a site in Fahy Bay, closer to the open sea. On balance, Gibbons, too, preferred the Fahy Bay site, but thought it best to leave the final decision to the Board before he would undertake a detailed survey. Graham wrote again to the Board in August, pointing out that failure of the potato crop locally would lead to dreadful starvation unless something was done soon by way of relief works. Gibbons straight away drew up a preliminary design and estimate of £863 for a suitable pier and slip at Graham's chosen site. Graham was concerned that work should be started immediately, as there was talk of a pier being built at Bundawlish, even further seaward than Fahy Bay. If the Board would give £450 for the Keelkyle site he would undertake to give £100, and the balance of £350 could, he proposed, be raised from the district. The Board agreed, Gibbons prepared the final detailed plan, and notices for a pier and slip at Keelkyle were posted throughout the district early in 1847.

A map and plan of the works, comprising a straight jetty pier with a slip alongside, are conserved in the National Archives of Ireland.[54] Work started under the local supervision of Robert Cassidy[55] and when completed, the pier was useful enough, although it was regularly inundated by high Spring tides. It had to be raised some feet higher in 1867 (chapter 5). It is still extant, but little used now, and in need of some repair.

Curhowna (Errislannan Harbour)

Errislannan is the name given to the whole peninsula between Clifden Bay and Mannin Bay.[56] According to the six-inch OS map of 1839, there were no manmade piers or quays existing anywhere on the peninsula at that early date. On 24 April 1846, Francis Burke of Eccles Street, Dublin, describing himself as the 'proprietor of the lands at Boat Harbour, Errislannan', applied to the OPW to 'build a pier for the better safety of the numerous fishing boats as well as for the coastguard, being their only place at present'.[57] According to Robinson,[58] the place locally called Boat Harbour (*Caladh an Bháid*) in the townland of Drimmeen did indeed belong to the Burkes and there was a small, sandy inlet there which gave the place its established local name. At the time of Burke's memorial, the Coastguard station was located in Drimmeen townland, much nearer to Boat Harbour than to the neighbouring townland of Carrownaught. A track leading from the Coastguard station to the boat harbour is clearly shown on the OS map.

Burke's application was not successful but John Lambert, another Errislannan proprietor, made a somewhat similar application about the same time. In a minute of 13 November 1846, Treasury sanctioned 'a cut from a lake to the sea at Errislannan, County Galway', the estimated cost of which was £800. Treasury approved a free grant of £535 for the project, and £265 would be levied on the district.[59] A month later, Lambert forwarded an abstract of his title to the 'lands of Errislannan upon which a grant has been made by the Treasury 'for the erection of a pier at Laughane Lea'. Laughaun Lea (there are many variants of the spelling) was situated in the townland of Carrownaught (Curhownagh; Curhowna today) and Lambert proposed 'to secure the advance signed by the OPW upon all the townlands (Kill, Keeraun, Carrownaught and Derryeighter) which will save trouble and increase the security'.[60]

It is clear from the documentary records that the separate applications of Burke and Lambert were for two completely different sites in Errislannan, and that Lambert's was the one sanctioned in 1846. The importance of correctly identifying the exact location of Laughaun Lea, and the confusion it causes in the records, will become clearer when we discuss Drimmeen Boat Harbour in the next chapter.

Focussing for now on Laughaun Lea, how did sanction for a 'cut from a lake', as stated in the Treasury minute, change into sanction for a pier and harbour? The solution may reside in the Irish name of the site' *Lochán Leá* (a name that is no longer in use). 'Laughaun Lea' is an English transliteration from the Irish, meaning 'small vanishing lake': *Lochán* is a small lake; *leá* means melting, dissolving or vanishing. The actual site was, at the time, a small rock-bound inlet closed off from the bay outside by an east-west rocky reef shown clearly in the OS map of 1839. Such an inlet would largely dry out (vanish) as the tide ebbed, and fill only late in the flood, hence a 'vanishing lake'. Boats could not enter it due to the rocky reef. Cutting a passage through that, thereby clearing the obstruction, would result in opening up the inlet to the sea and facilitating ready access for boats. The Laughaun Lea site as it existed then is shown in the six-inch OS map of 1839 and a small sketch plan of the site after the works were concluded was published in the 20th Report of the OPW.[61]

A poster advertising the proposed works went up in the locality on 24 December 1846 and preliminary plans were opened for examination in the offices of William Pierce in Clifden. Lambert wrote to the Board a month later, on 28 January 1847, expressing the hope that work would start soon in order to employ 'the wretches by whom we are surrounded'.[62] Within days Barry Gibbons provided specifications and a plan for a small harbour with a protecting breakwater and an inner quay.[63] First he proposed to excavate the west end of the reef of rocks to make a clear entrance into the harbour, 40 feet wide and 17 feet deep at high Spring tide. The sides of the excavation were cut to a slope of six inches horizontal to one foot vertical. The excavated material was deposited around the sides of the inner basin, and the unexcavated section of the reef was built up to form a breakwater, 3 feet wide at the top. Further excavation was carried out inside the basin clearing it of all obstructions. A quay wall, 140 feet long, 5.5 feet wide and 22 feet high was constructed of 'large flat bedded scantling stone, set flush with mortar, carried up in

FIGURE 4.4. Left: the stone steps at Tarrea. Note the plaque near the top left of the picture. Right: the original stone slip at Tarrea. The modern slip can be seen at the top right of the picture. Photos NPW.

regular, horizontal courses, 12 to 14 inches in height' on the west side of the new harbour. A set of stone steps was inserted into the quay wall at its southern end and a space 20 feet wide all around the harbour was coated with broken stone, four inches in depth, to form a roadway or secondary wharf.

Gibbons's drawings accompanying the specifications do not survive in the records, but the sketch map published in 1853 in the 20th Report of the OPW gives a general outline of the plan.

For all his concern in January 1847, Lambert seems to have lost momentum and failed to approve the plans until 20 August. Work commenced soon thereafter and would continue until 1851.[64] No other local contribution had been forthcoming in the meantime so Lambert had to take a loan to make up the £275 local contribution entirely on his own, which he did in 1850.[65] By then the project was being referred to in official documents as Errislannan harbour or Errislannan dock.

In October 1848, in a letter that manifests an incredibly detailed oversight of a very minor matter, Treasury sanctioned the appointment of George Gregory as temporary supervisor at Laughaun Lea. Progress on the works proved to be very slow, reflecting a serious underlying situation that involved Gregory, but not one for which he could be faulted. Twenty-two of the twenty-four men working on site were tenants of Lambert. Lambert complained to the OPW that Gregory was often absent and was 'addicted to drunkenness', thereby causing the works to suffer. In September 1849, Edward Hornsby, Secretary of the OPW, instructed the Board's engineer for the Galway district, Samuel Ussher Roberts, to visit the place and make a strict enquiry into the complaint.[66]

Roberts's report was devastating, but not in the way expected. He interviewed Gregory and, on hearing his side, found him innocent of all the allegations. Lambert was away at the time and, tellingly in his absence, some of the men 'stated that they each paid to Mr Lambert two shillings and sixpence out of each payment they received, some of them owing him arrears of rent'. One man, not a tenant of Lambert's, said he nevertheless paid him two shillings and sixpence 'for some grazing he got from him'.

It seems Lambert was exercising unwarranted patronage over the works; he had even asked Gregory to urge the men to pay the charge, and had asked Mrs Gregory to collect it! Gregory had demurred at this and it is possible that Lambert might have made the allegation against him because of that. Roberts had taken the sensible precaution of being accompanied in his investigation by Mr D'Arcy of Clifden, a Mr Morris of Ballinderrig and Mr Hart, the OPW paymaster, so the accuracy of his report could be relied upon. Roberts also remarked that Gregory had lost his mother and his recent wife from hardship in 'the wretched lodgings' in which he was housed.[67] Hornsby wrote to Gregory in October 1849, exonerating him of all the charges, and copied the letter to Lambert whose response is not on record. By then, Lambert's grasp on his Errislannan estate was coming under serious threat and it was sold soon after through the Encumbered Estates Court.[68]

Gregory was not the only one with problems at Errislannan. One William Campbell was a diver at the works. Anthony Gorham, a local shopkeeper, reportedly gave him groceries on credit, costing around £15, to be paid for when Campbell himself was paid. It was not an unusual practice in Connemara at that time to extend credit in that manner. But Gorham, thinking the works were nearing completion in January 1849, feared that Campbell would soon move away, leaving him, Gorham, unpaid. So he wrote to J. Scally and Co. in Dublin (presumably Campbell's employer under contract) requesting that they pay Campbell while he was still at Errislannan so that Gorham would get what was owed to him.[69] The records are silent on the outcome of this matter.

A total of £864 had been spent on Curhowna harbour by the time it was completed in 1851.[70] The 1853 sketch map published by the OPW was accompanied by a short description of the finished work: 'A small excavated basin, with a wharf and breakwater, affording excellent shelter for small vessels fishing in this important bay.'[71] It was listed in the schedule of the Grand Jury Act of 1853[72] as 'Errislannan or Loughawn Lea Dock', and transferred to the County by that Act.

In 1879, the Lord Lieutenant requested the OPW to provide reports and estimates for the repair and renovation of ten of the piers previously transferred to the counties and which the Fishery Inspectors had indicated were in need of repair; Curhowna dock was one of these.[73] The necessary works were sanctioned and commenced early in 1880. Nothing is known about the precise nature of these works beyond Robert Manning's terse comment that 'the reconstruction of a considerable part of the work' was required, at a cost of about £450.[74] It most likely involved renovation and extension of the breakwater; the Inspectors of Fisheries had pointed out that the breastwork was washed away in 1879 and since the harbour was important for the fisheries, the repairs needed to be attended to.[75] The money was advanced to the OPW by the Lord Lieutenant under the provisions of the 1853 Act, and everything was completed in 1882.[76] When Curhowna was surveyed in 1898, for the second edition of the six-inch OS map, the quay was shown and recorded; the breakwater was shown simply as a rocky projection with no indication that part of it was manmade.

FIGURE 4.5. Curhowna (Laughaun Lea; Errislannan Harbour). Left: the quay. Right: The quay wall is on the right and the breakwater on the left. The entrance cut is clearly shown leading out to the open bay. Photos NPW.

Today, the breakwater is a very substantial structure, much more massive than the quay/wharf and much larger and higher than originally constructed. It is clear that the whole structure was enhanced, widened and raised significantly above the original height; the stonework of the upper part differs greatly from that of the lower, and the whole has been extensively renovated in modern times. The harbour now serves the salmon-farming industry in the locality.

Rosroe Quay and Landing Slip

In 1846, Lieutenant General Alex Thomson of Salruck applied for a grant to build a quay with a landing slip alongside, together with a long approach road, at a place called Rosroe near the mouth of Killary Harbour. He was already distributing meal to his tenants and he wanted the work to start immediately in order to help them further with paid employment. He deposited £100 in the Bank of Ireland to the credit of the piers account of the Fishery Commissioners, probably to establish his *bona fides* and get things started.[77] It could take some time to have the plans and estimates drawn up and finally approved. George Roberts was the OPW district engineer for that locality and he advised on the plans, which he subsequently forwarded to Barry Gibbons for final approval.

The estimated cost of the proposed works was £800, for which the authorities agreed to give a free grant of £600, and Thomson took a loan of £200 from the OPW to make up the local contribution. Anxious to set men to work without any delay, Thomson decided to carry out the work with himself as the contractor, which was agreed with the OPW. When the work was inspected by William Pierce, in November 1847, construction of the road and pier was well in progress with seventy men employed; it proceeded to completion, although somewhat belatedly, in 1853. The quay, walled in good cut stone on three sides, projected outwards as a short, stub jetty, sheltering a small dock-like enclosure.

By 1880 it was in a damaged state; Thomas Brady confirmed that 'it was very useful for the fisheries. It should be put in repair.' The repairs were carried out in due course. Rosroe is still in use today for fishery and aquaculture activities, but much of its original masonry has been covered with concrete.

Some Unsuccessful Pier Applications during the Great Famine

General Thomson was only one of a number of landowners who responded positively and humanely to the crisis of the Great Famine, and he was fortunate in successfully obtaining a free grant of three quarters of the cost for Rosroe pier. Others were not quite so fortunate. Hyacinth D'Arcy of Clifden applied twice for £200 to build a pier at Coolakly (*Cúl an chlaí*) in Kingstown Bay.[78] He also applied to build a pier and slip at Gortdromagh and Falkeeragh in Clifden Bay,[79] and again for small piers at Gannoughs and Lackaneask in Rossadillisk townland;[80] none of these applications was sanctioned. Lord Wallscourt and others were similarly unsuccessful when they applied for a pier at Saleen, beside Ardfry House in inner

Galway Bay. Saleen would later become the site of Ireland's first government-funded oyster cultivation station when the Department of Agriculture and Technical Instruction took the site over in 1903.[81]

The agent of Baron de Basterot of Duras Demesne applied in April 1846 for piers at Ballybranigan, Crushua, Tawnagh, Bush Harbour and Kilturla, all located on the Baron's property on the western shore of Kinvara Bay. Each pier was estimated to cost £250, of which the Baron agreed to contribute one-fourth. At that time, nothing came of these requests either. Even the poor fishermen of Glenina in north Clare, who signed a memorial for a pier at that place, were unsuccessful, although they claimed that more than 120 fishermen were employed in the locality which had no pier at all.[82] They fished in currachs and sometimes sent their catches to Limerick and Ennis for sale. When they knew the Claddagh men were not fishing, they took their catches to market in Galway; whenever the Claddagh men were fishing they prevented the Glenina men from selling in Galway, possibly in order to keep fish prices artificially high.[83] At such times, the Galway buyers and curers would travel to the north Clare coast to buy up the Clare catches.[84] Nor was 1846 the first or the only time that Glenina was overlooked for a pier: in the famine of 1822, Bindon Blood, a local landowner, had requested a pier there, but the Commissioners of Fisheries replied that all the funds had already been used up.[85] That application was made in July 1822, and all new funding ceased entirely in August of that year, when the Government deemed the worst of the 1822 famine to be over.

On 4 April 1846, Thomas Nicholas Redington of Kilcornan, Clarinbridge, applied for a pier on Island Eddy, part of his estate in inner Galway Bay. A Catholic landowner, he was Liberal MP for Dundalk from 1837 to 1846, when he was replaced by Daniel O'Connell Jnr (son of the Liberator) as MP for that constituency. In July of the latter year, three months after sending in his memorial for Island Eddy, Redington was made Undersecretary of State for Ireland and a member of Sir John Burgoyne's Famine Relief Commission. Despite the high regard and respect in which he was obviously held, his application for Island Eddy was turned down. Barry Gibbons had examined it in May and deemed it not to be worth supporting. At the time, Island Eddy was inhabited (today it is not) and fishing was an important element of local life. Boats, usually *báid iomartha* – non-sailing craft of the hooker family up to about 21 feet in length – were essential for the normal life of the islanders and for their communication with the mainland. These boats are still remembered from the time the island was inhabited and at least one constructed on the island is still in use today.[86]

Island Eddy had then, and still has, one of the most unusual lying-up refuges for boats in Ireland. Being very low-lying (maximum elevation of the island is 24 feet), and open without shelter to the prevailing westerly winds, the problem of protecting the local fleet was solved in a unique manner. A series of about fifteen shallow indentations were excavated along a narrow inlet on the north side of the island, well in from the open sea; currachs and small boats could lie safely moored in these excavated indentations (technically termed 'nausts', but known locally as 'cloches') in any weather, and be launched easily when conditions ameliorated. The nausts of Island Eddy have been described and documented recently by Paul Gosling and his colleagues.[87]

That some memorials should be unsuccessful is hardly surprising as there were so many. One important reason for acknowledging them is the light they throw on the willingness of some landowners to meet the necessary one-fourth contribution from their own resources. Another is the fact that small piers now exist at many of the sites named, suggesting that either the full costs were obtained from sources other than Government, enabling the works to be carried out at the time, or that money was made available in later years from Government or from charities.

During the Great Famine, numerous charitable bodies were active in assisting distressed communities and they may have advanced, in an undocumented way, the necessary funds for some unrecorded piers. Some sites appeared again in memorials at later times and were funded in due course. Those that were built privately were generally small and rudimentary and for these there are no public records. Never having received public funds, they were not taken in charge subsequently by the counties under the 1853 Act[88] and for that reason they soon became neglected, were rarely repaired, and many were in virtual ruins by the year 2000. Generally speaking, the nineteenth-century piers were built when times were bad but were left to decay when times were better.

The Claddagh in Galway City

The works proposed for Claddagh in Galway city would bring the west bank of the Corrib estuary to the general state we recognise today. When Nimmo came in 1822, the harbour facilities were very defective. The western approach to the port was obstructed by a rocky outcrop called the Slate Rock or the 'Big Slate', which made entry to the port difficult for sailing boats, especially at low tide. A rock and gravel spit extended westwards from the Slate, sheltering Claddagh village from the open sea. Between that spit and the village was a marshy area, about twenty acres in extent, which was inundated at high tide, cutting the village off from the Slate Rock.

Within the estuary, the only commercial wharfage was along the east bank, where the Spanish Arch and Long Walk (known to generations of hooker boatmen as *an balla fada*, 'the long wall')[89] are today. A galleon is shown at the Spanish Arch quay in Thomas Phillips's MS survey sketch of 1685.[90] The old 'mud dock', properly called Eyre's Dock, existed at the seaward end of the Long Walk and is depicted in Bellin's map, dated 1764.[91] This dock still exists. Accommodation for fishing boats was confined to the Claddagh fishing village on the west bank of the estuary, where 1,000 families were said to live. There are no quays shown there in Bellin's map, reproduced in O'Dowd's *Down by the Claddagh*,[92] nor in the engraving of the Corrib estuary published in Hardiman's *History of Galway* (1820).[93] But two small jetties are shown in Logan's map, dated 1818,[94] reproduced in Woodman's *Safe and Commodious*.[95] These were low, rude, narrow groynes of loose stones, projecting out from the Claddagh strand where a small fleet of fishing boats would lie up. There was no substantial manmade quay there, the groynes extending out from the low, boggy, shoreline, and providing only the most rudimentary shelter from the main current of the river and the flow of the tide.

Nimmo immediately determined to erect a substantial breakwater at the Slate Rock in order to protect the outer harbour entrance; to raise the Claddagh jetties above high water and make a raised quay along the shoreline between them; and to excavate the shallow strand around the proposed new quay and jetties. His long-term plan included the construction of a causeway, which he estimated would cost more than £1,000, joining the proposed Slate breakwater to the proposed quay and jetties. There would be an opening in the causeway, either arched with stone, or crossed by a timber swivel bridge, permitting boats to enter into the marsh, which he intended to excavate and make into a proper sheltered dock, about five acres in extent. This plan was somewhat similar to the arrangement he successfully implemented on a much smaller scale at Killeenaran (St Kitts). In addition, raising the level of the spit that ran westwards from the Slate Rock would provide a protective wall for the south side of the proposed dock, and make for much better shelter within.

While devising his plan and drawing up his designs, he set men to work breaking up the Slate Rock and excavating the strand between the jetties. His design for the jetties and quay, which accompanied his report, dated November 1826, is reproduced in the 8th Annual Report of the Fishery Commissioners.[96] The plan and design for the Slate breakwater and causeway were never published by the Fishery Commissioners, because they did not contribute financially to that part of the works.

Around that time, the early 1820s, the local merchants were pressing for new, commercial floating docks to be built at an inlet (now the commercial docks) on the east bank of the estuary. This required parliamentary approval, and the Chamber of Commerce asked Nimmo to draw up a suitable plan for them, and to draft the necessary Bill.[97] This venture by Nimmo into designing commercial docks, when he was supposed to be erecting small fishing piers, seems not to have pleased the Fishery Commissioners, who took no immediate action when he submitted his plan for the Claddagh fishing jetties to them. Nimmo said their delay in approving the plan was due to the behaviour of the local boatmen: 'the Commissioners of Fisheries, from some insubordinate conduct of the boatmen, declined to give further assistance at that period'.[98] He certainly encountered early difficulties with construction workers in Galway,[99] but whether the delay can be attributed to these is not certain, especially in light of a comment by James Donnell quoted below. Noting the delay, the Lord Lieutenant, who had put the Chamber's dock memorial into Nimmo's hand in the first place, instructed him to proceed with the Slate breakwater, the Fishery Commissioners' reservations notwithstanding.

He commenced work in 1822, and the Slate pier, now known locally as Nimmo's pier, was completed in 1823. It stood entirely detached from the land because the causeway to it from Claddagh was not made at the time; access to it on foot was across the boggy mouth of the marsh, passable only at low tide. The original terminus of the pier can still be identified by examination of the present-day structure. Nothing further was done about the jetties and the quay, beyond the initial strand excavation works. When James Donnell surveyed all the fishery piers for the Commissioners in 1826, he made no mention of either the Slate breakwater or the Claddagh fishery quay and jetties.[100] Neither had the Commissioners' annual accounts ever made any mention of expenditure on the Slate breakwater up to then, so we can be certain that it had received no Fishery Commission approval or funds at the time.

But matters were soon to change. Donnell revisited the Claddagh on the Commissioners' behalf in 1828, finding only the poorly built jetties left in an unfinished state. He noted that 'Local contribution commensurate to an extensive work not being attainable, the Board have approved of a plan and specifications for rebuilding and completing the present imperfect piers and quay to the extent that the local contribution would warrant.'[101] By then, James O'Hara had offered a private subscription of £620 towards the jetties and quay, a small sum for the works originally contemplated but a positive step in providing the local contribution.[102]

In the Commissioners' final Annual Report (the eleventh, for the year to March 1830),[103] the Claddagh fishery piers were mentioned officially for the very first time. The Commissioners agreed to add £480 to O'Hara's £620, making a total of £1,100 available. It seems likely that absence of a local contribution in 1822, rather than Nimmo's explanation, was the true reason why the work had not gone ahead at the earlier date. Whatever the case, work finally commenced on the Claddagh quay and piers in 1829, funded by O'Hara's contribution and a loan of £300 raised by the Grand Jury which brought the total available from all sources up to £1,400.[104]

Construction continued through 1830 and was completed in 1831. Donnell's map and plan for the piers and quay, shown in the Commissioners' 10th Annual Report in 1829,[105] was virtually identical to Nimmo's, published in 1826. Both plans are useful in illustrating, by means of sectional drawings, the manner in which the small crude jetties were converted into substantial piers: they show the original low, narrow profile of the old jetties, with the proposed new superstructures superimposed on them. In effect, a new, larger stone shell was erected around each existing jetty, and the whole was then filled with rubble stone, so that the old jetties were covered, raised and fully enclosed within the new shells. This is akin to the modern encasement in concrete that has been effected on a number of the county's old stone piers since about 2001.

The specifications for the work were detailed and precise, drawn up by John Killaly (1776–1832), chief engineer of the OPW, who had approved Donnell's plan.[106] The foundation course was sunk 1 foot under low water of Spring tides, and the walls were raised 'to the level of the sill of Mr Rushe's door.'[107] The facings were limestone, rough hammer-dressed to fair face and beds, so that no filling spauls or small closures were needed. The coping was of 'thorough stones on edge, set archwise', at least eighteen inches deep and three feet across the wall, set in lime mortar. The surface of the longshore quay was extended forty-five feet inland from the face of the quay wall, and paved with rubble stone set on edge. The remaining space in front of the houses was covered twelve inches deep with strong, screened gravel. These specifications raised the shore to its present level, dramatically altering the site and forming the base of the roadway now named Claddagh Quay. The new piers were subsequently named Ballyknow pier (the northern pier, opposite the present Dominican church) and Claddagh pier, the more southerly structure.

In June 1832, the city took a loan of £300 from the OPW in order to extend the Claddagh Quay northwards from Ballyknow pier through the area known as *Garraí Glas*.[108] These modifications were done in 1833 and are indicated in a sketch of 'Part of the Town and Harbour of Galway' published by the OPW in 1833. The other works originally planned by Nimmo – the causeway from the Slate breakwater to the Claddagh piers, and a canal connection to Lough Corrib from the sea – remained unaddressed. A report made to the OPW in 1833 claimed that the Slate pier was then in a dilapidated condition and needed almost £1,400 worth of repairs, an inordinate sum for repairs alone. When works were, in fact, carried out on the quays and the Slate pier during the mid-1830s, they required only a small sum, nothing like £1,400, suggesting that some other purpose may have been in mind in 1833. It may, for example, have been that an attempt was being made to have the planned causeway started, but the small repairs were all that were carried out in the end.

The city's commercial docks were also built during the 1830s, largely to Nimmo's original design but modified by John Killaly and his son Hamilton Hartley Killaly (1800–74). The younger Killaly was engaged by the OPW as a superintending engineer under his father and a short time later, after the latter's death in 1832, as a principal engineer commissioned to finish his father's works.[109] Nimmo's original canal route, planned for the eastern side of the city (from Woodquay along present-day St Francis Street and Eglinton Street, along behind the Skeffington Arms Hotel to join the dock basin at Whitehall) was abandoned on John Killaly's advice. The sketch of the town and harbour of Galway, showing the improvements planned and in progress, published in the first Report of the OPW,[110] is almost certainly Nimmo's, rather than Killaly's; it shows Nimmo's proposed eastern canal route *in situ,* which would hardly have been shown if the sketch were really Killaly's, especially in light of the latter's outright dismissal of Nimmo's intended route. The sketch also shows the works in progress at *Garraí Glas* under the younger Killaly.

Hamilton Hartley Killaly emigrated to North America in 1834 and settled eventually in Ontario, Canada, in 1835. He became, in due course, engineer to the Welland Canal, Chairman of the Board of Works of Lower Canada, an elected member of the assembly, an Inspector of Railways and, in 1849, a founding member of the Canadian Institute. He was appointed to the Royal Commission on Fortification and Defence of Canada by Viscount Monck, whom we shall encounter in a different context in a later chapter. Buried in Toronto, Killaly is said to have been 'a vividly remembered personality, a genial gentleman and a superlative engineer.'[111] His works at the Galway commercial docks and at *Garraí Glas, Béal an Daingin*, Clifden and elsewhere make an interesting maritime connection between Galway and Canada that was further greatly enhanced through Viscount Monck in 1880 (chapter 6).

With Nimmo's proposed line of canal abandoned under the Killalys (although the Commissioners for the Galway Docks and Harbour preferred Nimmo's route),[112] the Harbour Board started to buy up land and property on the western side of the river for a new line of canal, applying to the OPW for loans for this purpose. When famine struck in 1846 there were, therefore, a number of projects in contemplation – the causeway to the Slate pier, the renovation of the Claddagh quays and the construction of a canal to Lough Corrib – that could be activated immediately as relief works.

On 11 July 1846, Treasury sanctioned works to upgrade the Claddagh piers and quay, estimated at £4,000.[113] Fr J.A. Rush OP of the Dominican friary in the Claddagh, together with some other unnamed benefactors, made up the requisite local contribution by taking a loan of £444 under the Piers and Harbours Act. The city Grand Jury also took a loan of £889 from the same source, and the Government made a free grant of £2,667. This made a grand total of £4,000, of which the free grant made up two thirds. This sum was allocated to build the causeway from the Slate breakwater to the piers, as well as upgrading the piers and quay from their 1831 state. Work commenced immediately, on contract by Mr W. Brady, and was completed by 1851.

Fr Rush also built the Claddagh Piscatory School close to the Dominican church in 1846, a novel venture to educate the Claddagh children in the skills of fishing and navigation. It opened in August 1847 with a reputed enrolment of almost 500 boys and girls, although it is difficult to envisage how that number could even fit into the school building, which is still extant. The curriculum ranged from simple writing to more esoteric subjects like trigonometry, bookkeeping, geography and navigation.[114]

The following year, 1848, when all the necessary land had been acquired, the OPW started work on excavating the new line of canal to Lough Corrib, a major operation carried out under the supervision of the engineer Samuel Ussher Roberts. It was completed by 1852. At the time, Roberts was the district engineer of the OPW in Galway. Later he would become County Surveyor of the town of Galway; he will appear many more times in this book, due to his eventual position as Chief Engineer of the OPW, and finally as a Commissioner for Public Works. The canal was officially opened in August of 1852 by the then Lord Lieutenant, the Earl of Eglinton, after whom it was named.[115]

From 1846 to 1852 therefore, the Claddagh and adjacent areas were busy places with plenty of work for stonemasons and general workers. The whole route from the Slate pier to the university campus and north to the Friar's Cut in the river was one continuous, linear construction site. Two of the separate projects – the causeway and the university – shared one practical, but not a financial, feature. Over 55,000 cubic yards of material had to be excavated from the bed of the canal route; some of this spoil was used to form the raised terrace at the front of University College, which was under construction at the time; more was used to make up the back-fill behind the wall of the new causeway joining the Slate breakwater to the Claddagh quay.

The new tidal basin at the entrance to the canal incorporated Ballyknow pier as its southern wall and Killaly's 1833 quay extension at *Garraí Glas* as its western wall. Both the pier and the quay were given new underpinnings in this transformation. The completely new east (riverside) wall of the canal basin was completed in 1851. The quay and the Claddagh pier south of Ballyknow pier were renovated and given new cut-stone ashlar facings to blend with those of the new canal and causeway. The causeway itself was 1,400 feet long, walled in ashlar on its river face, but it did not incorporate any entrance for boats into the marsh as Nimmo had originally intended. In fact, the new causeway completely closed off the marsh from the river and the sea. The marsh therefore filled up gradually over time to form a depressed, grassy area which flooded periodically. It became known as the Big Grass, and pictures of how it looked in the late nineteenth and early twentieth centuries are shown in O'Dowd's book on the Claddagh.[116] Eventually the Big Grass was filled in as a landfill site for the city, its southern wall (the old spit) was raised, and it was finally levelled and converted in the mid-twentieth century into a leisure park called South Park. It remains a public leisure park to this day. The later evolution of Galway's commercial dock does not concern us here and is described in detail by Woodman.[117]

Taken together, the totality of all the works represented an interesting consummation of Nimmo's vision: the Slate breakwater was now continuous by means of the causeway with the quays and piers of the Claddagh, the northern section of which was integrated into the canal basin; the canal to Lough Corrib, which Nimmo had first proposed in 1813 when working for the Bogs Commission,[118] was a reality, although it was not along the line he had proposed. A comprehensive atlas of Galway illustrating the Claddagh, the Slate pier, the river estuary and the dock area has recently been published.[119] A walk from Nimmo's pier (the Slate breakwater) along the causeway and on to Claddagh dock road, past the piers, the Piscatory School and the tidal basin, onwards along the canal bank to the university main building is therefore a walk through phases of Galway city's famine history from 1822 to 1852: the Slate breakwater was erected in the famine of 1822; the quay and piers in 1828 to 1852; the causeway, school, basin and canal during and just after the years of the Great Famine; and the old campus of the university was also a Great Famine construction. That the various structures should continue to grace the fabric of Galway city to this day is testimony to the skill and workmanship of men who were mostly starving, labouring under dreadful famine conditions of a severity not known since then.

Endnotes

1 16th Annual Report of the OPW, BPP 1847–48 [983], pp. 28–29.
2 13th Annual Report of the OPW, BPP 1845 [640].
3 3rd Annual Report of the Commissioners of Public Works, in re the Fisheries of Ireland. BPP 1845 (320).
4 57 Geo III C. 34.
5 1 Geo IV C. 81.
6 9 Vict. C. 3.

7 15th Annual Report of the OPW, BPP 1847 [847].
8 15th Annual Report of the OPW, BPP 1847 [847], appendix 28, p. 80.
9 15th Annual Report of the OPW, BPP 1847 [847], appendix 28, p. 81.
10 15th Annual Report of the OPW, BPP 1847 [847], p. 6.
11 10&11 Vict. C. 75, BPP 1847–48 [983].
12 17th Annual Report of the OPW, BPP 1849 [1098], p. 61.
13 19th Annual Report of the OPW, BPP 1851 [1414], appendix E, pp. 158–59.
14 17th Annual Report of the OPW, BPP 1849 [1098], appendix E1, p. 280.
15 16th, 17th, 18th Annual Reports of the OPW.
16 20th Annual Report of the OPW, BPP 1852–53 [1569].
17 21st Annual Report of the OPW, BPP 1852–53 [1651].
18 Return to Parliament, BPP 1859 (Sess 1) (119).
19 17th Annual Report of the OPW, BPP 1849 [1098].
20 17th Annual Report of the OPW, BPP 1849 [1098], Appendix E 2, p. 281.
21 19th Annual Report of the OPW, BPP 1851 [1414], p. 65.
22 16&17 Vict. C. 136.
23 16 & 17 Vict., C. 136, section VII. This quote is from the original version of the Act. The wording was later adapted to meet the requirements of the Local Government (Ireland) Act 1898 by the Local Government (Adaptation of Irish Enactments) Order 1899 to read 'shall from the date of transfer be held, maintained and preserved by the council of the county in which they are situate', which is the form printed in H.D. Conner, *Manual of Fisheries (Ireland) Acts* (Dublin: HMSO,1904).
24 22nd Annual Report of the OPW, BPP 1854[1820], appendix A8, p. 57.
25 23rd Annual Report of the OPW, BPP 1854–55 [1929], p. 25.
26 *An Act to amend the Act for transferring to Counties in Ireland certain Works constructed wholly or in part with the Public Money*, 19&20 Vict. C. 37.
27 25th Annual Report of the OPW, BPP 1857 session 2 [2228].
28 24th Annual Report of the OPW, BPP 1856 [2140], appendix A8.
29 NAI OPW/8/56.
30 Ibid.
31 Correspondence, July 1846 to January 1847, relating to the measures adopted for the relief of distress in Ireland and Scotland, BPP 1847 [765], p. 6.
32 19th Annual Report of the OPW, BPP 1851 [1414], p. 64.
33 T. Robinson, *Connemara. Part 1: Introduction and gazetteer. Part 2: A one-inch map* (Roundstone, Co. Galway: Folding Landscapes, 1990), p. 62.
34 NAI OPW/8/56.
35 Return to Parliament, BPP 1890 (276).
36 NAI OPW/8/11.
37 NAI OPW/8/56.
38 Petition of stonemasons to the Lord Lieutenant 18/1/1823. NAI CSORP 1823/5127.
39 Select Committee on Public Works in Ireland, BPP 1835 [573], query 3076.
40 S.C. Hall and A.M. Hall, *Ireland, its Scenery, Character etc.* (London: 1845), pp. 363–65.
41 *London Daily News* 21/10/1846, p. 3.
42 15th Annual Report of the OPW, BPP 1847 [847].
43 S. Jones to M. Blake, 16/11/1845. NAI OPW/8/383.
44 T. Kelly to OPW, 28/3/1846. NAI OPW/8/143.
45 B. Gibbons to OPW, 17/6/1846. NAI OPW/8/349.
46 NAI OPW8/349.

47 Correspondence, July 1846 to January 1847, relating to the measures adopted for the relief of distress in Ireland and Scotland, Fisheries Series, BPP 1847 [765], p. 1.
48 T. Kelly to P. Blake, 27/1/1846. NAI OPW/8/216.
49 NAI OPW/8/349.
50 17th Annual Report of the OPW, BPP 1849 [1098].
51 OPW to Gibbons. NAI OPW/8/349, item 26.
52 Rev. J. Griffin to OPW, 9/2/1846. NAI OPW/8/19.
53 Robinson, *Connemara, Part 1*, p. 34.
54 NAI OPW/8/36.
55 NAI OPW 8/19.
56 Robinson, *Connemara, Part 1*, p. 65.
57 Memorial of F. Burke dated 24/4/1846. NAI OPW8/139.
58 T. Robinson, *Connemara. The Last Pool of Darkness* (Dublin: Penguin, 2008).
59 Correspondence, July 1846 to January 1847, relating to the measures adopted for the relief of distress in Ireland and Scotland. Fisheries Series, BPP 1847 [765], p. 6.
60 Lambert to OPW, 18/12/1846. NAI OPW8/223.
61 20th Annual Report of the OPW, BPP 1852–53 [1569].
62 Lambert to OPW, 28/1/1847. NAI, OPW8/223.
63 B. Gibbons, Specifications for Works required to be executed in forming a Harbour and building a Quay Wall at Laughaun Lea, in the townland of Curhowna in the Parish of Ballindoon and county of Galway agreeably to the accompanying plans and sections. 2 February 1847. NAI OPW8/223.
64 17th Annual Report of the OPW, BPP 1849 [1098], appendix E1, p. 278.
65 19th Annual Report of the OPW, BPP 1851 [1414], appendix A, p. 32.
66 E. Hornsby to S. U. Roberts, 11/9/1849. NAI OPW8/139.
67 Report of S. U. Roberts to the OPW. NAI OPW8/139.
68 Melvin, *Estates*, p. 252n.
69 A. Gorham to Mssrs J. Scally and Co., 8/1/1849. NAI OPW8/139.
70 19th Annual Report of the OPW, BPP 1851 [1414], appendix E, p. 158.
71 20th Annual Report of the OPW, BPP 1852–53 [1569], p. 49.
72 16&17 Vict. C. 136.
73 48th Annual Report of the OPW, BPP 1880 [C. 2646], p. 64.
74 49th Annual Report of the OPW, BPP 1881 [C. 2958], p. 68.
75 Report of the Inspectors of Irish Fisheries on the Sea and Inland Fisheries of Ireland for the year 1879, BPP 1880 [C. 2627], p. 10.
76 50th Annual Report of the OPW, BPP 1882 [C. 3261], p. 25.
77 NAI OPW/8/84.
78 NAI OPW/8/214.
79 NAI OPW/8/164.
80 NAI OPW/8/221.
81 N.P. Wilkins, *Squires, Spalpeens and Spats: Oysters and Oystering in Galway Bay* (Galway: the Author, 2001).
82 NAI OPW/8/158
83 1st Report of the Commission of Inquiry into the Irish Fisheries 1836, p. 113.
84 R.P. MacDonnell, in First Report of the Commissioners of Inquiry into the State of the Irish Fisheries 1836, evidence, p. 105
85 NAI CSORP 22/2237.
86 P. de Bhaldraithe, personal communication, 2015.

87 P. Gosling, B. MacMahon and C. Roden. 'Nausts, púcáns and 'mallúirs', *Archaeology Ireland*, 24, 3 (Autumn 2010), pp. 30–34.
88 16&17 Vict. C. 136.
89 P. de Bhaldraithe, personal communication, 2015.
90 Thomas Phillip's sketch is reproduced in P. O'Dowd, *Down by the Claddagh* (Galway: Kenny's Bookshop, 1993).
91 J.N. Bellin, (Paris 1764). Plan de Galloway. Reproduced in O'Dowd, *Down by the Claddagh.*
92 O'Dowd, *Down by the Claddagh.*
93 Hardiman, *A History.*
94 M. Logan, *Map of Galway, 1818*. Reproduced in K. Woodman, *Safe and Commodious. The Annals of the Galway Harbour Commissioners* (Galway: The Galway Harbour Company, 2000).
95 K. Woodman, *Safe and Commodious.*
96 8th Report of the Commissioners of Fisheries, BPP 1827 (487), plate 11, figure 2.
97 2nd Report, Royal Commission on Tidal Harbours, BPP 1846 [692], p. 58a.
98 8th Report of the Commissioners of Fisheries, BPP 1827 (487), appendix 10, p. 42.
99 NAI CSORP 1823/5127; Wilkins, *Alexander Nimmo*, pp. 199–200.
100 J. Donnell, Report to the Commissioners in 8th Report of the Commissioners of Fisheries, BPP 1827 (487).
101 J. Donnell, Report to the Commissioners in 10th Report of the Commissioners of Fisheries, BPP 1829 (329), appendix 10, p. 22.
102 11th Report of the Commissioners of Fisheries, BPP 1830 (491), appendix 7, p. 10.
103 11th Report of the Commissioners of Fisheries, BPP 1830 (491).
104 1st Annual Report of the OPW, BPP 1833 (75), p. 25.
105 10th Report of the Commissioners of Fisheries, BPP 1829 (329), plate 12.
106 1st Annual Report of the OPW, BPP 1833 (75), p. 8.
107 10th Report of the Commissioners of Fisheries, BPP 1829 (329), appendix 10, p. 40.
108 17th Annual Report of the OPW, BPP 1849 [1098], appendix A, p. 70. See also endnote 112.
109 21st Report of the Commissioners for Auditing Public Accounts in Ireland, p. 175.
110 1st Annual Report of the OPW, BPP 133 (75), sketch of part of the Town and Harbour of Galway, p. 25
111 Full information on Killaly's career in Canada is given in G. Mainer, *Dictionary of Canadian Biography, vol. 10* (Toronto: University of Toronto/Université Laval, 1972).
112 2nd Report of the Royal Commission on Tidal Harbours, BPP 1846 (629), appendix B, p. 174.
113 16th Annual Report of the OPW, BPP 1847–48 [983], Table 1P, p. 290.
114 A. MacLochlainn, 'The Claddagh Piscatory School', in Anonymous, *Two Galway Schools* (Galway: Labour History Group, 1993).
115 M. Semple, *Reflections on Lough Corrib* (Galway: the Author, 1974).
116 O'Dowd, *Down by the Claddagh.*
117 Woodman, *Safe and Commodious.*
118 A. Nimmo in 4th Report of the Commissioners for the Bogs of Ireland, BPP 1813/14 (131), appendix 12.
119 J. Prunty and P. Walsh, *Galway/Gaillimh. Irish Historic Towns Atlas No. 28* (Dublin: Royal Irish Academy, 2016).

CHAPTER 5

From a Species of Delusion to a New Fishery Inspectorate

1857–1879

I have at all times considered that our Department was a species of delusion. We were supposed to be the persons having charge of the sea fisheries of Ireland. Our duties ... were mistaken by the public. It was believed that we were to be the encouragers of the fisheries and many persons ... looked to us as the parties whose duty it was to do that which we thought we had no right to do ... I think it is rather unfortunate that there existed any Department at all ...

Inspecting Commissioner of Fisheries
J. Redmond Barry, 1866.[1]

The Aftermath of the Great Famine

With the piers transferred to the counties and their maintenance resting with the Grand Juries, the engagement of the OPW with fishery piers and quays waned significantly. In the five years from 1847 to 1851, it had spent £106,000 on such works; this reduced to £21,868 in the seven years from 1852 to 1858, and in the six years to the end of 1864, the Board spent just over £1,000, spending nothing at all in four of those years.[2] In truth, the OPW was exhausted; in the words of a modern commentator, 'the department became so overheated that it burnt itself out'.[3] This was because of the impossible demands made on it by the Great Famine, and the predictable criticism it had to face in its aftermath: many of its actions were deemed not to have been 'useful', and its many failures to act were regarded as inexcusable.

William T. Mulvany took the brunt of the opprobrium for this, especially in the matter of drainage, and this precipitated his premature resignation and eventual departure to Germany in 1854. There he developed the 'Hibernia' and 'Shamrock' mines, starting the serious industrial development of the Ruhr valley. Going on from there he initiated canal and railway networks that were critical in the industrial development of Westphalia for which, among other achievements, he received the Kaiser's Medal in 1875. He is commemorated in Germany to this day as the father of its industrial revolution.[4]

The OPW, on the other hand, entered a period of serious decline that lasted for the ensuing two decades. Fishery piers and quays disappeared entirely from its Annual Reports, from the 25th Report in 1857 to the 35th in 1867. During that time, the Board underwent a number of personnel changes, which resulted in its membership declining to only two active Commissioners in 1864, and continuing with that complement for a number of years thereafter.

During this hiatus, the Grand Juries, having had the fishery piers and quays foisted on them, were understandably slow to undertake their maintenance, since that required a charge to be imposed on the cess-payers. By and large, therefore, they ignored this duty. Even the most recent piers, not to mention the older ones, were beginning to deteriorate and nothing was being done. A habit had long become established among users of the piers of anchoring their boats by inserting grappling hooks or anchors into the pier fabric, thereby loosening the stones of the coping and walls, which they or others would then take away as ballast. This, naturally enough, caused the edifices to fall to pieces.[5] Despite that, the offending practice continued long into the twentieth century.[6]

Observing the situation, Inspecting Commissioner of Fisheries J.R. Barry, took what action he could in the matter, although piers and quays were not officially within his fisheries remit. Several times he remonstrated with the Grand Juries regarding their duty, but to no avail. Inland proprietors on the Juries claimed to have no interest in fishery piers on the coast and did not see why the county as a whole should be burdened with their upkeep.[7] As to the fishermen on the coast, they were usually so poor that they could not contribute to the requisite one-fourth of the cost of repairs that the law required.[8] Barry was adamant that the OPW could not legitimately undertake any repair or maintenance on its own initiative, unless requested to do so by an interested party and then only on the expressed order of the Lord Lieutenant:

> from the time of the passing of the Act which transferred to the Grand Juries the management of the piers in the first instance, and the supervision of them, the Board of Works have not felt themselves at liberty to expend any money in the improvement or in the repair of piers ... The process is that some interested parties shall, in the first instance, apply to the Grand Jury to get them repaired, and that if the Grand Jury do not consent to do so they shall then have recourse to the Board of Works ... The Lord Lieutenant shall signify [the sum needed] to the Treasury [who may] direct that any such sum or sums of money required be advanced and paid ... There must be somebody taking the initiative in the first instance ... The Board of Works is a mere engine.[9]

He was scathing in regard to the transfer of the fishery piers to the counties in the first place: 'I think that the fatal Act of all with regard to those small structures which were erected upon the coast, was that Act which transferred them to the Grand Juries of Ireland.'[10]

In 1859, there was still more than £15,000 remaining unspent from the £90,000 made available for piers during the Great Famine.[11] All the time, money was coming in from works done earlier (interest payments on loans, and repayments of loan capital), so that the fund was in a reasonably healthy state and growing. Despite this, there were, it would appear, no memorials seeking the construction of anything new. Indeed, no new works would be undertaken and very little done for the next five years.[12]

New Stirrings of Activity

After the Famine, there was considerable conflict between the coastal and the inland proprietors regarding, on the one side, the number and location of stake weirs erected on the coast to catch salmon before they entered freshwater and, on the other, the abundance of illegal fishing weirs and mill dams in rivers. These involved matters of law and property rights best left to special commissioners to adjudicate.

In 1863, new legislation, laid down in *An Act to amend the Laws relating to Fisheries in Ireland*,[13] transferred all the duties of the OPW and the Inspecting Commissioners of Fisheries concerning the salmon fisheries and vested them in three new Commissioners called the Special Commissioners for Irish Fisheries. In this way, the Act relieved the OPW of all involvement in the perennially contentious salmon fisheries, effectively the only important inland fisheries. Their powers and duties relating to oysters and sea fish were specifically excluded from the transfer (section 15 of the Act), and nothing in it referred to fishery piers and quays either. In separating off the inland fisheries in this way, the Act, for the first time ever, severed them from the deep-sea and coast fisheries and from all other aspects of the Irish fisheries, which remained the responsibility of the OPW and its Inspecting Commissioner. The Special Commissioners were to hold office for two years initially, and thereafter until the end of the next session of Parliament. After that, all their powers and duties were to transfer to, and vest in, two permanent Inspectors of Fisheries who would be appointed by, and subject to the control of, the Lord Lieutenant of Ireland.[14]

The 1863 Act therefore constituted a first step in a major administrative change: whereas the OPW answered directly to Treasury in Whitehall, the proposed new Inspectors of Fisheries, once appointed, would answer directly, as a separate department, to the Lord Lieutenant in Dublin. In effect, responsibility for the regulation of inland fisheries was being repatriated by statute to the Irish administration for the first time since 1831. The deep sea, the coastal and the shell fisheries remained under the OPW with Barry as the only Inspecting Commissioner. The title of the annual sea fishery reports issued by the OPW changed thereafter from the *Reports of the Commissioners of Fisheries, Ireland* (1853 to 1862) to the *Reports of the Deep-Sea and Coast Fishery Commissioners, Ireland* (1863 to 1868). The separate reports on the inland fisheries were issued from the Office of the Special Commissioners under the title *Report of the Special Commissioners for Irish Fisheries* (from 1864 to 1867).

Commission of Enquiry into the Sea Fisheries of the United Kingdom, 1865

On a wider scale, the Government in Westminster resolved to undertake an investigation into the sea fisheries of Great Britain and Ireland, resulting in the setting-up of the Royal Commission to Enquire into the Sea Fisheries of the United Kingdom.[15] Its brief was to determine whether the sea fish stocks were in decline, whether fishing methods were detrimental to them and whether the existing legislation was injurious to them. While these questions were not specifically directed at the Irish fisheries, which were only a minute part of the total UK fisheries and enjoyed entirely separate legislation, the Royal Commission nevertheless held oral hearings in the autumn of 1864 at no less than twenty-one Irish venues. These hearings heightened awareness of the problems and difficulties commonly held to be retarding the Irish fisheries. When asked, for example, to describe his duties first as Inspector, and now as Inspecting Commissioner of Fisheries, Barry damned the OPW with faint praise: 'We had generally to carry out the directions of the Board. Their policy was to do as little as possible, but to preserve the peace, if possible.'[16] The decrepit state of the fishery piers and the deficiencies in their management also came to greater public notice, particularly in Barry's oral evidence.

The renewal of interest in piers precipitated by the Royal Commission caused questions to be raised in the House of Commons about the way in which the fishery piers were funded. Parliamentary Returns of Expenditure were requested and provided, showing not only how the Famine period sums of £50,000 and £40,000 had been dispensed, but giving details of all the expenditure incurred on all the fishery piers going back to 1820, i.e. back to the original funding by the old Commissioners for Irish Fisheries from 1820 to 1830, including the legacy funds transferred to the OPW in 1831.[17] Because of small differences in the way the information is presented in the records, and the merging of the legacy funds with the later Fishery Piers and Harbours funds, it is difficult to reconcile with full precision the sums detailed in the various Returns. Nevertheless, the information is sufficiently consistent and clear to show that almost £200,000 had been spent since 1820, of which about £137,000 (69 per cent) was given in free grants, less than but close enough to 75 per cent of the overall expenditure.

By March 1864, there remained an unappropriated balance of about £16,000 in the piers fund. However, an accountant's footnote attached to this balance in a Return presented in 1865 stated that 'of this sum about £11,500 has been allocated to piers, leaving £3,800 available [in April 1865] for further sanctions.'[18]

In 1864 also, the Board of the OPW underwent major change: William Le Fanu was appointed a Commissioner to replace John Radcliff, who had retired in 1862. The doughty, eighty-year-old Sir Richard Griffith retired from public office, having spent over fifty years in public service, seventeen of them with the OPW (fifteen as chairman). He was not replaced, but was made an Honorary Commissioner. Colonel John G. McKerlie, the new chairman, and William Le Fanu, served as the only active, paid Commissioners for Public Works for the next sixteen years.

A New Stimulus for Fishery Piers

Since the Report of the Commission on the British Fisheries had drawn attention to the continuing failure to build new fishery piers in Ireland, the Irish executive moved to take remedial action when the Commission's report was published. In April 1866, a new Piers and Harbours Bill was prepared and brought to Parliament by Mr Hugh Childers MP, financial secretary to the Treasury, Mr Chichester Fortescue MP, and the Attorney General for Ireland, Mr John Lawson MP. It was enacted and passed into law in June 1866 as *An Act to extend the Provisions for the Encouragement of the Sea Fisheries of Ireland, by promoting and aiding with Grants of Public Money, the Construction of Piers, Harbours and other Works.*[19]

This Act made two very important changes. First, it increased the limit of £5,000 that the OPW could give as a free grant for any individual fishery pier to a new limit of £7,500, with loans continuing to be available as advances from the Board's loan funds to help the locality or individual contributors to make up the one-fourth local contribution. The result was that the total allowed for each pier was raised to a maximum of £10,000 (£7,500 free grant plus £2,500 local contribution). Second, the money could be used for the extension, enlargement or improvement of any existing structure built under the earlier Acts. This was a real step forward, since renovations, modifications and extensions were precisely what were needed in many cases, rather than entirely new constructions.

The Act had an instant beneficial effect: reports on fishery piers and harbours, which had disappeared entirely from the OPW Annual Reports, reappeared in the 35th Annual Report for 1866, published in 1867.[20] By then, construction was reported to be ongoing at four locations, memorials had come in for new piers at three more, and for improvements at one other; none of these memorials concerned Co. Galway or north Clare.

By the next report, in 1868, and out of the blue, so to speak, the Board reported that works were ongoing in Co. Galway at *Spidéal* and Clifden, had been completed at Keelkyle (*Bearnadearg* in Ballinakill Bay) and at Leenaun, and a memorial for works at Drimmeen (Errislannan) was among six others the Board had actively in hand![21] All are listed in Table 5.1. Michael Morris (later Lord Killanin) of *Spidéal*, who was the Liberal MP for Galway, had succeeded John Lawson as Attorney General in November 1866 and that fact may, perhaps, have influenced the Board's sudden interest in the Co. Galway works, especially those at *Spidéal.*

TABLE 5.1

Nine piers and harbours erected or modified in Co. Galway and north Clare,1867–79

Site	RH No.	Date	Work done
Spidéal (New)	144	1871	New pier erected
Leenaun	002X2	1867	Western jetty pier erected. Old east jetty raised above H.W.
Keelkyle (Ballinakill)	013	1868	Existing pier raised above H.W. level
Drimmeen (Errislannan)	022	1869/70	Roadway, Quay and Pier with backwall erected.
Clifden	021	1869	Wharf wall extended and quay perfected. Scouring wall erected.
Inishbofin	164	1876	New pier and quay erected.
Inishshark	164B	1876	Slip excavated and inclined plane of oak skids added.
Bournapeaka	CS027	1879	New pier erected.
Ballyvaughan	CS031	1877	Existing pier repaired.

RH No. = Pier number in the Ryan Hanley survey. CS = Pier number in County Clare Coastal Architectural Heritage Survey.

The Sea Coast Fisheries (Ireland) Bill 1867

While the various pier works were underway, events were moving rapidly on the general fisheries front. The Royal Commission had delivered its report in 1866, recommending wholesale repeal of all restrictive sea fisheries legislation and the complete deregulation and liberalisation of the fishing industry.[22] Many of its recommendations were implemented in an Act passed in 1868 that applied to Britain alone.

A completely separate Bill to amend the Irish sea fisheries laws, called the Sea Coast Fisheries (Ireland) Bill, was introduced in Westminster in February 1867 by John A. Blake MP, Col. Tottenham MP and Dr Brady MP.[23] Among its provisions, it proposed that the direction and control of all the Irish fisheries, i.e. the inland, the deep-sea and the coastal fisheries, should in future be vested in the Lord Lieutenant. He, in turn, would appoint three Commissioners to execute the Act, one of whom was to be an engineer. If passed into law, this Bill would have transferred responsibility for the entire Irish fisheries from the OPW to the Lord Lieutenant, as the 1863 Act had already done with the inland fisheries. In a later amendment added by a Select Committee, the following was proposed, among other provisions:[24]

> All Acts relating to the building, making or improvement of piers and harbours shall be incorporated with this Act; and after the passing of this Act any works to be undertaken and carried out thereunder shall be undertaken and carried out by the Commissioners appointed under this Act.

That provision would have removed the last remaining aspect of fisheries – the piers and quays – from the remit of the OPW, and would have combined responsibility for fisheries and fishery piers in one single authority, acting under the Lord Lieutenant. Section 33 went on to propose: 'The Chief and other Commissioners to be appointed under this Act shall be Commissioners for executing the laws affecting both the Inland and Sea Fisheries in Ireland, and all powers now vested in the Board of Works with regard to the Fisheries shall be transferred to them'. That section, if passed into law, would expressly re-unite the regulation of the sea and the freshwater fisheries, but remove them entirely from the OPW. These were very radical proposals which needed much further and wider debate. For instance, the possible benefit to be derived from the transfer of all the engineering and construction aspects of piers and harbours from the OPW to a new Fishery Commission, even one with an engineer on it, seemed to be particularly in need of more detailed consideration.

Unsurprisingly, the original Bill was referred in May (1867) to a Select Committee chaired by Blake, who was the MP for Waterford, one of the Bill's proposers and a staunch advocate of the Irish fisheries.[25] Among the twenty-one Committee members was Mr George Shaw Lefevre MP, who had chaired the Commission on the Sea Fisheries of the UK. It is little wonder, therefore, that the Committee on the Bill concentrated on matters cognate with those already addressed by the Royal Commission: the state of the fisheries (the Irish fisheries, in this case) and whether they could be developed; whether restrictions on particular modes of fishing and regulations regarding owners and crews were advisable; and the desirability of a general consolidation of the Irish fisheries legislation. However, the Select Committee paid special attention to the most radical, uniquely Irish issue: should the administration of the Irish fisheries be vested in the Lord Lieutenant, with their day-to-day superintendence being placed under a new, dedicated Board or Commission of Fisheries?

Sixteen witnesses were examined, including Col. John G. McKerlie and Edward Hornsby, Chairman and Secretary, respectively, of the OPW; Inspecting Commissioner of Fisheries James R. Barry; Thomas F. Brady, previously an acting Inspecting Commissioner of Fisheries in the OPW; Sir James Dombrain, late Inspector General of the Irish Coastguard from 1819 to 1849; Professor Thomas H. Huxley, the eminent British scientist and a dominant member of the UK Commission; and ten others.

It was almost inevitable that the Committee's deliberations would embrace the matter of the fishery piers and harbours also. In his evidence, Barry was adamant that the fishery piers should never have been transferred to the counties in the first place and he was just as trenchant regarding their ongoing maintenance: 'I am of opinion that the right of the Commissioners for Public Works to look after those piers constructed under their direction should never have ceased.'[26] When pressed, he widened his answer to include all the Irish fishery piers and harbours that had been constructed in part or entirely with public money, in any place, at any time. This was consistent with his overall favourable opinion of the OPW. He was entirely happy with the *status quo,* in which the OPW had full responsibility for the deep-sea and coast fisheries, and he did not think a new body was necessary for their regulation:[27]

> I declare and I think that it would take a great many years to train any new department to the knowledge which is at present possessed by the Board of Works ... and further I think that the expense of such an establishment would be very materially increased ... I do not think that what you call an independent Board would be better than the arrangement which I propose.

When asked if the sea and inland fisheries should be re-combined he replied: 'I do not think them so closely connected, nor do I think them so much of the same nature as to make it desirable that they should be under the same supervision'.[28] He certainly did not favour bringing the inland fisheries back under the Board of Works: they were entirely separate at that time, and in his view, should remain so.[29]

Thomas F. Brady, who had been with the fisheries section of the OPW from 1846 to 1863,[30] first as a Clerk, then as an Acting Inspector of Fisheries from 1861, was now the Secretary of the Special Commissioners for Irish Fisheries (the body that took over control of the inland fisheries of Ireland under the 1863 Act); he was also Secretary to the English Commission on Inland Fisheries. He held a diametrically opposite, and far more jaundiced, opinion of the role of the OPW in Irish fisheries: 'I think it was rather an unfortunate thing that the fisheries were ever placed in the hands of the Board of Works'.[31] He had not the slightest doubt that the Board had failed to give the fisheries their due attention.[32] On many occasions, he and Barry had had to wait days and even weeks before they could get a Board meeting for fishery purposes. He now favoured recombining the inland and the sea fisheries responsibilities in one department, dissociated altogether from the Board of Works:

> I think both the inland and the sea fisheries could be very well combined, to the great advantage of the public and in an economical point of view also ... I know that it would be advantageous to both services. I am perfectly convinced of it, and I state that from my knowledge of both subjects ... I think that there cannot be the least doubt that it would be most desirable to have the salmon and inland fisheries and the sea fisheries dissociated from the Board of Works altogether.[33]

As for the fishery piers, he said that most were decaying and many of them were located in places where they were of no use to fishermen. This was because the proprietors who provided the one-fourth local contributions were principally concerned with their own advantage, and not with the needs of fishermen, when choosing the precise locations of the piers. In this opinion, at least, he was in full agreement with Barry.[34] Brady also believed that the Board of Works should have the power to maintain the piers on its own initiative, if the Grand Juries failed to do so. However, in a supplementary written report that he submitted after his oral evidence, he changed his mind, preferring that the proposed new Fishery Board should have responsibility for the maintenance of the fishery piers and harbours, which they should address by enlisting the engineering expertise of the County Surveyors rather than that of the OPW. Brady's evident antipathy to the involvement of the OPW in the fisheries and the piers could hardly have been made more obvious. It was to grow more virulent as the years went on.

The Select Committee could not have failed to detect a simmering animosity between Barry and Brady, which went back to the time of the Great Famine, when both men worked as fishery officials in the OPW. In his evidence, Barry alleged that the Board used to delete portions of the Reports of the Fishery Inspectors, of whom he was one, prior to their publication: 'it was done, I think, by Sir Richard Griffith, with the aid of a gentleman in whom he placed great confidence, namely, Mr Brady'.[35] Brady was only a clerk at the time, so the perceived slight had rankled with Barry until now. It was further inflamed by similar perceived, but undescribed, slights that allegedly occurred in 1863, at the time Brady took up his present secretarial post with the Special Commissioners of Inland Fisheries. Brady denied Barry's allegation, although he had no hesitation in claiming that the Board had been dysfunctional in fisheries matters during and since the Famine. In fairness to the OPW, he said, he had not the slightest doubt that it was its attention to other activities that had prevented the Board from giving the fisheries the attention they deserved. But the fact remained that the fishery business had not been attended to as it ought to have been, and 'Mr Barry knows that perfectly well'.[36] When asked whether the existing fishery staff in the OPW was quite insufficient for all practical purposes, Brady's reply might be interpreted a number of ways: 'The present fishery staff consists of Mr Barry, and, I think, that it is positively absurd to imagine that one gentleman can do the business. I, who am a much younger man, could not do it, and would not attempt to do it.'[37] Barry, the sole fisheries official in the OPW at the time, was then almost seventy-seven years old and the state of his health was suspect.

Sir James Dombrain's evidence was most depressing. In his opinion, none of the older piers had ever been any good:

> With respect to piers that have been built nominally for the fisheries, they are of no more use than this table, and some have never been used ... Of course I am speaking of very bye-gone days; but I am satisfied that everything which is undertaken by the Board of Works will be properly done ... Nothing which they have done has been improperly done.[38]

His criticism was directed at the piers dating from Nimmo's time and hand, but he made no allowance for the fact that those early piers (then over forty years old) had been envisaged, not just for fisheries but for trade, communication and general use at a time when steam ships had yet to come into vogue.

In the very first Annual Report of the OPW, Colonel Burgoyne, the Chairman, had acknowledged that, and praised them for their general usefulness.[39] While denigrating the pre-1831 piers, Dombrain had high praise for the OPW's piers, an opinion he may have proffered to cast his own contribution as Chief of the Coastguard in a more flattering light: before erecting piers it was the practice of the Board to consult the Coastguard regarding the advantages of each individual site, and to consult their own Fishery Inspectors only to know the extent of the fisheries in the general vicinity. If there were no problems with the precise locations of the piers, the credit would surely have lain as much with the Coastguard as with the OPW. Dombrain went on to declare that no fishery piers, not even small piers in sheltered creeks, were currently wanting anywhere in Ireland. Barry read this statement 'with much surprise' and when asked if Dombrain's views tallied with his own thoughts, replied: 'I do not think anything of the kind'.[40]

The Fisheries (Ireland) Act 1869

From the point of view of those hoping for change, the deliberations of the Select Committee were a disaster, and its Final Report in July 1867 indicated a serious failure to arrive at general agreement, especially on the issues of management and control.[41] Rather weakly, the Committee found it merely 'desirable' to amalgamate the sea and inland fisheries under one board, but it made no mention of any future role for the OPW in fisheries, nor of its possible relationship with any new fishery board that might be established. Neither did it make any recommendation on the future responsibility for erecting or maintaining fishery piers. Naturally enough in the absence of general agreement, the Bill did not progress successfully through Parliament and it was July 1869 before an entirely new Bill on Irish fisheries, brought in by Mr Fortescue and the Attorney General for Ireland, was enacted into law as *An Act to amend the Laws relating to the Fisheries of Ireland*.[42]

Considerable effort and extensive change of mind had gone into the formulation of this new Act, which was much shorter and less complex than the original Bill of 1867. It combined the posts of the two Inspectors of Salmon Fisheries (these had been created only months earlier by an amending Act of 1869)[43] with that of the long-established Inspecting Commissioner of Fisheries (J.R. Barry's post). These three posts were consolidated and newly constituted by this Act (32 Vict. C. 9), and their new holders were styled the 'Inspectors of Irish Fisheries'. All the powers, authorities and jurisdictions granted by the Act were vested in these three Inspectors, acting under the direction of the Lord Lieutenant. But the Act went much further than might have been anticipated. Section 8 stated:

> From and after the commencement of this Act, all powers, rights, privileges, authorities and jurisdictions vested in or exercised by and all duties imposed upon the said Inspectors of Fisheries under the said Salmon Fishery (Ireland) Act, 1869, and the Commissioners of Public Works in Ireland and the Inspecting Commissioner of Fisheries, or any of them, by any Act relating to the Oyster and White Sea-fisheries in Ireland, shall be transferred to, vested in, and exercised and discharged

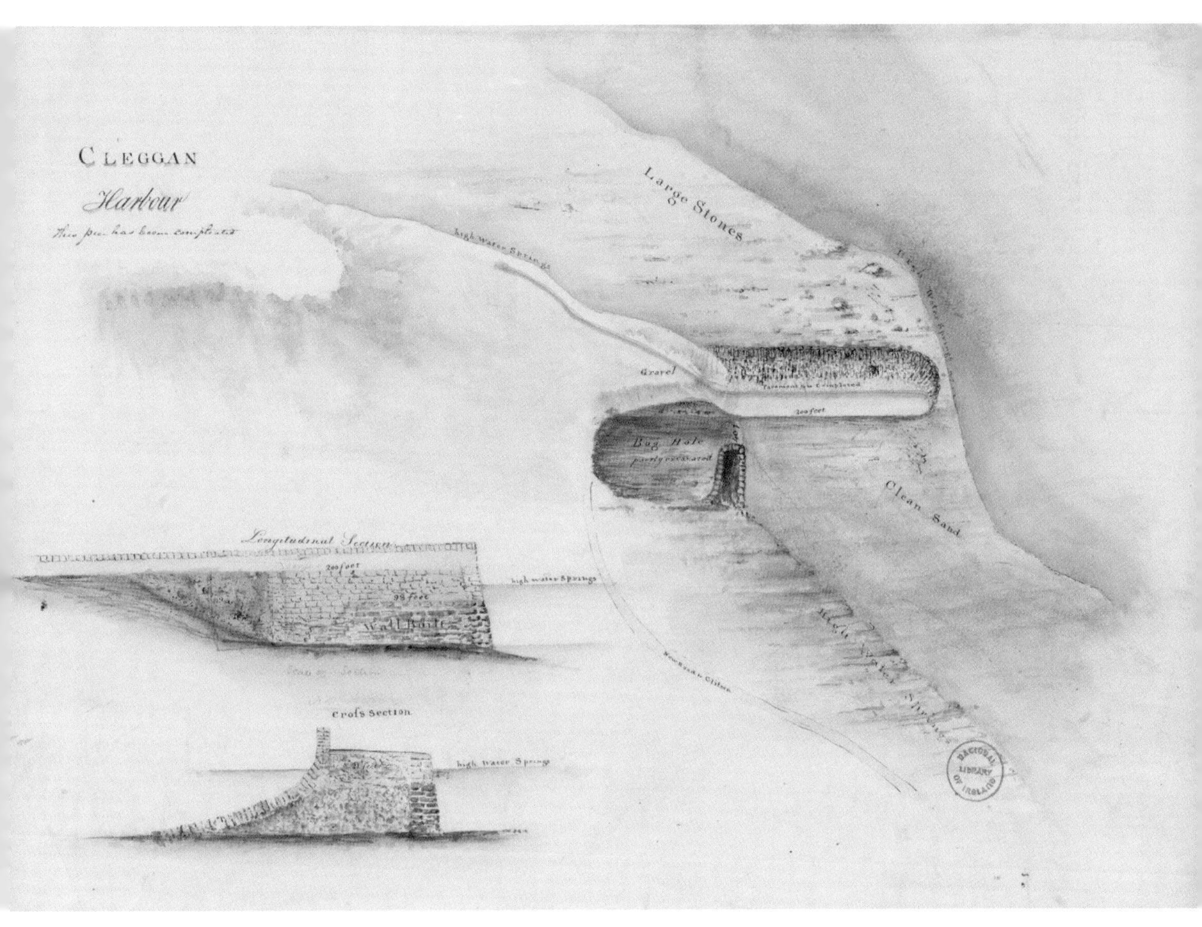

Nimmo's plan for a dock and pier at Cleggan Harbour in 1827 (detail). Reproduced from Map No. 61 H 5 (5), National Library of Ireland. Reproduced courtesy of the National Library of Ireland.

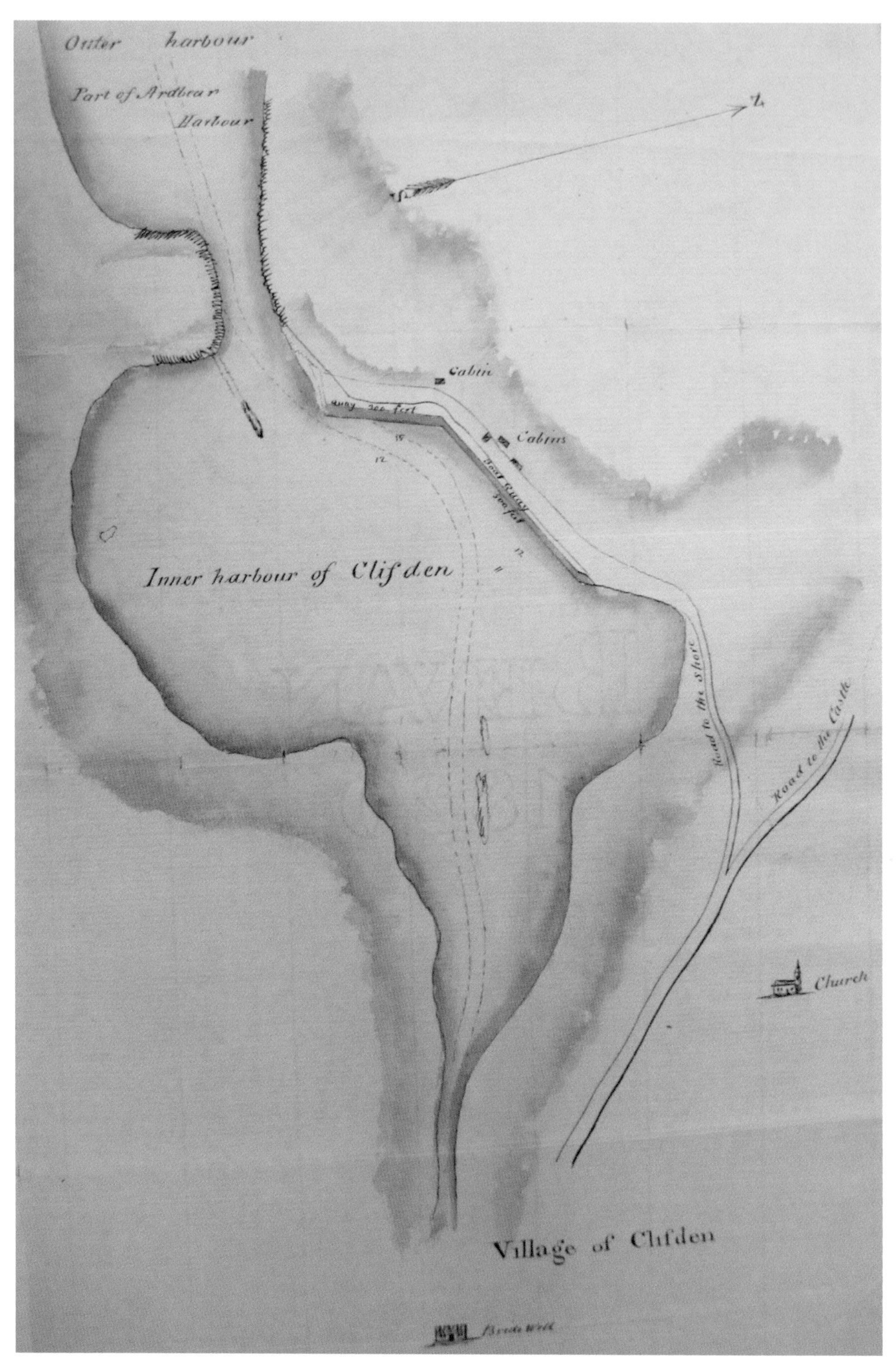

Alexander Nimmo's map of Clifden Harbour, 1823, showing his site for the proposed quay. Reproduced from OPW/8/84 with the permission of the Director, National Archives of Ireland.

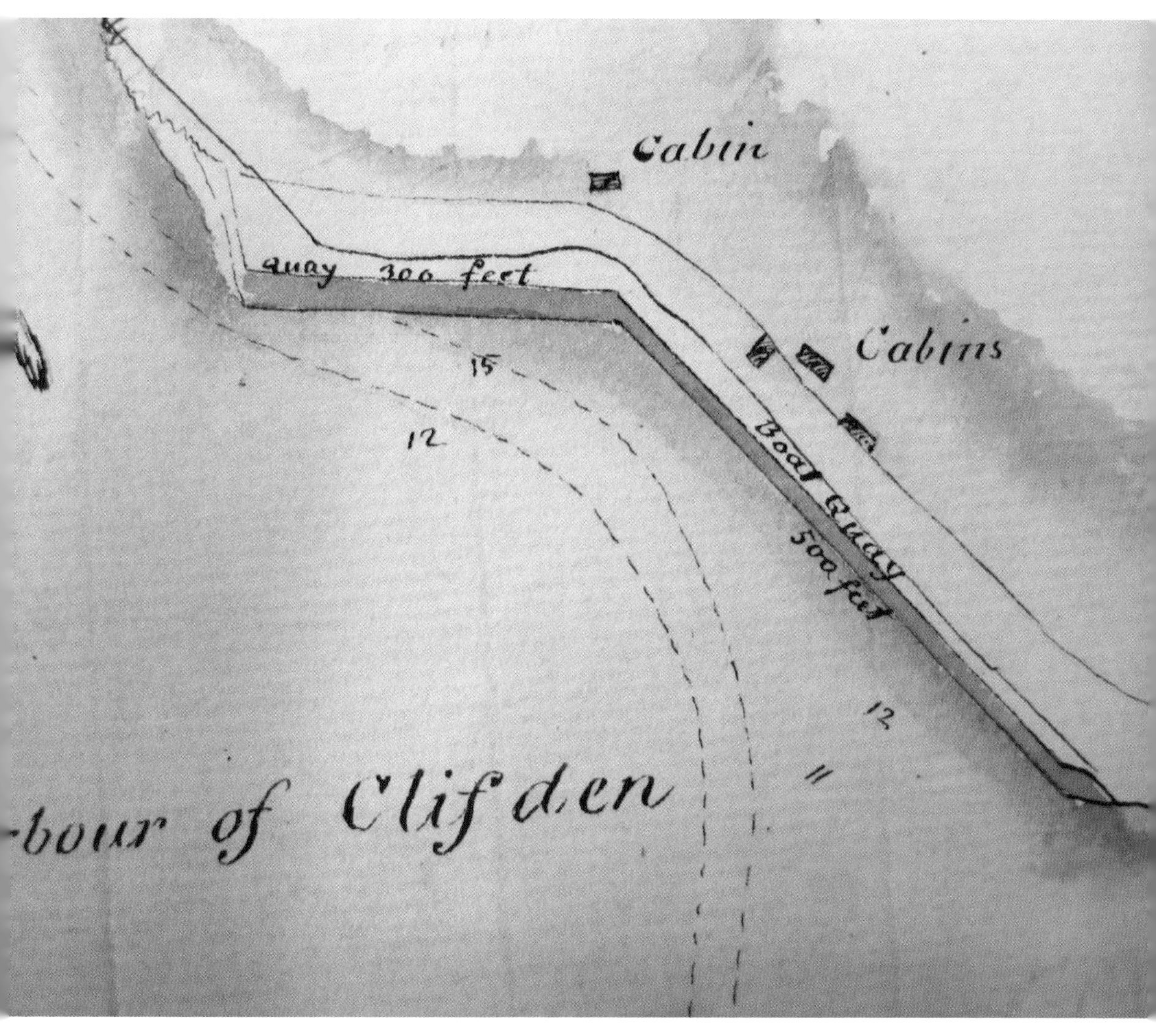

Alexander Nimmo's plan for Clifden quay, 1823 (detail). Reproduced from OPW/8/84 with the permission of the Director, National Archives of Ireland.

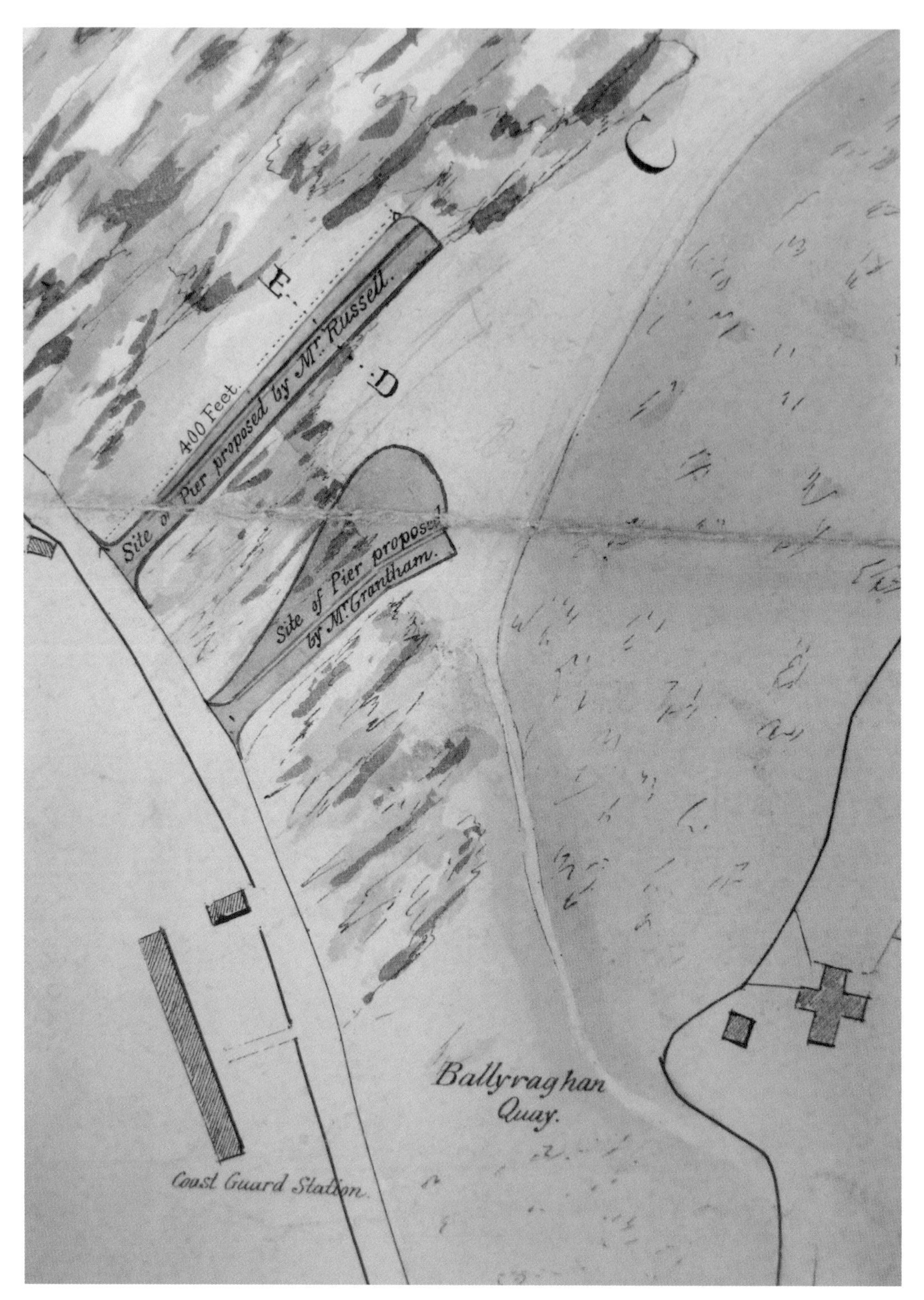

Edward Russell's 1844 map illustrating his own proposed site for a new pier in Ballyvaughan, together with Grantham's earlier proposed site. Also indicated is the original site of 'Ballyvaughan quay' near the Coastguard station. The original quay location shown here has no manmade structures and is a grassy, low shore today, and there is an old pier ruin (CS030) located directly opposite the small house shown beside the Coastguard station. Reproduced from OPW/5/10225/77 with the permission of the Director, National Archives of Ireland.

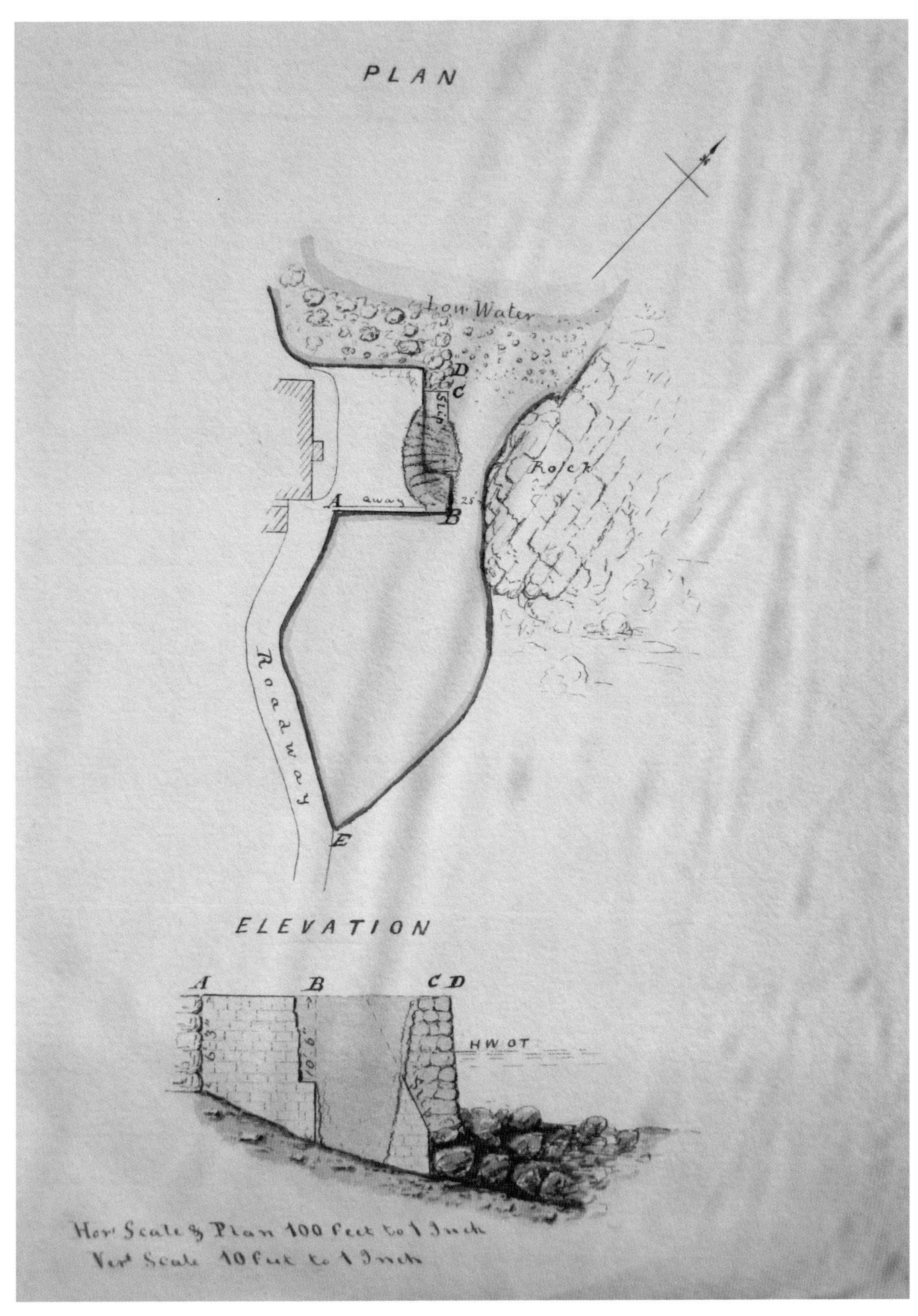

Plan of repairs to Rosroe pier carried out in 1880. Reproduced from OPW/59570/80 with the permission of the Director, National Archives of Ireland.

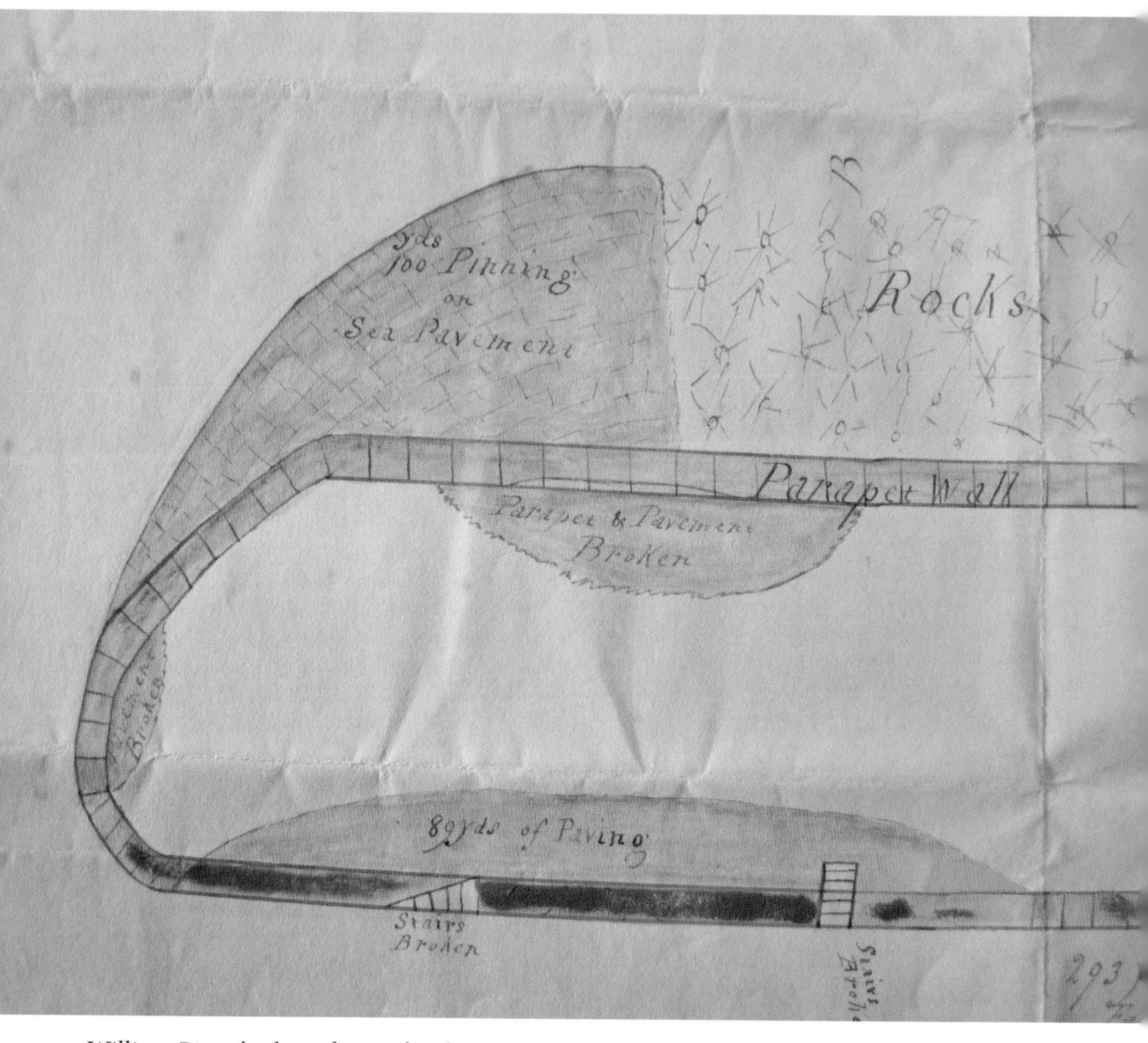

William Pierce's plan of 1835 for the restoration of Nimmo's old pier at *Spidéal*. Parts requiring repair are shown in yellow. Reproduced from OPW/8/341 with the permission of the Director, National Archives of Ireland.

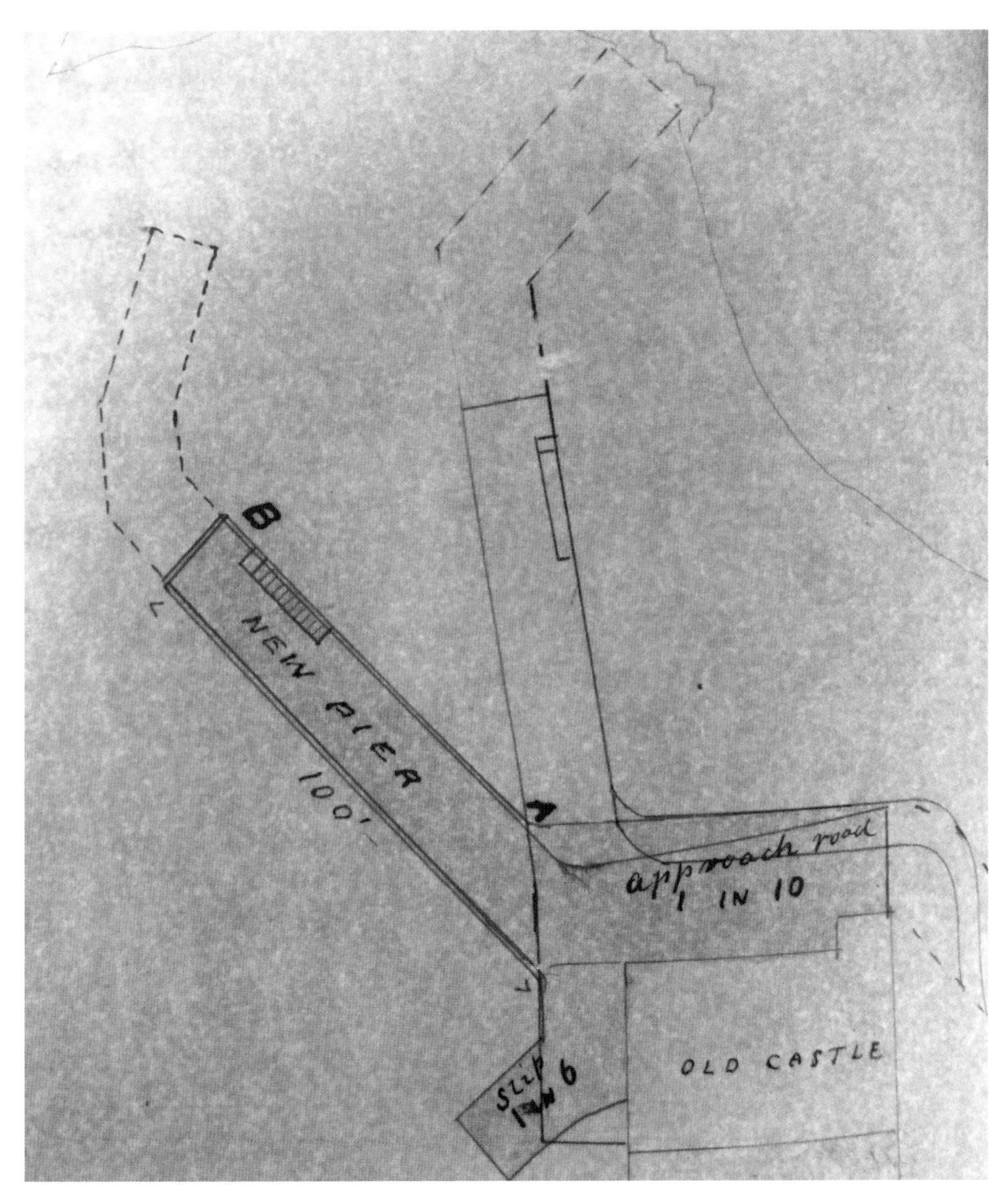

Plans for the pier at Oranmore Castle. The original OPW plan for the pier is shown in outline. The later, modified plan is shown in colour. Note the angle of the pier to the castle site (hatched). Reproduced from NAI OPW5/4188/14 with the permission of the Director, National Archives of Ireland.

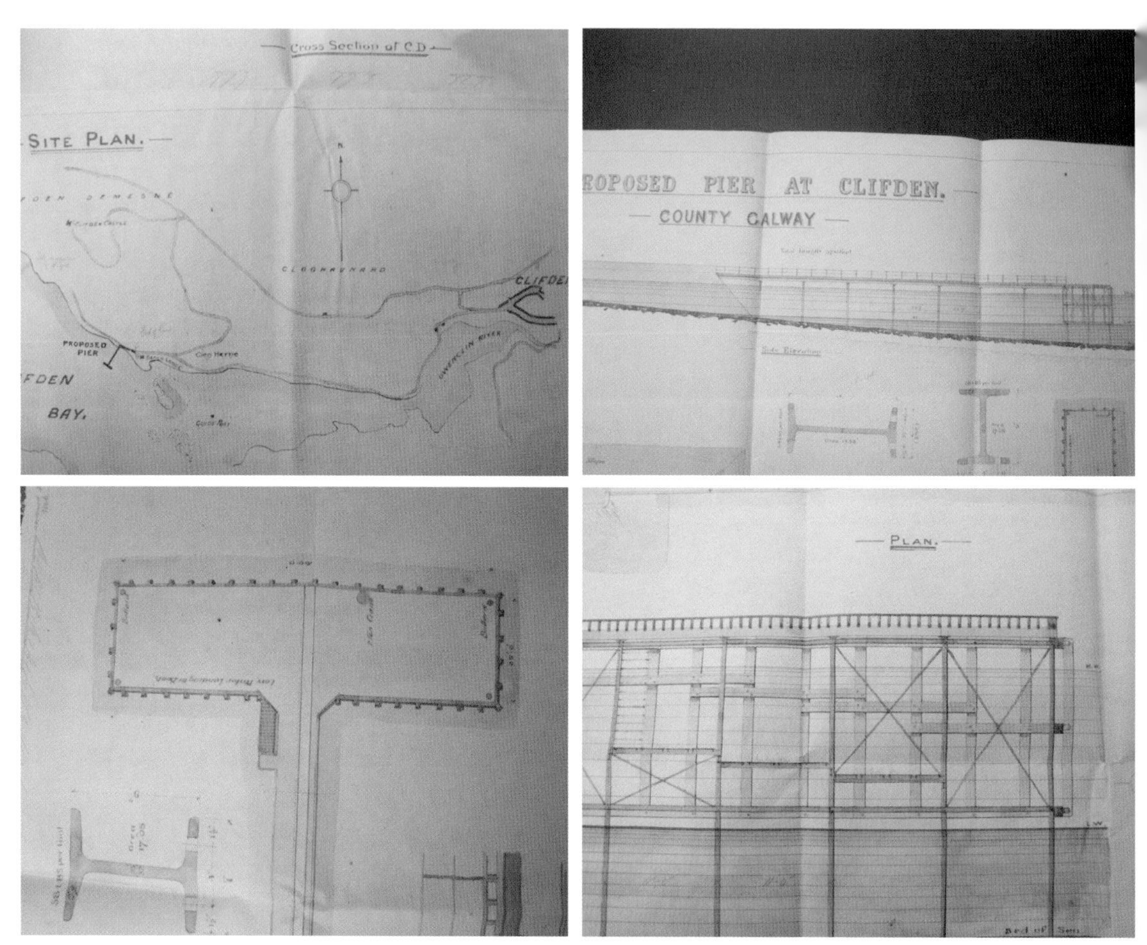

Details from James Gilmour's plan for a pier at Clifden. Top left: the site chosen. Top right: longitudinal elevation of the pier. Bottom left: plan of the pier head. Bottom right: section showing proposed bracing of the joints of the joists and piles. Reproduced from NAI OPW/5/4076/5 with permission of the Director, the National Archives of Ireland.

> by the Inspectors of Irish Fisheries acting in the execution of this Act during their continuance in office.

In one sweep, responsibility for the sea, the coast, the shellfish and the inland fisheries was combined once more in one single authority, the fishery Inspectorate, which was entirely separate from the OPW. For the first time since 1831, the Board of the OPW no longer had any responsibility for, or input into, the fisheries *per se*. This meant that the whole of the fisheries portfolio was repatriated from Treasury control (acting through the OPW) and placed under the control of the Irish executive (acting through the Lord Lieutenant and the new Inspectors). The office of the new fisheries department was located initially at 12, Ely Place, Dublin, and subsequently moved to Dublin Castle, the seat of the Irish administration.

Crucially for our purposes, the powers and duties of the OPW under the Piers and Harbours Acts remained unaffected, except for the provision of section 18:

> The Inspectors of Irish Fisheries shall, when so directed by the Lord Lieutenant, report for the information of the Commissioners of Public Works in Ireland in regard to the necessity for, and the advantage to be derived by the fisheries from any work which may be proposed to be carried out under the provisions of the Acts enumerated in the Schedule B to this Act annexed. [The annexed Acts were the Piers and Harbours Acts, 9 & 10 Vict. C. 9; 10 & 11 Vict. C. 75; 29 & 30 Vict. C. 45.]

Now, for the first time since 1842, an independent authority exercising direct responsibility for the fisheries (the new Inspectorate) was given by statute a formal, if only advisory, role in regard to fishery piers and harbours, small as that role would prove to be. The Act represented a major victory for Brady, Blake and other reformers who had striven to have the sea and inland fisheries combined and removed from the OPW. In addition, it gave them a second, although only partial, victory by reducing the OPW's exclusive control over the fishery piers. Their third and final victory was a satisfyingly practical one: Blake and Brady became two of the three new Inspectors of Irish Fisheries created by the Act; the third Inspector appointed was Major Joseph Hayes. James R. Barry was retired and took no more part in fisheries administration. He was the major loser in the reorganisation, a sad end to a long and dedicated service. He died in 1879, aged eighty-seven years, and is buried in Dundrum cemetery.[44]

The OPW Comes Under Scrutiny: the Lansdowne and Crichton Enquiries

Transferring fisheries out of the OPW was to prove a fortuitous blessing in disguise for the Board, whose performance had been under sustained criticism since the Great Famine. In 1872, Lord Lansdowne (Henry Petty-Fitzmaurice), was appointed by Treasury specifically to enquire into the matter. He produced an unflattering confidential report on the Board's reputedly poor administration.[45] Despite alleged attempts by Col. McKerlie to keep the report confidential (within the OPW, he alone was said to have a copy,[46] a claim he subsequently denied),[47] observations made in it leaked out and fuelled furious debate in which the Board was generally alleged to be almost totally ineffective.

Previous to 1860, Board meetings were held only 'with tolerable regularity', the normal business being carried out entirely by informal consultations between the Commissioners.[48] Formal meetings of the Commissioners petered out after that, and none at all had taken place since 1864.[49] Once this unsatisfactory state of affairs was aired in Parliament, Treasury decided it was past time for a fuller enquiry, and in November 1877, it set up a Departmental Committee, called The Board of Works (Ireland) Enquiry Committee, or simply the Crichton Committee, to consider the matter. Chaired by Viscount Crichton, MP for Enniskillen, the Committee comprised Arthur MacMurrough Kavanagh, MP for Co. Carlow, Mitchell Henry, MP for Galway, Charles Fremantle, Deputy Master of the Mint and Herbert Murray, Treasurer Remembrancer in Ireland. They were to enquire into the burden of statutes that the OPW was charged to administer; how the Board implemented these, and the extent to which it made a fair interpretation of the law in accordance with the spirit and intention of the legislature; whether, in its interpretations or its actions, the Board had been directed or advised by others; and to what extent the principles, practices and rules of the Board should be altered to achieve the aims of the legislation. This remit involved consideration of the division of duties among the Commissioners, and between them and their permanent officials; the supervision and control exercised over those officials; and whether there existed any reasonable grounds for complaint on the part of the public against the OPW on the score of undue delay or unnecessary expenditure in the transaction of its business.

However, long before the Crichton Committee was ever set up, the Lansdowne report had exerted its influence on the Board: one of the main outcomes from Lansdowne was the appointment, on 19 September 1873, of Samuel Ussher Roberts, a talented and eminent engineer, as an Assistant Commissioner of the OPW. He had been the District Engineer of the OPW in Galway from 1848 to 1855, at which time he was made County Surveyor to the County and City of Galway.[50] During his early years with the OPW, one of his assistants was the engineer, Robert Manning; the latter was made Chief Engineer of the OPW one year after Roberts's appointment as Assistant Commissioner and, from then on, under the influence largely of these two men, the OPW reports on the piers and harbours would have a fuller, more professional and much improved character. Since piers remained virtually the exclusive responsibility of the OPW, the new Inspectors of Fisheries quite properly made no mention of them at all in their first three Annual Reports published from 1869 to 1871.[51]

The Crichton Enquiry

The Crichton Committee took evidence at the OPW office in Dublin from 21 November to 8 December 1877, but there was very little public interest in the proceedings. Had it been a

Royal Commission, its hearings would have been open to the public and, in the Committee's own declared opinion, the resultant daily reports in the newspapers might have elicited a greater amount of interest and a larger number of complaints for investigation.[52] As it was, the sittings were closed-door affairs. Notice of them was given in the Dublin papers, stating that the Committee was prepared to hear evidence from anyone who had complaints or statements to make, on condition that due notice of the questions they wished to raise was given in advance. However, the advertisement made clear 'that persons who intimate their wish to give evidence will not attend sittings of the Committee except on receiving a notice to that effect.'[53] Special invitations were issued to two members of Parliament who had been vocal on the matter in the House during the previous session; one ignored it entirely, and the other declined the offer to attend and give evidence. The Board and its officers, however, gave exemplary assistance to the Committee.

No doubt the fisheries, had they still been under the OPW's control, would have constituted a stout stick with which to thrash the Board. After all, Thomas Brady – once an acting Inspector of Irish Fisheries in the OPW, well experienced as a fishery administrator and hardly a firebrand – was on public record as having said that he thought it unfortunate that the fisheries were ever placed under the Board of Works, and he had doubted that any of the Commissioners, with a few exceptions, had ever really understood or known anything about fisheries.[54]

Fortunately for the Board, it did not have to face such long-standing complaints, since it was no longer the responsible authority. However, fishery piers were still within its portfolio and, although only a very minor aspect of the investigation, the Crichton Committee successfully elicited (as we shall discuss later in chapter 9) interesting information on how the Board operated with regard to them. But from the very outset, the Committee exonerated Treasury from all implications attending any comments they were about to make:[55]

> So far as we have been able to judge, the instructions issued by the Treasury for the guidance of the Board of Works (and it is the Treasury to whom they look for instructions), have, as a rule, been conceived and framed in a spirit of liberality, and have had for their object the fulfilment of the intentions of the legislature. Whatever complaints there have been, or may be, do not, we think, find justification from any action of the executive government.

This judgement is little surprise, coming from a Committee directly selected and appointed by Treasury and taking its evidence *in camera*. But in fairness, there appears to have been no serious complaints raised at the hearings about the works done under the Piers and Harbours Acts, or the manner in which those Acts were implemented, possibly because the matter formed such a relatively minor aspect of the overall deliberations. Altogether, the Crichton Committee said very little about fisheries or fishery piers. Its main significance for our purpose is that it failed utterly to detect the simmering lack of understanding that was growing between the Board and the Inspectors of Fisheries regarding piers – or if it did, it chose not to acknowledge the problem openly, because of its possible implication both for Treasury and the Irish executive. As we shall see in chapter 9, a Royal Commission on Irish Public Works that sat in 1887, the third serious investigation into the operation of the OPW in the space of fifteen years, would not be so reticent in criticising the behaviour of all parties – Treasury, the executive, the OPW and the Fishery Inspectors – in a scathing report.[56]

How the Fishery Piers Fared after 1869

After the 1869 Sea Fisheries (Ireland) Act came into effect in October of that year, memorials for fishery piers were submitted in small but steady numbers, and the decade from 1869 to 1879 was one of two distinct halves with respect to them. Up to 1875, the inflow of memorials was slow, only thirty-four being submitted for the entire country. Twelve of these were rejected by Treasury or were not pursued further by the memorialists; thirteen were awaiting Treasury sanction; work was ongoing at five; and just four had been taken to completion.[57] That was hardly a spectacular achievement and it reflected unanticipated problems arising from the 1869 Act.

In the new dispensation, a memorial sent to the OPW for a new pier or quay was forwarded by them to the Fishery Inspectors for their opinion as to the pier's necessity and its likely benefit to the fisheries. The Inspectors would visit the proposed site before formulating their advice. If that were positive, the OPW had its own engineer examine the matter, if necessary visiting the site first, and had him draw up a preliminary plan and rough estimate. The Board would then determine whether the proposal was within regulations – whether the total estimated cost was under £10,000, the local one-fourth contribution was in place or was repayable by loan, and so on – and would then indicate the amount of the grant and loan that they were prepared to recommend to Treasury. Only then would the Board's engineer draw up a definitive plan and a final estimate, a process that often required yet another visit to the site. Finally, in every single case, the memorial and all the relevant surveys, plans, estimates, documents and approvals were forwarded to Treasury for final sanction or otherwise.

It was a slow, cumbersome process, which neither facilitated an early decision, nor guaranteed a successful outcome. In addition, from 1 January 1867, a further requirement was introduced: under the *Crown Lands Act* of 1866,[58] works on the foreshore had to be approved by the Board of Trade, either by the applicant or the OPW purchasing or leasing the Crown rights in the foreshore, or by paying a fee in recognition of those rights and interests of the Crown. The OPW, for example, paid a fee of £1 for a 'licence by deed' for Leenaun pier and for *Spidéal* pier in August 1867[59] and for Inishbofin and Inishshark in 1874.[60] Similar licences covering piers built under the Relief of Distress Act would be paid in 1880[61] and 1885.[62] In due course the fee, which was of a peppercorn nature, would be paid for all the piers and quays and for all other marine structures that were erected, or impacted, on the foreshore.

Matters Go from Bad to Worse

Throughout the relatively lean period for new piers in the first half of the 1870s, the Inspectors of Fisheries repeatedly raised the difficulty presented by the one-fourth local contributions. 'It is a matter of great difficulty', they wrote in 1873, 'to induce local parties to subscribe, or ratepayers to charge the Barony, with the required quota',[63] and in 1874: 'The one fourth contribution is an insurmountable obstacle.'[64] Even when the contribution was available locally, there remained an added obstacle: should no tender be received for the work and it had to be carried out directly by OPW staff, the memorialists were to be held liable for any excess cost incurred above the engineer's original estimate – an uncertain and unpredictable charge that no memorialist would readily accede to.[65]

Although, in the end, no such extra charge was ever demanded by Treasury in any instance known to the Board's engineer, it most likely did have some inhibitory effect on the submission of memorials.

On 3 December 1875, the Board received a minute from Treasury: as the Government intended to advance money in the coming year for the improvement of the important fishery harbours of Arklow and Ardglass, which would involve large expenditure, the Board was directed, for the present, not to bring forward applications for grants in aid of fishery piers from any other place, with the sole exception of Kinsale, Co. Cork.[66] Proposals for piers at these three named places were supported by special Bills introduced in Parliament in 1876 for Arklow[67] and Ardglass,[68] and in 1880 for Kinsale.[69] These piers would be constructed on a much larger scale than the fishery piers erected elsewhere in Ireland. The estimated cost of the fortunate few greatly exceeded £10,000 each (Arklow, for instance, would require £26,000 and Ardglass £20,000) and that placed them well beyond the limit imposed by the Piers and Harbours Acts; that is why they needed special local Acts of Parliament before they could go ahead.[70] They were the places where the British and foreign fleets used to gather when fishing the Irish Sea and the south coast.

Because they required so much, the allocation of money to piers elsewhere was suspended 'for so long a period as may be necessary to reduce the average annual expenditure of the next few years (for this service) to a level with the average expenditure of the past five years [i.e. about £5,000 per annum]'.[71] By putting funding on the long finger in this way, applicants for small fishery piers elsewhere in Ireland were being made to suffer for developments at the 'important' sites. But it also indicates that a coherent policy for Irish fishery piers, long wanting, was beginning to emerge. It was driven, not entirely by a rational strategy directed at developing the native Irish fishing industry, but by the pragmatic allocation of funds to improve facilities for commercial fishing enterprises, predominantly non-Irish, that worked out of the east and south coast ports of Ireland.

Because of the suspension of funding, the OPW took no further action on the existing memorials it had in hand, which stood in abeyance until Treasury should change its instructions. The wider implication of all this was that the Irish fishing industry, admitted by all to be seriously underdeveloped, was being treated in like manner to the British – specifically the Scottish – industry: larger boats for offshore and pelagic fishing were being facilitated and encouraged with bigger, better-equipped ports in the east and south, while coastal inshore fishing, which was the mainstay of the indigenous fisheries of the west of Ireland was being largely made to pay. It was, arguably, the most appropriate and necessary strategy for the time and in the circumstances.

If the first half of the decade was lean with respect to new fishery piers, the second half (1875 to 1879) was to be famished. In consequence of the suspension, the OPW reported that it received only a single new memorial (for Ray in Lough Swilly). No new piers were approved between 1875 and 1879, although in its Annual Reports each year it continued to list the thirty-four memorials it had in hand since 1870. These comprised ten that had been completed and thirteen that were in abeyance; the rest had been refused sanction, or the memorialists had not followed up on their applications.[72]

The Inspectors of Fisheries, who had actively championed the case for Arklow, began to regret that their support had contributed unintentionally to the suspension of all other memorials.[73] They were not at all happy with the number stalled in the suspension, but they felt that there was little they could do but ask for updates on their status in 1878[74] and 1879, expressing again their disappointment and reiterating the potential benefit to the country at large of providing safe and convenient shelters for small fishing boats.[75] The upshot of it all was that 1875 marked a temporary end to almost three decades of pier building: twenty-seven had been erected between 1845 and 1853 (eight of them in Co. Galway and north Clare) and a total of sixty-seven had been erected nationally by 1875.[76]

Maintaining the Existing Piers

Normal ongoing maintenance of piers and quays had been a recurring problem from 1853. If the legislation governing the construction of new piers was deficient to a degree, that governing their maintenance by the counties was, in the words of William Lane Joynt, Solicitor to the Treasury in Ireland, 'utterly defective.'[77] Section 8 of the Act that transferred them to the counties made it lawful for the Grand Juries to raise from the county at large, or from any district or barony, the sums necessary for maintenance.[78] But the Act did not expressly oblige the counties to maintain them, or even to monitor their condition. Section 9 made it lawful, but not mandatory, for them to appoint harbour constables who, among other duties, would 'settle all disputes which may arise with respect to ... the due and proper care and preservation of the works'.

In the event that the piers were not maintained in good and proper repair, the Lord Lieutenant was to inform Treasury of the fact; Treasury could then instruct the OPW to carry out the necessary repairs, eventually charging the county with the cost. But who was to inform the Lord Lieutenant when the Grand Juries failed, or refused, to act? On this, the

Act was silent. Neither the Board of Works nor the Inspectors of Fisheries were empowered to do so on their own initiative. Some persons thought the task should devolve on the County Surveyors, who were the professional officials most likely to be aware of the state of publicly owned structures in their own respective counties. William Lane Joynt considered any intervention by them to be highly improbable:[79]

> The county surveyor who must put the thing in motion is very seldom anxious to do so, and to fly in the face of his employer. His employers refuse to improve, or to mend, or to alter the pier, and then he has to memorialise His Excellency the Lord Lieutenant to do the very thing that his employers refuse to do. Some have done it, and I have known a county surveyor do it, but it is very seldom resorted to, and it is too much to expect that county officers will generally do it; the public never dream of doing it.

Although Galway and north Clare did not fare badly (Table 5.1), the period from about 1855 to 1879, all in all, had been a relatively poor one for fishery piers and quays both new and existing. It was about to be followed by the greatest decade of fishery pier construction the country has ever experienced: pier-building activity that had been seriously curtailed after one Great Famine was about to explode with the anticipated onset of yet another such catastrophe.

The Piers Built in Co. Galway and North Clare in the 1860s and 1870s

The Old and the New at Spidéal

In 1868, Nimmo's original old pier and harbour at *Spidéal*, known today as *an tSeancéibh* ('the old quay'), was reported to be almost useless. Built originally in 1823, it suffered recurrent episodes of siltation and had been badly damaged by the storms of 1830–31. It was said to be useless for fisheries by 1835 when it was inspected by Mr Owen, chief architect of the OPW (or one of his many engineering sons) and William Pierce, the engineer from Clifden. At first Owen seems to have been dismissive of the old harbour until Pierce explained that 'the country people' used it extensively to land seaweed and to load turf for sale in Galway city. That persuaded Owen to change his mind and declare that 'as it had been made, it was a pity it should be allowed to go to ruin'.[80] Grasping the moment, Pierce wrote immediately to the OPW explaining the situation and he was soon directed to forward a plan of the pier (still extant) and an estimate of the cost of repairs, which came to £89. 1*s*. 3*d*.[81]

Repairs and renovations were duly carried out but by 1839 the harbour had completely silted up again, putting it beyond service, and beyond the patience of the OPW and that of potential users. The repeated siltation and the recurrent structural damage by storms testify to the inclemency of the swell and the mobility of the seabed at the particular location.

A memorial 'to enlarge and repair the present very insufficient and ill-constructed pier or harbour at *Spiddal*' was submitted to the OPW again in March 1846.[82] Barry Gibbons examined it in October and found it a very rude structure, in his opinion not worth the cost of renovation, which he estimated would require £200 to £300. (One suspects that Gibbons may have mistakenly confused Furbo pier, a really rude and derelict structure located about four miles to the East, with *Spidéal* old pier; there is no hard evidence for this conjecture except the continued survival of *an tSeancéibh* as a useable harbour for almost 200 years.)

Be that as it may, nothing was done until *Spidéal* reappeared unexpectedly in the 36th Annual Report of the OPW in 1868[83] as a place where work was already in progress on building a new pier. An entirely new site at the mouth of the local river, a small distance west of *an tSeancéibh*, had been chosen for a completely new quay and pier, 950 feet in total length, which was being built under the supervision of James J. Boylan, Assistant Engineer of the OPW. Local lore has it that the fishermen wanted the new pier to be built at a place called *Aill Fhinn* near *An Cnoc* (Knock) a few miles further west from the village. The records confirm that a memorial for a pier at Knockallia (*Cnoc Aille*) had indeed been received by the OPW some time before 1880 but was never sanctioned.[84] According to O'Tuarisg, it was not sanctioned because Michael Morris (later Lord Killanin) wanted the pier built at the present site, closer to his own residence.[85] Morris had, in fact, contributed £300 to the cost of the new pier, which entitled him to select the exact site.[86]

Work commenced at the present location in June 1867 by forming a roadway 1,600 feet long to the site from the main public road and erecting workshops and a store yard. Wagons, cranes and other machinery were brought on site, and a rail track was laid from a local quarry to facilitate the delivery of stone to the works. A rough mole of rubble, 250 feet in length and 18 feet high, on average, faced on one side with a battered quay wall of dressed masonry, was then constructed along the shoreline leading from the new road to the root of the proposed jetty pier.[87] Construction progressed rapidly over the next fifteen months, despite the fact that operations were largely suspended in the springtime for want of labourers; the usual workers were local men who attended to their holdings in March and April in order to get their crops sown.[88] Agriculture was going through a prosperous phase, so it was to their advantage to leave paid public work for a short while to plant and sow, in order to ensure a harvest in the autumn.

The new site proved to be even more exposed to the elements than Nimmo's old one. By January 1869, when the jetty pier was well underway, it suffered serious damage in a storm. A 100-foot section of the outer, unfinished part was completely flattened and had to be rebuilt in its entirety. That work was almost finished by the end of the year, when the pier head suffered once again in storms during December. A large section of the sea pavement was also washed away.[89] Robert Manning was the OPW's chief engineer by then, and he moved quickly to have the damage repaired. All the works were substantially finished by early 1871 and brought to completion by June of that year, when the pier was handed over to the county.[90]

FIGURE 5.1. The massive aspect of the wharf face of *Spidéal* new pier and its enormous storm wall, are evident when compared with the boats lying alongside. Photo NPW.

Spidéal new pier would once again prove to be a real problem a decade later, when it was severely damaged by storms in November 1881. Repairs commenced in March 1882, but when almost finished in October, the pier was hit again by a big storm that tore up an unfinished part of the sea slope and breached the parapet. Little could be done during the winter period and more trouble was to follow later. On 12 February 1883, yet another storm, greater than the previous, tore up more of the sea pavement and washed stones weighing up to two tons off the wharf and into the harbour.[91]

Repairing this damage was not finalised until the autumn and it is little wonder that the final cost came to £2,826,[92] or that the pier has such a massive aspect today. Clearly, care was taken to ensure that such repeated failures would never happen again, by using massive blocks in the reconstruction; the sheer size of some of those making up the wharf face of the pier, laid in uneven courses without dressing or pointing, is notable. They and the enormous, terraced storm wall along its entire length, reminiscent of the defences of a mediaeval fortress, deprive the structure of all charm. Despite this, the existing paved sea slope is an attractive feature, not always noticed by visitors to the area. While not as distinctive as the curved stonework of the sea slope at Nimmo's pier in Galway – also only rarely noticed by casual visitors, since sea slopes are always located at the 'back', or sea face, of piers – it is worthy of remark nevertheless. The initial cost of construction, together with the multiple renovations that *Spidéal* new pier underwent throughout the whole of the century, amounted ultimately to around £20,000, more than any other single pier or harbour in the West had received.

Nimmo's old pier and harbour went into decline when the new pier was built, although it never entirely lost its attraction for boatmen. It has been greatly altered in modern times and has lost its original ambience as a pier of the 1822 famine and a work of Nimmo's design. When referring to *Spidéal* pier, the public documents do not distinguish between the old (1823) and the new (1871) structures, an omission that may confuse researchers not familiar with the detailed history. *An tSeancéibh* is now in use again to an even greater extent than the 'new' pier. That huge pier – the single most massive, manmade maritime structure in the county – does not attract the amount of marine traffic that its size might have warranted. Recently, a plaque has been erected on it, correctly recording its date of construction as 1871, but making no mention of the later extensive rebuilds and renovations that had given rise to severe criticism at the Select Committee on Harbour Accommodation in 1883.[93]

Keelkyle Raised Up

The repair works that commenced in July 1867 at Keelkyle (Bearnadearg) in Ballinakill Bay, involved remedial improvements to the existing pier, which had been erected during the Famine, rather than anything entirely new. The whole surface of the old pier, which was inundated by high water of Spring tides, was raised two and a half feet. This involved removing the centre paving and the coping, raising the walls, filling up the core to the desired height and then re-setting the coping and paving.[94] Some rocks were also removed and a rough shingle slip was made at the east side of the pier root. Everything was completed by March 1868. Today the pier is much as it was when completed in 1868 but it is rarely used.

Clifden Quay Resurrected

When work had finished at Clifden quay in 1830, it comprised only one half of the plan originally drawn up by Nimmo. About 500 feet of the boat quay had been made, and in his modification of Nimmo's plan, Donnell had considered that sufficient; he had abandoned Nimmo's proposed ship quay and finished off the boat quay with a short return head. The entire quay was therefore a single, straight longshore structure, with a short return head at each end.[95] At no stage was there ever a jetty pier at Clifden; it was always, and still is, simply a longshore quay. The OPW surveyed the place in 1847 with a view to undertaking famine relief works, but nothing came of that immediately.[96] Clifden quay did, however, vest in the OPW in 1847[97] and transferred to the county under the 1853 Grand Jury Act.

Its next appearance in the public records is in the 36th Report of the OPW for the year 1867, when the quay was reported to be 'in course of construction'. Work had commenced in July on a 200-foot extension. Donnell's return head at the seaward end was demolished and Nimmo's original planned ship quay was constructed. The angle originally envisaged between the upper boat quay and the lower ship quay was incorporated to make the whole structure follow the natural line of the shore. Within the year, the new ship quay was completed for its entire length to within two feet of the top, along with a new twenty-foot return at its seaward end. The entire works needed 900 cubic yards of dressed masonry. The space enclosed behind, to a width of forty feet, was filled with 3,350 cubic yards of well-packed gravel forming what is now the bed of the roadway along the quay. Rough stone pitching was added at the seaward return as a batter to give protection from surges of the tide. A flight of steps was constructed of hand-set stones near the angle. A completely novel development was the construction of a so-called 'half-tide barrier' in order to increase the scour in the navigation channel during the ebb and flow of the tide.[98] This was an inter-tidal rock wall, like a long, submerged, disengaged jetty, designed to deflect the late ebbing and early flowing tide closer to the quay wall, thereby increasing the scour and helping to keep the bottom clear of silt. Its construction was undertaken in the summer of 1868. Forty-five cubic yards of rock were also removed from the bottom of the entrance to the new channel between the ship quay and the barrier wall before work was suspended for the winter.

Resuming in February 1869, the half-tide barrier was completed, the quay was brought to its planned height, the surface was finished and everything was bought to finality in November. The entire work had taken a total of 1,300 cubic yards of dressed masonry for the quay wall and 579 cubic feet of ashlar for steps and coping.[99] It was a major undertaking, initially estimated to cost £1,600 and completed for £1,565. The structures had been designed by the then OPW chief engineer, William Forsyth (1798–1868) and completed after his death under the direction of James J. Boylan, assistant engineer. The sum of £1,174 was given as a free grant by government and £391 as a local contribution by presentment of the Grand Jury.[100]

The work brought Clifden quay to the form envisaged originally by Nimmo in the 1820s and that still broadly exists today. Vessels of 100 to 200 tons traded occasionally with Glasgow, Liverpool and other British ports, and large hookers were regular visitors, trading with Galway and Westport. No fishing boats were said to enter there because of the absence of facilities that could be used at all stages of the tide.[101] The tidal scouring wall can still be seen in the harbour at low tide, giving the approach to the ship quay the appearance of an artificial channel, which in fact it really is. Fr Lynsky, PP of Clifden, whom we will encounter again at Aughrus Beg, proposed at one stage that the channel should be fitted with a set of dock gates, as a means of retaining the water level at the upper quay during low tide, but nothing ever came of that idea.[102] In 1895 the Congested Districts Board installed navigation beacons in the harbour, erecting one at the innermost end of the scouring wall, and others on the extended shoreline below the ship quay. They are still there today but do not have operational lights on them. We will deal in a later chapter with proposals made to construct a genuine jetty pier at Clifden at the turn of the twentieth century and why they came to nothing.

Drimmeen Boat Harbour (Errislannan) – A Case of Bilocation?

As we saw earlier, Francis Burke of Boat Harbour in Drimmeen townland had applied for a pier at this place in 1846 but had not received sanction for his proposal. The Burke family subsequently sold the land in 1865 to John Byrne who, according to Robinson,[103] resurrected Francis Burke's proposal. The OPW certainly recorded a memorial for a grant in aid of a pier there in 1867, without indicating who had submitted it.[104] It was sanctioned in 1868 for support under the Piers and Harbours Act – a free grant of £450, and a loan of £100 from the public works loan fund – and work commenced in 1869.[105]

The construction, which was carried out under the Board's officers, comprised a low quay wall extending around the west, south and east sides of the inlet, with an L-shaped jetty pier attached at its western end. The total expended by 1870 was £650 and Manning reported that the only tasks that remained to be completed were the filling-in of the quay, which would require 380 cubic yards of stone and gravel, and the erection of a back wall on the pier, 179 feet long and 2.5 feet high.[106] By March 1871 everything was fully finished, the outstanding balance of the grant and loan were paid over and the harbour transferred to the county.[107]

The pier and quay wall are sturdily constructed of local stone of varying sizes and shapes, well laid to fit together, but not in any clearly defined courses. Like *Spidéal* new

FIGURE 5.2. Left: Drimmeen (Errislannan) Boat Harbour (RH 022). Note the rocks on the shore that make it a difficult berth for boats. The incised stone is located in the wall leading out to the sea. Photo NPW. Right: The incised stone in the wall of the pier at the entrance to the harbour. The incision has been highlighted temporarily with soft chalk that does not damage the incision and is easily removed by the tide. Photo Michael Taylor, reproduced with permission. .

pier, erected around the same time, this kind of construction gives the appearance of being of a much earlier date than, for example, the quay wall at nearby Errislannan (Curhowna) Harbour, which was built with greater finesse almost twenty years earlier. The harbour at Drimmeen was never properly cleared of rock and boat access to the low wharf on the south and west section is difficult.

An inscription incised roughly on a stone in the sea face of the pier seems to be contemporaneous and to record the date of construction. It says: 'V-LYIDEN S AUG 10 1869', perhaps a record by one of the men who built the structure.[108]

In 1882, application was made for repairs at Drimmeen[109] which were not addressed until 1886.

Curhowna and Drimmeen, both located in Errislannan, are widely confused in the public records. Curhowna, as we saw, started life as 'Loughaun Lea', and became 'Errislannan Harbour' on completion of the quay and breakwater in 1852, but is mostly referred to as 'Errislannan Dock' or 'Errislannan Pier' in the records; it is often referred to by the different names even within a single source record. The name 'Laughaun Lea' gradually disappeared from the published records and it is not known in the locality nowadays. The Drimmeen site started out being called 'Boat Harbour' and became known as 'Errislannan', or 'Errislannan Harbour', or 'Errislannan Boat Harbour' after the 1869–71 works. It is shown simply as 'Boat Harbour' in the second edition of the six-inch OS map of 1898, whereas Curhowna is shown only as 'quay' on that map. The interchanging of names reflects real conflation of the two sites in the records, even by the OPW, who built both of them: the maps published by the OPW and the Inspectors of Fisheries only ever record a single site on the Errislannan peninsula.

Although both sites, named respectively 'Errislannan (Laughawn Lea)' and 'Errislannan', are correctly recorded with financial information and dates of construction in the 53rd Report of the OPW in 1885,[110] only one site named 'Errislannan' is recorded in the schedule of piers given in the 54th Report in the following year.[111] A different version of the latter schedule was submitted by General Sankey Chairman of the OPW, to the Allport Commission in 1888,[112] which further increased the confusion. In this, only one Errislannan structure is recorded, called 'Errislannan (Drimmeen)', described as 'a quay 225 feet by 16 feet, a groin [*sic*], 60 feet by 6 feet, a roadway and retaining wall.' The final confusion occurs in the 55th Report of the Board in which it is recorded: 'During the year the pier at Errislannan (Drimmeen) in the county of Mayo [*sic*] has been repaired by Your Lordships' authority at a cost of £203.14s.3d.'![113]

Today, Curhowna serves the local salmon-farming industry and Drimmeen has only a few small local boats, mainly concerned with the lobster fishery.

Bournapeaka Pier, north Clare

A memorial for a pier at Bournapeaka in north Clare was submitted by the proprietors and ratepayers of the district in 1874. The application was approved by the Inspectors and

FIGURE 5.3. Bournapeaka pier near Ballyvaughan, Co. Clare. Photo NPW.

sanctioned by Treasury just in time to avoid the impending suspension of pier building. The memorial, dated 21 February 1874, was accompanied by a report from an engineer, John Hill of Ennis, in which he stated that the pier at nearby Ballyvaughan (Russell/Gibbons's pier of 1846–47) was approachable only from half tide onwards and therefore not convenient for fishermen. He drew up specifications for a structure to be erected at Bournapeaka just outside the village, and followed it up with a plan and drawing on 4 June.[114]

The Grand Jury, agreeing to make the local contribution, approved the proposal in July and Fishery Inspector Thomas Brady also approved it, advising a free grant be given in October 1874. The OPW estimate was £4,350; the free grant was £3,000 and the county agreed to take a loan of the remaining £1350, £500 to be levied on the county at large and £850 on the Barony of Burrin. Some opposition must have arisen because the Fishery Inspectors called a public meeting in August 1875 which was attended by a large number of ratepayers, including J.B. MacNamara, the owner of the land involved, and William Lane Joynt of Clareville. Unlike the objection to Ballyvaughan pier made in 1845 by the Duke of Buckingham (which also involved the MacNamara family and the Clareville estate), this new proposal received general acclaim at the meeting and final public notices of the intention to proceed with the proposed work were posted in September.

A contract was quickly drawn up with Messrs Mannix and Slade to undertake the construction.[115] They set to work right away and spent the period up to July 1876 fencing the approach road and assembling materials and equipment. On 20 July they started setting concrete, the first time this material was mentioned in relation to piers in the Annual Reports of the OPW. Large concrete blocks, representing a total quantity of 286 cubic yards, were cast on site and about 280 linear feet of the pier was laid by 1877.[116] These dimensions suggest that the size of the blocks made was one cubic yard (3x3x3 feet). Everything was completed by 1878[117] and the pier was handed over to the county on 6 June 1879.[118]

Nearby at Ballyvaughan, the pier dating from 1846–47 was in some disrepair, having been damaged by storms in 1876. On application to the Lord Lieutenant, a grant was advanced to the OPW to make good the damage. Repairs to a breach in the sea wall were commenced on 1 January 1877 while work was still underway at Bournapeaka.[119] The repairs, which can still be seen today in the sea face of Ballyvaughan pier, were finished within a few months and other small renovations to the pier head were expected to be completed later that summer.

A Fishery Harbour Proposed at Ballyloughane near Galway City

Seven of the thirty-four memorials received up to 1875 were from Co. Galway. One of these, for a small pier, quay and landing slip in Mannin Bay, was never pursued subsequently by the memorialists. Another, for the repair of a quay at Ballyloughane near Galway city, put forward in a memorial from J. Wilson Lynch, was refused final sanction by Treasury. This refusal prompted questions at the Crichton enquiry in 1877.[120] Lynch may have thought that repairs to an existing pier were more likely to be grant-aided than the construction of an entirely new structure, a view that other information seems to support. For example, later in the early twentieth century, when Ernest Holt, then Chief Inspector of Fisheries, was under examination at the Select Committee on Transport, he explained what used to happen:[121]

> Query 4318: If you saw a work that you thought ought to be done, and which would cost £5,000, what would you do then?
>
> A: I am afraid under present circumstances we should have to leave it alone. Of course it is often possible to say 'This is a work of reconstruction'. There might be some little tiny harbour to start with. If it is an entirely new harbour we are bound by that rule [no expenditure above a certain sum].
>
> Query 4319: So if you find a place where there is very good fishing going on, and you thought it would be very useful and necessary to expend £5,000 on it, you would leave it alone?

A: Yes, but where there is good fishing there is nearly always some scrap of a harbour in existence.

Q 4320: There is the loop hole?

A: That is the loop hole.

When the Ballyloughane site was examined by the Inspectors of Fisheries on 19 September 1870, the only evidence for a pier existing at the site was a few blocks of stone scattered around here and there. However, the claim, whether accurate or not (and there is no documentary evidence for any pre-existing structure), that a pier had existed there previously had acted to ensure that the Inspectors actually visited and examined the locality. On assessing the situation, they decided that the memorial should be treated as one for an entirely new pier, rather than for the repair of a supposedly pre-existing one, and that was what they recommended. The OPW therefore progressed the application by instructing their engineer, William Forsythe at the time, to examine the site. He visited Ballyloughane on 9 November and proposed that a new breakwater, 300 feet long, could be built along the gravel spit protecting the beach. It would be detached from the land except for a timber causeway, 120 feet long, supported by five pairs of piles, joining it to the south side of the inlet. That way, the fishermen could have easy access on foot to their boats lying in the shelter of the proposed breakwater.

The OPW agreed to grant £400 of the estimated cost of £540, and forwarded that recommendation to Treasury on 10 December. In a letter of 8 February 1871, Treasury refused to sanction the proposal, without giving any reason. That was all the information that Mitchell Henry MP had been able to elicit during the Crichton enquiry; it confirms the nuances inherent in the system and the final uncertainty facing any memorial. Even the Inspectors of Fisheries expressed regret that Ballyloughane was refused sanction; they were convinced the local contribution would indeed be paid, whereas 'in poor localities it is found impossible in many instances to comply with the conditions to obtain a grant-in-aid for the construction of a pier or harbour'.[122] The memorialist in Ballyloughane's case, J. Wilson Lynch of Renmore, was one time MP for Galway and member of a well-known local merchant family that had built Moneen pier in Lough Atalia and Duras pier in south Galway in the 1820s.

Even that progressive pedigree was no help in Ballyloughane's case. With the failure of this memorial, it dropped off the list of sites for potential fishery piers and nothing was ever built there subsequently. Ballyloughane had been the chosen location for a massive commercial harbour proposed for Galway city in a report of 1858 commissioned by the Admiralty from Captain Washington, RN, Captain Vetch, RE, and Barry Gibbons CE (accompanied by Samuel Ussher Roberts CE, then the Galway County Surveyor).[123] That proposal was a far bigger, more sophisticated and more ambitious enterprise than anything proposed for Galway to this very day, but it was not sanctioned by the Government of the time. Treasury in 1871 may have been more than happy to let the proposal for a fishery pier lapse, and let Ballyloughane return to obscurity, for fear of resurrecting that immensely more grandiose and elaborate plan, which would have elevated Galway port to a maritime status eclipsing that of Liverpool. *Sic transit gloria mundi.*

Inishshark and Inishbofin

In contrast to Ballyloughane, both Inishshark and Inishbofin applied successfully for works, not once but twice each, during the 1870–75 period. The Inishshark memorial was for a boat slip that required the excavation of rock and gravel from the site to make an inclined plane. It would have fifty-nine oak skids fixed to it, set on stone blocks, and fastened with oak tree nails. For the work, 800 cubic yards of loose stone were excavated and dumped into the sea, along with 400 cubic yards of rock that were excavated by blasting. Boats could be drawn up the new slip by means of a crab-and-chain that was set on two large limestone blocks at the top of the shore.[124] On Inishbofin, a new pier

FIGURE 5.4. Inishbofin pier undergoing renovation in recent times. The original stonework is evident in the wharf face. Photo NPW.

was started in January 1874 with Patrick Duffy as overseer. First, an excavation into the harbour was opened and foundations laid for a pier with a return head, and a longshore quay wall, 226 feet in total length. Stone was drawn from a nearby quarry using a specially laid railed wagonway across the beach.

Remnants of this wagonway have recently been located on the island by Dr J.P. Mercer.[125] Dressed ashlar of 7,570 cubic feet was used for the facings of the pier and the wall, and 600 cubic feet of coping for the pier itself; four mooring posts were set in masonry and one set of steps was made. The peculiar difficulty associated with the work can be gauged from Patrick Duffy's comments:[126]

> This pier would have been finished long since if I could have got masons to come in to work here – not one would come. I had to train up the men of the island to dress and set the stones on the pier, in the use of quarry tools, and in blasting rocks. The men on the works now are indeed very handy men, well fit to go into any public work.

That final comment echoed those made by Thomas Telford and Alexander Nimmo over fifty years earlier, when they pointed out the potential value of public works in creating a skilled workforce.[127] To have completed the work at all, under the prevailing adverse conditions, was quite an achievement, especially when Duffy's first report had indicated the distressed state of the islanders at the time: 'I further state that there are many families sick in this island and in great distress, the men seeking employment and cannot be employed on a small job like this.'[128] Work at both sites was completed and the works handed over to the county by June 1876.[129] The second application from Inishbofin and Inishshark, made in 1878, was to clear a channel to the pier and make improvements to the slip, respectively, works that should probably have been done originally. The engineer reported positively on these applications but they became caught up in the suspension that was imposed on new marine works.

Endnotes

1 Commission of Enquiry into the Sea Fisheries of the UK, part 2, BPP 1866 [3596-I], minutes, query 37050.
2 Further Return to Parliament, BPP 1865 Sess.1 [207–1].
3 A.R.G. Griffiths, *The Irish Board of Works.*
4 J.J. O'Sullivan, *Breaking Ground. The Story of William T. Mulvany* (Cork: Mercier Press, *c.* 2004).
5 J. Burgoyne in Report of the Select Committee on Public Works in Ireland, BPP 1835 [573], query 717.
6 See, for example, letter from OPW to Secretary, Galway County Council 23/8/1905 in Galway Library Archives, GC/CSO/2/56.
7 J.R. Barry, in Report of the Select Committee on Sea Coast Fisheries (Ireland) Bill, BPP 1867 (443), minutes, query 3813.
8 E. Hornsby, in Report of the Select Committee on Sea Coast Fisheries (Ireland) Bill, BPP 1867 (443), minutes, query 234.

9 J.R. Barry, in Report of the Select Committee on Sea Coast Fisheries (Ireland) Bill, BPP 1867 (443), minutes, queries 3977–3983.
10 J.R. Barry, in Report of the Select Committee on Sea Coast Fisheries (Ireland) Bill, BPP 1867 (443), minutes, query 3772.
11 Return to Parliament, BPP 1859 (Sess 1) (119).
12 Further Return to Parliament, BPP 1865 Sess. I [207], p. 5.
13 *An Act to amend the Laws relating to Fisheries in Ireland (The Salmon Fisheries (Ireland) Act 1863),* 26 & 27 Vict. C. 114.
14 Section 42 of the 1863 Act.
15 Report of the Commission of Enquiry into the Sea Fisheries of the UK, BPP 1866 [3596-I].
16 J.R. Barry, in Minutes of the Commission of Enquiry into the Sea Fisheries of the UK, BPP 1866 [3596-I], query37014.
17 Returns to Parliament, BPP 1859 [sess 1] [119]; BPP 1864 [563]; BPP 1865 [sess 1] [207]; BPP 1865 [491].
18 Return to Parliament, BPP 1865 [sess 1] [207].
19 29&30 Vict. C. 45.
20 35th Annual Report of the OPW, BPP 1867 [3875].
21 36th Annual Report of the OPW, BPP 1867-68 [4043], p. 15.
22 Report of the Commission of Enquiry into the Sea Fisheries of the UK, BPP 1866 [3596 I].
23 *A Bill to amend the law of Ireland as to Sea Coast Fisheries*, BPP 1867 (50).
24 *A Bill [as amended by the Select committee] to amend the Law of Ireland as to Sea Coast Fisheries,* BPP 1867 (268), Section 26.
25 Report of the Select Committee on the Sea Coast Fisheries (Ireland) Bill, BPP 1867 (443).
26 J.R. Barry, in Report of the Select Committee on the Sea Coast Fisheries (Ireland) Bill, BPP 1867 (443), minutes, query 4405.
27 J.R. Barry, in Report of the Select Committee on the Sea Coast Fisheries (Ireland) Bill, BPP 1867 (443), minutes, queries 3702–3713.
28 J.R. Barry, in Report of the Select Committee on the Sea Coast Fisheries (Ireland) Bill, BPP 1867 (443), minutes, query 3805.
29 J.R. Barry, in Report of the Select Committee on the Sea Coast Fisheries (Ireland) Bill, BPP 1867 (443), minutes, queries 3803–3808.
30 T.F. Brady, in Report of the Select Committee on the Sea Coast Fisheries (Ireland) Bill, BPP 1867 (443), minutes, queries 5280–5283.
31 T.F. Brady, in Report of the Select Committee on the Sea Coast Fisheries (Ireland) Bill, BPP 1867 (443), minutes, query 5399.
32 T.F. Brady, in Report of the Select Committee on the Sea Coast Fisheries (Ireland) Bill, BPP 1867 (443), minutes, query 5287.
33 T.F. Brady, in Report of the Select Committee on the Sea Coast Fisheries (Ireland) Bill, BPP 1867 (443), minutes, queries 5301–5303.
34 J.R. Barry, in Report of the Select Committee on the Sea Coast Fisheries (Ireland) Bill, BPP 1867 (443), minutes, query 3361.
35 J.R. Barry, in Report of the Select Committee on the Sea Coast Fisheries (Ireland) Bill, BPP 1867 (443), minutes, query 3383.
36 T.F. Brady, in Report of the Select Committee on the Sea Coast Fisheries (Ireland) Bill, BPP 1867 (443), minutes, query 5287.
37 T.F. Brady, in Report of the Select Committee on the Sea Coast Fisheries (Ireland) Bill, BPP 1867 (443), minutes, query 5415.

38 J. Dombrain, in Report of the Select Committee on the Sea Coast Fisheries (Ireland) Bill, BPP 1867 (443), queries 1126–1128.
39 1st Annual Report of the OPW, BPP 133 (75), p. 3.
40 J.R. Barry, in Report of the Select Committee on the Sea Coast Fisheries (Ireland) Bill, BPP 1867 (443), minutes, query 3778.
41 Report of the Select Committee on the Sea Coast Fisheries (Ireland) Bill, BPP 1867 (443).
42 32&33 Vict. C. 92.
43 *An Act to amend 'The Salmon Fisheries (Ireland) Act 1863' and the Acts continuing the temporary provision of the same*, 32 Vict. C. 9.
44 Coombes, *Utopia in Glandore.*
45 A.R.G. Griffiths, *The Irish Board of Works,* p. 159.
46 Report of the Committee appointed to enquire into the Board of Works, Ireland [The Crichton Committee], BPP 1878 [C. 2060], paragraph 278, p. lx. The questioning of McKerlie concerning the Report is given in the evidence at paragraphs 388–422, pp. 15–16.
47 Col. J.G. McKerlie, Statement on the Report of the Committee appointed to enquire into the Board of Works, Ireland, BPP 1878 [C. 2080], pp. 6–7.
48 Report of the Committee appointed to enquire into the Board of Works, Ireland [The Crichton Committee], BPP 1878 [C. 2060], para 315.
49 Report of the Committee appointed to enquire into the Board of Works, Ireland [The Crichton Committee], BPP 1878 [C. 2060], p. lxviii.
50 O'Donoghue, *County Surveyors.*
51 Report of the Inspectors of Irish Fisheries for 1868, BPP 1868–69 [4177]; Report for 1869, BPP 1870 [C. 225]; Report for 1870, BPP 1871.
52 Report of the Committee appointed to enquire into the Board of Works, Ireland [The Crichton Committee], BPP 1878 [C. 2060], para 13, pp. ix–x.
53 Report of the Committee appointed to enquire into the Board of Works, Ireland [The Crichton Committee], BPP 1878 [C. 2060], p. xi.
54 T.F. Brady, in Report of the Select Committee on the Sea Coast Fisheries (Ireland) Bill, BPP 1867 (443), minutes, queries 5399–5410.
55 Report of the Committee appointed to enquire into the Board of Works, Ireland [The Crichton Committee], BPP 1878 [C. 2060], para 15, p. xi.
56 Second Report of the Royal Commission on Irish Public Works [The Allport Commission], BPP 1888 [C. 5264–I], p. 11.
57 44th Annual Report of the OPW, BPP 1876 [C. 1509], p. 22.
58 29&30 Vict. C. 62
59 Statement under Crown Lands Act, BPP 1872 (61).
60 Statement under Crown Lands Act, BPP 1877 (127).
61 Statement under Crown Lands Act, BPP 1882 (122).
62 Statement under Crown Lands Act, BPP 1887 (100).
63 Report of the Inspectors of Irish Fisheries for 1873, BPP 1874 [C. 980], p. 7.
64 Report of the Inspectors of Irish Fisheries for 1874, BPP 1875 [C. 1176].
65 Report of the Inspectors of Irish Fisheries for 1882, BPP 1882 [C. 3248].
66 42nd Annual Report of the OPW, BPP 1874 [C. 1001], p. 22.
67 BPP 1876 Bill 199; Act 45&46 Vict. C. 13.
68 BPP 1876 Bill 200; BPP 1878 Bill 159.
69 BPP 1880 Bill 266; Act 43&44 Vict. C. Clxxiv [A public Act of local character].
70 Report of the Committee appointed to enquire into the Board of Works, Ireland [The Crichton Committee], BPP 1878 [C. 2060], query 1718, p. 104.

71 Report of the Committee appointed to enquire into the Board of Works, Ireland [The Crichton Committee], BPP 1878 [C. 2060], paragraph 201, p. xlvi.
72 47th Annual Report of the OPW, BPP 1878–79 [C. 2336], p. 22.
73 Report of the Inspectors of Irish Fisheries for 1875, BPP 1876 [C. 1467].
74 Report of the Inspectors of Irish Fisheries for 1877, BPP 1878 [C. 2041].
75 Report of the Inspectors of Irish Fisheries for 1878, BPP 1878 [C. 1467].
76 Data from the 19th Annual Report of the OPW, BPP 1851 [1414]; See also Return to Parliament dated 1859 [BPP 1859 (Sess 1) (119)].
77 W.L. Joynt, in Report and Minutes of the Select Committee on Harbour Accommodation, BPP 1883 (255), query 2833.
78 16 & 17 Vict. C. 136, section 8.
79 Report and Minutes of the Select Committee on Harbour Accommodation, BPP 1883 (255), query 2899, p. 196.
80 William Pierce to Henry Paine, 30 July 1835. NAI OPW8/341.
81 William Pierce to Henry Paine, 31 August 1835. NAI OPW8/341.
82 NAI OPW/8/341, item 11.
83 36th Annual Report of the OPW, BPP 1867–68 [4043], p. 15.
84 48th Annual Report of the OPW, BPP 1880 [C. 2646], p. 26.
85 L. O'Tuarisg, 'Stair Chois Fharraige', *Biseach: Iris Chumann Forbartha Chois Fharraige* (Galway, Chumann Forbartha Chois Fharraige, 1999).
86 W. Le Fanu, in Report of the Committee appointed to enquire into the Board of Works, Ireland [The Crichton Committee], BPP 1878 [C. 2060], query 1713.
87 36th Annual Report of the OPW, BPP 1867–68 [4043], appendix D, p. 55.
88 37th Annual Report of the OPW, BPP 1868–69 [4174], appendix D, p. 54.
89 38th Annual Report of the OPW, BPP 1870 [C. 154].
90 40th Annual Report of the OPW, BPP 1872 [C. 584], appendix B, pp. 50–51.
91 51st Annual Report of the OPW, BPP 1883 [C. 3649], appendix B, p. 64.
92 52nd Annual Report of the OPW, BPP 1884 [C. 4068].
93 Report and Minutes of Evidence of the Select Committee on Harbour Accommodation, BPP 1883 (255), queries 2810–2821.
94 36th Annual Report of the OPW, BPP 1867–68 [4043], appendix D, p. 55.
95 Plan and estimate by James Donnell for works at Clifden, 9 January 1827. NAI OPW8/84.
96 15th Annual Report of the OPW, BPP 1847 [847].
97 17th Annual Report of the OPW, BPP 1849 [1098], Appendix E2.
98 36th Annual Report of the OPW, BPP 1867–68 [4043], appendix D, p. 55.
99 37th Annual Report of the OPW, BPP 1868–69 [4174], appendix D, p. 54.
100 Return to Parliament, BPP 1884–85 (167).
101 Return to Parliament, BPP 1883 (313), p. 679.
102 R. Ruttledge-Fair, Report on the Clifden District to the Congested Districts Board, 26 August 1892. In J. Morrissey (compiler and ed.), *On the Verge of Want. The base line reports of the Inspectors of the Congested Districts Board, 1892* (Dublin: Crannóg Books, 2001), p. 154.
103 Robinson, *Connemara, Part 1*, p. 257.
104 36th Annual Report of the OPW, BPP 1867–68 [4043], p. 15.
105 37th Annual Report of the OPW, BPP 1868–69 [4174], p. 15.
106 38th Annual Report of the OPW, BPP 1870 [C. 154], appendix D, p. 55.
107 39th Annual Report of the OPW, BPP 1871 [C. 383], pp. 17 and 35.
108 I am grateful to Michael Taylor, Errislannan, who brought this incised stone to my attention and for his permission to use his photograph of it.

109 50th Annual Report of the OPW, BPP 1882 [C. 3261], appendix B, p. 68.
110 53rd Annual Report of the OPW, BPP 1884–85 [C. 4475], appendix D, p. 68.
111 54th Annual Report of the OPW, BPP 1886 [C. 4774], appendix E, schedule no. 2, pier no. 100, p. 69.
112 Report of the Royal Commission on Irish Public Works (The Allport Commission), BPP 1888 [5264], list no. 2, handed in by Gen Sankey, appendix, p. 706.
113 55th Annual Report of the OPW, BPP 1887 [C. 5142], p. 26.
114 Specifications dated 2/2/1874 and plan dated 4/6/1874 by John Hill. NAI, OPW5/31019 /81.
115 44th Annual Report of the OPW, BPP 1876 [C. 1509], p. 51.
116 45th Annual Report of the OPW, BPP 1877 [C. 1762], p. 53.
117 47th Annual Report of the OPW, BPP 1878–79 [C. 2336], p. 53.
118 48th Annual Report of the OPW, BPP 1880 [C. 2646], p. 24.
119 45th Annual Report of the OPW, BPP 1877 [C. 1762], p. 53.
120 Report of the Committee appointed to enquire into the Board of Works, Ireland [The Crichton Committee], BPP 1878 [C. 2060], queries 1779–1780, pp. 107–09.
121 First and Second Reports from the Select Committee on Transport with the Proceedings of the Committee and Minutes of Evidence, BPP 1918 (130, 136), p. 200.
122 Report of the Inspectors of Irish Fisheries for 1871, BPP 1872 [C. 642].
123 Report to the Admiralty on the Port and Harbour of Galway, BPP 1859 [107].
124 43rd Annual Report of the OPW, BPP 1875 [C. 1223], p. 53.
125 J.P. Mercer, personal communication, 2012.
126 43rd Annual Report of the OPW, BPP 1875 [C. 1223].
127 2nd Report from the Select Committee on the State of Disease, and Condition of the Labouring Poor in Ireland, BPP 1819 [409].
128 42nd Annual Report of the OPW, BPP 1874 [C. 1001], p. 51.
129 Report of the Inspectors of Irish Fisheries for 1875, BPP 1876 [C. 1467].

CHAPTER 6

The Kindness of Strangers and the Relief of Distress

1879–1884

The landlords, as a rule, care very little about encouraging the fisheries. Those who follow them are usually holders of only very small plots of land, and as their success in fishing would add little, if anything, to the landlords' profits, they are not sufficiently interested to contribute much to harbours for purely fishing purposes. The same feeling operates with the ratepayers not engaged in fishing. It frequently happens, therefore, that where a harbour is much required as a shelter for fishing boats, but would serve no other object, that it is left unmade ... Nothing is more essential for the success of the fisheries than good harbours.

Fishery Inspector John A. Blake, 1875[1]

Yet Another Famine Looms

The timing of the directive suspending new pier construction in 1875 was particularly unfortunate for the western counties. The twenty previous years had been ones of relative prosperity, even though the rural population and the number of fishermen in particular had continued to decline due to emigration. Starting in 1875, local agricultural prices began to collapse and a series of poor harvests made things worse for poor Irish tenants. Agricultural output stagnated and rural incomes declined sharply.[2]

The move away from tillage to pasture in Britain resulted in a shortage of work for Irish seasonal migrants, which exacerbated the difficulty in those parts of the West that relied on the earnings of migrant workers. Not since the Great Famine was paid work so necessary to ensure the subsistence of the rural poor. Harvests went from bad to disastrous in 1879 and the Government, fearful that a famine was looming, tried to initiate a rapid response – almost a pre-emptive strike – to ameliorate the worst anticipated scenario. It sought advice from the OPW as to what measures, by way of public works, might be initiated to meet the imminent threat. On 31 October and 7 November 1879, the Board submitted memoranda outlining projects on land drainage and reclamation, the construction of new fishery piers and harbours, and major arterial drainage.[3] The OPW had, in fact, been well ahead of the Government: on 23 July 1879, it had already submitted for Treasury sanction, despite the suspension, a list of twenty-three memorials for piers previously recommended by the Inspectors of Fisheries.[4] Three of these were in Co. Galway, Smeerogue (Renvyle), Errislannan (Curhowna) and Inishbofin/Inishshark, and one, Ballyvaughan, was in north Clare.[5] On instruction from Treasury, the Board directed Robert Manning to prepare rough estimates of the likely costs of constructions and renovations, without actually visiting all the sites in advance. His estimate for the twenty-three came to a total of £61,250.[6]

By November 1879, therefore, Treasury had to hand all the relevant reports, information and estimates regarding these twenty-three projects. Yet, when the letter of sanction issued on 26 December, it named only six piers 'as those in respect of which they should be ready to ask parliament for grants, or grants on account, for the next year, provided the localities concerned are ready to contribute the fourth part of the total cost which the Act requires'.[7] The piers selected were: Goleen Creek in Co. Clare; Burtonport, Ballysaggart, Porturrin and Bunatruthan in Co. Donegal; and Smeerogue (Renvyle) in Co. Galway. The total estimated cost of these six was £9,200. In light of the concern it professed earlier, this limited sanction was hardly an overwhelmingly positive response by Treasury.

Treasury did, however, add some Christmas cheer. The suspension of pier building had been imposed partly to offset the large sum of £26,000 allocated to the proposed works at Arklow harbour. Very late in the Arklow process, after a contractor had been appointed, plant and materials had been delivered, and rails laid to the quarry for stone, the Bill seeking parliamentary approval for the work was withdrawn from Parliament.[8] This happened because the Wicklow Copper Mining Company, the owners of the harbour, did not agree to its being handed over to Harbour Commissioners when completed. As a result, the earmarked money became available again, and Treasury decided to lift the suspension on further memorials, perhaps little expecting the huge reaction that would trigger. In fact it opened the flood gates: the OPW claimed it received seventy-four new memorials in the year 1879 to 1880, swelling the number on its list of memorials to 110, about ninety of which were 'in abeyance'.[9] The surge suggests a release of pent-up demand; or the possibility that the OPW had been receiving and retaining, but not publicly declaring, memorials since 1875; or that an awareness had gripped the country that additional funds were about to be made available; or maybe all three. In addition, Treasury authorised the OPW to advance loans, and the Grand Juries to hold extraordinary presentment sessions, in order to initiate immediate relief works, actions that were legitimised retrospectively by the first *Relief of Distress Act (1880)*,[10] passed by Parliament on 15 March 1880. The construction of fishery piers was being cast, once again, as a relief works solution to an anticipated famine crisis.

A Canadian Gift of Great Generosity

The crisis anticipated in the country was brought to the attention of the world in letters and newspaper reports circulating widely in Europe and North America. In response, early in 1880, the Government of Canada magnanimously gave a gift of $100,000 (more than £20,000 at the time, equivalent to millions of euros today, depending on the method of calculation) for the relief of distress in Ireland.[11] On the advice of the Secretary for the Colonies, Sir M.E. Hicks-Beach, this sum was to be administered by a special committee of six persons, drawn equally from the Mansion House Committee for the Relief of Distress (set up and chaired by Edmund Dwyer Gray (1845–88), Lord Mayor of Dublin and proprietor of the *Freeman's Journal* newspaper) and the Relief Committee of the Duchess of Marlborough (wife of the Lord Lieutenant).

While the Canadian gift was unfettered by pre-conditions, Hicks-Beach suggested that some appropriate uses for the money might be to assist fishermen in purchasing boats and nets and for 'grants providing the contributions required from the localities interested in order to secure the construction of fishing piers and harbours', i.e. the one-fourth local contribution.[12] He and Treasury had no intention of breaching the absolute requirement for the local contribution, no matter how distressed the condition of the people might become. A kinder interpretation might be that this proposed use of the Canadian fund was a stratagem devised by Hicks-Beach to leverage funds already voted for piers but not hitherto accessible due to the lack of local contributions, thereby maximising the beneficial effect of the gift. That, at least, was the diplomatic interpretation of the special committee that administered the fund.[13] Whatever the motivation, the special committee, formally entitled the Canadian Committee for the Relief of Distress in Ireland, but more often referred to as the Canadian Fund Committee, was quickly established and held its first meeting on 17 April 1880 in the Shelbourne Hotel, Dublin. On the retirement of the Duke of Marlborough as Lord Lieutenant around that time, the Duchess resigned from the Committee and was succeeded by a most appropriate replacement, Viscount Charles Stanley Monck (1819–94), the Tipperary-born first Governor General of Canada (by then retired) and previously a Lord of the Treasury under Palmerston.

After full consideration at its first meeting, the Committee agreed that the Canadian fund should be limited to the objects suggested by Hicks-Beach. The members then acquainted themselves with the existing situation regarding the piers. They learned that, owing partly to the insufficiency of the annual grant, and partly to the difficulty of obtaining the local one-fourth contributions, the construction of piers had proceeded only very slowly, some languishing for several years without any action being taken. Matters could not be allowed to remain like that as far as the Committee was concerned and, at their next meeting on 26 April 1880, the members adopted a resolution proposed by Lord Monck: 'That all grants made, or to be made, by the committee in aid of the erection of piers and harbours, be subject to the condition, that the work be actually commenced within a period of three months, from the 1st day of May next, as a maximum.'[14]

They wrote to Treasury on 5 May 1880, a detailed letter indicating their deadline and describing their proposed awards.[15] For once, it was a letter of determination and strength, couched in terms entirely unequivocal. The Committee had found that, at the time, only one pier in Ireland had received Treasury sanction for three fourths of the cost, but even that one was not going ahead because it still lacked the local one-fourth contribution. This was Smeerogue (Renvyle) pier in Co. Galway, approved by the Inspectors since 1874[16] but lacking the local contribution since then. The Committee immediately granted the sum required, £566, to provide the entire one-fourth of the cost so that the work could be started at once by the OPW. It also noted, from returns received, that although the Inspectors of Fisheries had recommended grants in twenty-eight other cases, no steps had yet been taken by the OPW to have the preliminary surveys, plans and estimates drawn up due to the suspension of funding. They therefore allocated various amounts to assist with making up the contributions in all of these cases too. Having every reason to believe that the residual balance of the contributions would be fully made up locally, the Committee urged Treasury that instructions be given to the OPW to have the necessary surveys and plans prepared without any further delay.

Of the twenty-nine piers supported by the Canadian Committee, eight were in Co. Galway and one in north Clare: Smeerogue, Pollnadhu (Ardmore), Inishlacken, Bush Harbour, Roundstone, Doleen, Leenaun, Glengevlagh, and Glenina in north Clare (Table 6.1). Some of them benefitted from other charitable donations as well, and these are indicated in the table. In due course, work started at all of them. Two other Galway sites, Scrahalia and Callowfeenish (*Caladh Fínis*) were recommended by the Inspectors but they did not feature in the list supported by the Canadian fund; years later, piers would eventually be erected at these places under other schemes.

Allocations to Oranmore Castle pier and to five more sites in other counties were made possible at the last moment by the release of £1,000, originally allocated to Killala, Co. Mayo, which was not taken up there.[17] Learning from the OPW that its engineers could make only 'two surveys in a fortnight', the Committee urged Treasury to authorise the Board to employ additional staff in order that the surveys and plans could be made with greater expedition 'or otherwise the committee will reluctantly be obliged to abandon the idea of endeavouring to assist in the accomplishment of these works'. That was indeed a serious threat. The letter continued:

> As the funds of this committee are to be given in aid of grants to be sanctioned by the Lords of Treasury, the committee further hope that you [the Secretary of the Treasury] will be good enough to move their Lordships, to convey to the Board of Works their sanction of the grants in the cases referred to, as speedily as possible.[18]

The letter ended with: 'unless the works can be immediately commenced, no aid can be afforded by the Committee and they would beg their Lordships' help in the manner mentioned'.

TABLE 6.1

Fishery piers and quays in Co. Galway and north Clare that benefitted from the kindness of strangers between 1879 and 1883.

Location	Est. cost	Govt Grant	Canada Fund	Liver-pool Fund	*New York Herald* Fund	Land League Fund	Private funds
Ardmore (New pier & quay)	£1,400	£1,050	£350				
Bush Harbour (New pier)	£3,130	£2,348	£583	£100			£100[1]
Doleen (New quay)	£2,000	£1,500	£350			£20	£130[2]
Glengevlagh (New pier)	£650	£488	£63				
Inishlacken (New pier)	£600	£450	£150				
Leenaun (Pier extension)	£800	£600	£200				
Oranmore (New pier)	£580	£435	£100	£50			£50[3]
Roundstone (North pier rebuilt)	£2,000	£1,500	£150				
Smeerogue (New pier & slip)	£4,000	£2,821	£566	£250	£189		
Glenina, Co. Clare (New pier)	£1,200	£900	£300	£100			

[1]Count de Basterot; [2] Fr Flannery, PP; [3] Fr Quinn, PP. Other small contributions are not recorded separately, e.g. J.H. Tuke gave £5 to Glengevlagh as he passed the pier site.

Were their Lordships ever before admonished in this manner 'to move as speedily as possible', or given such an ultimatum, one wonders? One wonders even more what the fate of Ireland might have been had they been admonished in this forthright way at other troubled times in our history, especially by persons of the calibre and repute of Viscount Monck. We will see later (chapter 10) that a Select Committee of the House of Commons in connection with the Irish Board of Works expressed its dismay and dissatisfaction at the excessively timid attitude of the Board in its normal dealings with Treasury. Receiving no immediate reply, the Canadian Committee hastily met the new Chief Secretary for Ireland,

W.E. Forster, on 12 May, to press the case, urging him to impress on Treasury the necessity for 'energetic action with regard to grants towards fishery piers and harbours.'[19] So their admonition was not only firm, but persistent, too.

The Lords of the Treasury replied to Forster on 25 May proposing certain courses of action for his consideration.[20] Treasury would present to Parliament a supplementary estimate for £30,000 to be applied exclusively in the distressed Poor Law Unions. It would set up a special committee of three persons to advise the Lord Lieutenant, for his approval and ultimately that of Treasury, of the fishery stations where a total of £40,000 (made up of the supplementary £30,000 vote, plus the associated one-fourth contributions of £10,000) could be expended 'with the best advantage for the double purpose of providing employment for unskilled labour, where it is urgently needed, in the construction of the works and of permanently improving and extending the fisheries of the west of Ireland.' When appointed, this new body, later to be called the Fishery Piers Committee (FPC), could employ whatever extra assistance it judged necessary for the purpose of enquiries, surveys and plans, in addition to the assistance already available to it from their respective Boards (the members of the FPC were all drawn from Government departments). Having sugared the pill in this way, their Lordships launched into a justification of their prescribed medicine: the law was the law, and the provisions of the Piers and Harbours Acts had to be respected, cumbersome as they might be, especially where the Acts permitted extensive interference with private property.

In addition, it was not at all clear to Treasury who was to pay for the maintenance of the piers as required by law, or how local consent for that expense was to be ascertained, or how the necessary funds could be guaranteed before construction was even started. All of this temporisation was topped off with a barbed observation: the Lords of the Treasury 'recognise the wisdom of the Canadian Fund Committee in confining their action, apparently, to the provision of the necessary fourth, and after that leaving the Government to settle with the local representatives as to the application of it' – perhaps a veiled caution that the Committee should 'butt out' of what was clearly a Government responsibility. Treasury had no evidence, the letter continued, that the sum of almost £200,000 spent on the fishery piers and harbours of Ireland in the previous thirty years had succeeded in the establishment of a fishing trade; if it had seen such evidence, results as well as hopes could have been appealed to, in support of such works at the present time.

It was a long and detailed reply, directed at the Canadian Fund Committee through the Chief Secretary. The latter added his own missive to the Committee on 4 June: the Viceroy, he wrote, was glad to hear that the Canadian Committee would work along with the new Fishery Piers Committee, but, he indicated, the Lords of the Treasury 'are of opinion that the [time] limit should be extended at least to 31 August, and possibly to 30 September ... His Excellency [the Viceroy] concurs in the opinion, and hopes that the Canadian Relief Fund Committee will be kind enough to accede to the proposal.'[21]

To its great credit and ultimate effectiveness, the Canadian Fund Committee was in no mood to back down or acquiesce when replying to the Chief Secretary on 7 June, three days later. The members declared:

> their public statements, as well as their relations with the Canadian Government, render it difficult to make any relaxation in the rule ... The conditions upon which they voted the money would be complied with, provided the sum which they voted should be in course of expenditure before the date named [31 July] ... They trust that the Committee appointed to administer the Treasury grant [the FPC] will be able to expedite matters so as to have the works commenced ... The Canadian committee cannot forget that the intentions of the Canadian Government in voting the money were, that it should be expended in the relief of existing distress, and they think, therefore, that they are bound to see to its expenditure within the time named.[22]

The letter ended by pointing out, that since the total of the grants being provided by Treasury (£30,000) for the proposed works did not equal three fourths of the estimated costs of those already approved, it appeared that the choice of which piers would be proceeded with must be made at once, 'unless a larger vote be obtained'. The Committee already had available a sum that was more than one-fourth of the £40,000 total that Treasury proposed. The Committee's response was, without a shadow of doubt, very far from the meek acquiescence that so often resulted when Treasury 'requested' compliance with its proposals.

At its next meeting, on 11 June, the Committee adopted a formal resolution which it forwarded to the Viceroy and the Chief Secretary, urging the Government to obtain an increase in the proposed supplementary estimate. By then the combined Canadian Fund allocations, along with the local and other private contributions already secured, stood at £15,000, so that Government could, if it so wished, advance £45,000 as its three-fourths share. Remarkably, the Lords of the Treasury indicated on 30 June that they were 'willing, on consideration of the Resolution of 11 June forwarded to them, to increase the amount of the grant to be proposed to Parliament, from £30,000 to £45,000'.[23]

True to its word, Treasury immediately sponsored a second, amended Relief of Distress (1880) Act that was passed by Parliament on 2 August, granting the promised £45,000 for piers.[24] Separately, Alan Hornsby, the Secretary of the FPC that was set up to determine which sites were to benefit from the Act, indicated that he hoped to have working plans ready within a week and that great endeavours had been made on all sides to ensure that the funds of the Canadian Committee should not be lost. Without any delay, a list of thirty sites was drawn up by the FPC, and the Canadian Fund agreed immediately to support twenty-nine of them.

All the piers built with relief of distress funds between 1879 and 1884 are listed, with the sums allocated, in Appendix E. The Canadian Fund Committee also 'executed works under its own superintendence' at Carntullagh in Co. Donegal (£55), Ballygarry in Co. Mayo (£65) and Seafield in Co. Clare (£203).

Initially referred to as the Warrant Committee (it had been appointed under warrant of 14 June 1880), the Fishery Piers Committee (FPC) comprised Thomas F. Brady of the Fisheries Inspectorate, William Le Fanu of the OPW and Henry Robinson, Vice-President of the Local Government Board. Staff Commander Charles Langdon RN was added to it later by Treasury. It had, as a first step, drawn up a list of twenty-four sites to be supported, which were named in a Return to Parliament in 1880.[25] Later, it added the remaining six to these, bringing the total to thirty, published in a Return of 1881.[26] The OPW listed all thirty sites in its 49th Report in 1882 and showed their locations in a large-scale coloured map.[27] Only piers in the western half of the country were included, naturally enough since that was where the distress prevailed; the list was never meant to be a random selection from the large number of memorials then in the hands of the OPW. Indeed, a few years later, the effect of this selectivity in favour of the West would be rebalanced by favouring sites on the east and south coasts under the Sea Fisheries (Ireland) Act of 1883, to be discussed in chapter 7.

Matters had indeed moved with remarkable rapidity and considerable success since the Canadian Committee's first meeting on 17 April, so much so that the members congratulated themselves 'for having contributed to accomplish, in a comparatively short time, more than would have been accomplished for many years to come if they had not directed their attention to the matter'.[28] It was well-merited satisfaction, by any measure.

The fund had contributed to a total twenty-nine works, all erected between 1879 and 1881. In one section of its report,[29] the Committee indicated that it had supported only twenty-three, excluding Killala; but an addendum to the report[30] confirms the final total as twenty-nine sites (including Smeerogue), consequent upon the abandonment of Killala. The Committee's staunch determination demonstrably helped to break the bureaucratic and financial deadlock that had been holding back new constructions. Its support not only assisted materially with paying the one-fourth contributions, it also exposed dramatically the fiction that the contributions were necessarily 'local', in the sense of donated by local interests. Finally, without the Committee's trenchant stand, Treasury was unlikely ever to have requested Parliament to increase its allocation for funds from £30,000 to £45,000, especially when the impending threat of famine was receding, or at least not proving as severe as first feared.

In the first two weeks of August 1880, the Committee paid £8,790 into the bank to the credit of the OPW, and work commenced immediately at the chosen locations. Adding-in the £566 it had already contributed for Smeerogue, and a small sum given separately for a pier in south Clare, the total contributed to the piers by the Canadian fund alone came to £9,704. At the most basic level of the gross number of men employed, 490 were reported to have been engaged at twenty-three of the sites, making a very significant contribution to the local communities involved.[31] As discussed below, relief work was so badly needed in most places that desperate men were forcing themselves on to the sites, leading in some instances to riotous behaviour and serious conflict.

By the time it was wound up in February 1881, the Canadian Fund had expended a grand total of £21,000 on various relief measures, having given more than £11,100 for the purchase of boats and nets for poor fishermen in addition to its contribution to the piers. It was altogether a most munificent gift, magnanimously given, by the Parliament and people of Canada. It deserves to be better remembered, acknowledged and suitably commemorated today.

Other Charitable Benefactors also Helped

Some private individuals and other charitable funds also helped in making up the one-fourth contributions, but with more modest amounts. Thomas Brady listed these donors in the Annual Report of the Inspectors of Fisheries in 1881.[32] The smallest amount placed directly in his hands was an anonymous donation of five shillings. Other sums ranged from ten shillings to £150 given, in a number of cases, for particular named sites. For example, Baroness Burdett-Coutts, better remembered for her support of the fisheries and the Fishery School in Baltimore, Co. Cork, gave two donations totalling £85 for marine works in Co. Clare. The humanitarian, James Hack Tuke, gave £100, the Archbishop of Tuam £150, and other churchmen and MPs gave sundry amounts, again for named sites. Fr Thomas Flannery PP of Carna gave a generous £130 towards Doleen quay (Ballyconneely). The Liverpool Relief Committee fund stepped in with almost £2,000 for piers in counties Donegal, Mayo, Galway, Clare and Cork. The Relief of Distress fund of the Irish Land League gave £200 for piers in the same counties, and the *New York Herald* contributed £188 towards Smeerogue pier from funds collected from its readers. From elsewhere abroad, and as examples of the wide extent of public generosity, the Municipal Council of Rouen, France voted 1,000 francs towards food for the poor in Ireland, and the bishops of Versailles and Nimes authorised collections at Sunday Masses throughout their respective dioceses for Irish relief purposes.[33]

The Duchess of Marlborough Fund and Major Gaskell

The Canadian Fund was the largest of a number of relief funds operating in the West during the early 1880s. The Duchess of Marlborough Relief Committee, for instance, operated a number of relief schemes on its own account during the summer of 1880. Like other charities, it worked independently of the State-run schemes and at a more localised level. In Connemara, the fund's representative was Major Gaskell. According to Captain George Morant, 2,200 men and women were employed making roads and building small piers under Gaskell, being paid at the rate of one stone of meal per person per day for five working days.[34] Staff Commander Charles Langdon of the FPC described these works as follows:

> he [Gaskell] was building in Ardmore and in that direction west of Ardmore; I went one day and saw the little piers being built by the fishermen themselves, who received wages, and I thought that was a very good way of doing it, because it relieved their distress, and improved their port, whereas in a great many fishery pier cases, the contractors employed their own men, and not men in the localities.[35]

For obvious reasons, Gaskell's works would have been small and rudimentary. Piers of that artisanal kind near Ardmore include *Aill an Eachrais* (No. 067), *Caladh Feenish* (No. 069), *Aill na mBrón* (No. 071), Rossduggan (No. 072A), *Sruthán na mBrácaí* (No. 073), *Aill Uaithne* (No. 074) and Derryrush (No. 078). Further away, small additions at *Clais na nUan* (No. 120) and *Caladh Thaidhg* (No. 129) were of similar, vernacular form. It remains to be confirmed which of these artisanal piers were indeed Gaskell's, although we know he was certainly engaged in the Ardmore/Kilkieran area in 1880. Morant states it and sketches by Charles W. Cooke archived in the National Library of Ireland[36] confirm his presence at Loch Conaortha near Kilkieran at the time. Gaskell's own reference to Teernea (*Clais na nUan*)[37] suggests that he did some work there, and a minute of the OPW[38] confirms that he erected a small pier at *Caladh Thaidhg* that was later dismantled by the Piers and Roads Commission. Some of his works would remain largely unfinished, as Morant explained with some dismay, 'owing to the time fixed for stopping relief work, viz. the 1st of August, having arrived before they were completed'. Little seems to have changed since work on Maumeen pier was similarly abandoned prematurely in the famine of 1822!

Eaten Bread Is Soon Forgotten

Sadly, the generosity of the various benefactors of that time is seldom, if ever, recalled now and there was nothing to acknowledge them at any of the piers that were built with their generous support. The Canadian Committee had expressed the hope that securing the erection of the piers 'will be of lasting advantage to this country, and a permanent record of the generosity of the Canadian people'.[39] The absence of a commemorative plaque at Smeerogue (Renvyle) is particularly regrettable: the Canadian support for that pier was the catalyst that broke the funding deadlock, and swayed Parliament to increase to £45,000 the sum granted by the second *Relief of Distress Act*. The deficiency was, to a very small extent, rectified in August 2015 with the unveiling by the Canadian ambassador, the Hon. Kevin Vickers, of a plaque at Parkmore pier (Bush Harbour) acknowledging the generosity of the Canadian gift. The plaque, which revives the traditional name, *Caladh na Sceiche*, was inscribed by the sculptor Tom Glendon and erected through the personal efforts of Dr Michael Brogan. Parkmore is the only one of the assisted piers and quays on the Irish coast that carries such an acknowledgement. Strangely, even the very names of three of the sites in County Galway that benefitted – Smeerogue, Doleen and Bush Harbour – seem to have been erased from memory: they are now known as Tully (or Renvyle) pier, Ballyconneely quay and Parkmore pier, respectively. Another beneficiary, Glenagimlagh pier in Killary Harbour, is now known as Glengevlagh pier, or Aasleigh pier, or sometimes (incorrectly and most confusingly) even as Leenaun pier.

The Two Relief of Distress Acts of 1880

The first Relief of Distress (Ireland) Act 1880 was enacted in March 1880.[40] It allocated the sum of £750,000 to the relief of distress, principally by means of outdoor relief administered by the Poor Law Unions. It also legitimated, retrospectively, certain urgent relief actions that had been initiated on Treasury instruction before the actual passing of the Act. Sections 10 and 20 prescribed:[41]

FIGURE 6.1. Parkmore (Bush Harbour; *Caladh na sceiche*) pier as it is today. The commemorative plaque unveiled by the Ambassador of Canada in 2015 is inserted into the central wall where the box-like structure is shown in this picture. The stone dated 1881 is at the seaward end of the wall. Photo NPW.

> Sect. 10. All presentments made or to be made at extraordinary presentment sessions in accordance with instructions (already issued) and which have been, or shall be, approved by the Commissioners for Public Works and all works done in execution thereof shall be ratified and confirmed as if they had been presented, made and done strictly in accordance with the statutes
>
> Sect. 20. All persons who have acted in making loans etc ... shall be released and indemnified from and against all penalties in consequence thereof ...

The second Relief Act, *An Act to amend the Relief of Distress (Ireland) Act 1880 and for other Purposes relating thereto,*[42] was enacted in August of 1880. It granted the OPW, in lieu of the earlier £750,000, a new sum of £1.5 million, with £45,000 of that amount specifically earmarked for the purposes of the Fishery Piers and Harbours Act of 1846, exactly as

Treasury had promised in its letter of 30 June to the Viceroy.[43] It was from this source that the OPW allocated grants for the piers selected by the FPC, ones for which memorials had been submitted before the passing of the August 1880 Act, or submitted before 30 September immediately after its passing. The funds made available were fully absorbed straight away by the chosen sites, leaving no money at all for the approximately 130 other memorials still on file with the OPW.[44]

The 1880 Relief Acts required the promoters of any pier to contribute the usual one-fourth of the estimated cost, but once that was paid to the OPW, or secured in some other way from whatever source, the work could go ahead without any of the preliminary surveys, enquiries and reports to Treasury prescribed in the 1846 Piers and Harbours Act. This provision helped to reduce the procedural hurdles surrounding the process, and, as such, was a very welcome improvement. However, once any pier was completed it remained subject to all the other provisions of the 1846 Act, which meant, for example, that it 'would be deemed and be taken to be, public property absolutely vested in the Commissioners' as specified in section 4(2).

While construction work made rapid progress at many of the sites, nothing had commenced at six of them one year later, according to a Return to Parliament dated 12 May 1881.[45] This inordinate delay was raised in the House of Commons by Arthur O'Connor MP. Lord Frederick Cavendish, the new Financial Secretary to the Treasury responded in explanation:[46]

FIGURE 6.2. The commemorative plaque erected at Parkmore pier and unveiled by the Ambassador of Canada to Ireland in August 2015. Photo NPW.

> In addition to the various provisions of the Fishery Piers and Harbours Act which cause delay, such as the section providing for preliminary surveys and inquiries, a Report to the Treasury, the publication of notices, and security for the due maintenance of the pier when completed, there was the difficulty of selecting a limited number out of a very large number of applications for grants.

His reply explicitly acknowledged the cumbersome nature of the whole process from memorial to construction and the difficulty caused by the unexpectedly large number of memorials. However, he appears not to have realised that section 14 of the amended Act of 1880 had actually swept away many of these procedural hurdles, and he went on to state that 'selection had to be made not simply on the ground of greatest benefit to the fisheries, but also for the purposes of relief of distress'. Once again, it was the need for relief work as much as the needs of the fisheries that coloured the Government's attitude to fishery pier construction. Yet, O'Connor's complaint did have some substance, if not in the slow rate of completions, at least in highlighting the failure to deal effectively with the very large number of memorials that remained on file with little realistic hope of being addressed. Reports on the constructions underway, given in the Annual Reports,[47] confirm the OPW's determined application to the very challenging task it faced.

New memorials were coming in faster than works were being sanctioned and undertaken, and since the whole sum of £60,000 (£45,000 plus £15,000 local contributions) was fully committed from the very outset, it was abundantly clear that something more needed to be done if any serious inroad were ever to be made into the growing list of memorials. In 1883, therefore, Parliament voted an additional, although meagre, £4,000 'in aid of the construction of Fishery Piers and Harbours in Ireland'[48] and Treasury sanctioned the application of this sum to eight of the sites still remaining unaddressed. These were in Donegal (Downies pier; Rannagh pier); Kerry (Portmagee slip); Sligo (Pollnadivva harbour); Mayo (Tontanavally pier extension); Wexford (Carnsore pier and slip); and two (Ballyhees and Carrowmore) in Galway. Four were grants in aid of entirely new structures; the rest were supplementary grants to places that had already received funds but needed some more for completion. Five of the eight were completed by June 1885, the remaining three being almost finished by then. The benefits to Co. Galway from this small sum were a new landing place at *Trácht Each* near Ballyhees (*Baile Thíos*) on *Inis Oirr*, the smallest of the Aran Islands, and the construction of a pier at Carrowmore (Lynch's Pier), a small natural landing place in Ballinacourty, in inner Galway Bay.

What Did the Relief of Distress Acts Achieve?

The total amount allocated to fishery piers between 1879 and 1883, inclusive of all donations and local contributions, was £76,760: £3,500 spent on Cape Clear; £9,200 allocated to six piers in December 1879; £60,000 allocated to thirty piers under the *Relief of Distress Acts 1880*; and £2,060 expended of the £4,000 voted specially in 1883 for the final eight.[49] In all, thirty-seven works were fully completed with these amounts. The savings made below the estimates were used for other unnamed piers and quays and a small residue was left in the account. The Commissioners of Public Works summarised briefly, and with a degree of satisfaction, the operation of the various schemes.[50]

Generally, their procedure had been to seek tenders for the works by public advertisement and where no tender was received, or where no contract could be agreed within budget, the OPW did the work with day labourers under the direction of its own engineers. At the sixteen sites completed using day labour, the works were estimated to cost £26,350 and, at the other twenty-one, done by contract, estimated to cost £48,510. Taken together, the overall estimated cost of the thirty-seven works was £74,860 and the actual gross expenditure, including administrative costs, was £74,765, a remarkably precise and accurate outturn. It was just as well they got things right: with so much expenditure underway, Treasury was careful to ensure that it should not run out of control. As the OPW put it, 'it was obvious that without additional funds being provided by contributions, which under the circumstances could not be relied upon at the time, we were directed to curtail the works, should any necessity arise for doing so, in order to avoid an excess of expenditure over estimate for which no provision could be made'. As it turned out, only one project, at Teelin in County Donegal, had to be cut back and the dependence of other sites on the charitable donations was amply confirmed. Overall, seven counties, on the southwest, west and northwest coasts, benefitted from the employment created by the *Relief of Distress* pier works as recorded in the report of the Chief Engineer, Robert Manning, and shown in Table 6.2.[51]

The information in the table needs careful evaluation: the total number of men (188,938) reported as employed in pier construction during the years 1881 to 1882, is unbelievably large. Information given in a more detailed Return to Parliament in 1881[52] indicates, by way of comparison, that the numbers employed daily at eight named Galway sites ranged from nineteen to thirty-eight men, the aggregate for the eight sites being 203. Working a six-day week for fifty weeks a year, the number of man-days would be 203 x 6 x 50 = 60,900 man-days. That figure is rather close to Manning's figure (58,135) given for the 'number of men employed' at nine Galway sites. Alternatively, if Manning's total figure of 188,938 men allegedly employed at all the sites during that year is divided by fifty (approximate number of weeks worked) and then by six (days worked per week) the result is 630, a figure close to 621, the 'total number of men employed daily' according to the Return to Parliament of 1881. Therefore it is most likely that Manning's figures in Table 6.2 for the 'number of men employed' should be interpreted more correctly as 'number of man-days', rather than as the absolute number of men employed. That puts a far less flattering gloss on the employment information.

As to costs and wages, seventeen of thirty works recorded in a Return to Parliament of 1882[53] were done by contractors at an aggregate contract price of £22,680 and they engaged an average of nineteen men per site (an aggregate of 323 men at the seventeen sites). The other thirteen works were carried out by the OPW by direct labour at a total cost

TABLE 6.2

Works executed 1881–82 on 35 piers in 7 counties where works were carried out under Treasury sanction of October 1879 and the Fishery Piers Committee of 1880.

County	No. of works done			Estimated Expenditure			No. of men employed in year 1881–82		
	Total	Contract	OPW	Total	Contract	OPW	Total	Contract	OPW
Clare	4	0	4	£2,350	-	£2,350	17,650	-	17,650
Cork	4	3	1	£5,590	£5,370	£220	7,590	6,982	608
Donegal	9	8	1	£22,880	£21,900	£980	38,566	34,272	4,294
Galway	9	3	6	£15,160	£4,360	£10,800	58,135	15,152	42,983
Kerry	1	1	0	£1,350	£1,350	-	10,728	10,728	-
Mayo	6	3	3	£14,560	£6,420	£8,140	38,946	11,256	27,690
Sligo	2	2	0	£7,310	£7,310	-	17,323	17,323	-
	35	**20**	**15**	**£69,200**	**£46,710**	**£22,490**	**188,938**	**95,713**	**93,225**

Source: Manning's report in the 50th Report of the Commissioners for Public Works.

of £16,475, employing an average of twenty-two men per site (an aggregate of 286 men). Calculations from these figures show that contractors provided one job for every £70.22 expended, whereas the OPW provided one job for every £57.60 expended.

Rule-of-thumb calculations based on these figures, and other estimates of the number of days worked at each site,[54] yield an estimated maximum rate of pay under the OPW of about two shillings per man per day. This is very similar to the daily wage calculated below for the men working at Glenina and at Inishlacken. It could, however, be an over-estimate: two years later, in 1885–87, the Piers and Roads Commissioners would pay unskilled labourers only one shilling and four pence per day in summer and one shilling per day in the shorter days of the winter months.[55]

While the overall amount of work generated by the *Relief of Distress* funds from 1880 to 1883 appears therefore to have been less than a casual glance at the published official figures might suggest – an average of 630 men employed daily over seven counties (ninety per county) is fairly meagre in the circumstances – it nevertheless must have made an important contribution to the economy of those families lucky enough to benefit from it.

Many more would have welcomed any opportunity to work. At Glengevlagh, for instance, the contractor 'was much harassed by men forcing themselves on the work in greater numbers than he could profitably employ', sure evidence of the sheer desperation of men in that district for work.[56] The same occurred at Lacken pier in Co. Mayo: 'From the commencement there has been great difficulty in maintaining order, in consequence of men crowding on the works in larger numbers than it was possible to employ. In consequence of the insubordinate conduct of the men, the works had to be suspended on several occasions.'[57] Things were little better at Inishlacken pier in Co. Galway where there were 'several strikes ... but on the superintendent obtaining a conviction against one of the men for using threatening language, all annoyance ceased'.[58] Somewhat similar occurrences were reported from Poolnadhu (Ardmore in Co. Galway), but after an 'aggravated assault on the superintendent' there were no further outrages at the works.[59] Occurrences such as these indicate real desperation for work, and dismay at the small amount of it that was available, possibly aggravated by partial hiring practices engaged in by some contractors. As Captain Morant pointed out, contractors tended generally to bring in their own men rather than take on local labour.[60]

Many more memorials remained unaddressed over the whole country. According to the Inspectors of Fisheries, they had received a total of no less than 264 memorials by 1884, which they listed in their Annual Report for that year.[61] Galway (55), Mayo (42), Donegal (42) and west Cork (33) were the counties submitting the largest numbers of memorials, and more were coming in all the time. The Annual Report of the OPW for 1883–84 stated that 173 were received in that one year alone.[62] Not all memorials were necessarily valid and, in due course, some would be rejected or not taken any further. But the total number indicates the volume and distribution of demand for fishery piers at that time. The demand may have been stimulated by the parliamentary debate leading to, and the eventual passage of, the Sea Fisheries (Ireland) Act of 1883, which would, as we shall see in the next chapter, go a long way towards providing the very large sum of money clearly needed.

What Did It All Mean for Co. Galway and North Clare?

Co. Galway had done well from the operations of the *Relief of Distress* Acts. Overall, it had received the second largest amount of money, and had the greatest number of sites serviced: approximately £14,000 in Government grants had been spent on nine marine works in the county by the spring of 1882. North Clare's only benefit was the single new pier at Glenina.

Of the nine sites funded in Galway, seven were for completely new piers and two (Roundstone and Leenaun) were for major renovations. The sites were widely distributed throughout the whole region, from the head of Killary Harbour to Black Head on the south shore of Galway Bay. Overall, they constituted rather a mixed bag. Some were clearly of potential benefit to the fisheries and would appear for a number of years in the records of the Coastguard and the Fishery Inspectors as places where fish were landed. These include Leenaun, Smeerogue (occasional records only), Roundstone, Ardmore, Bush Harbour and Glenina. Most of these are located facing deep open water and did not require the boats to undertake the long inbound passage necessary in order to land catches at piers further inland. Ardmore, Bush Harbour and Glenina were particularly useful, the latter two for the easy access they gave to the traditional fishing grounds in Galway Bay. Other piers, such as Doleen and Oranmore Castle pier were erected too far from the fishing grounds, or were too landlocked or difficult to access at most states of the tide, to be of much value to fishermen. Inishlacken, while useful for the fisheries, was much more important for communications and the general convenience of the island population.

Why were certain sites selected for piers above others equally needful? A major reason derives from the various requirements of the Relief Acts, which, while aiming to provide relief work, limited the application of funds exclusively to places from which memorials were already, or very soon to be, in hand.[63] Moreover, the works were to be subject expressly and strictly to most of the prescriptions of the *Fishery Piers and Harbours Act* of 1846; therefore, neither the Fishery Piers Committee (FPC), which made the selection, nor the OPW which did the work, could vary or alter the precise locations chosen by the memorialists.

For their part, the memorialists had still to provide or secure the local one-fourth contributions prescribed by the 1846 Act – unless they were supplied by charitable donors. Only those places that met the memorial and the contribution criteria, and were at the same time 'distressed', were eligible for support. Doleen and Oranmore, for example, were successful but Scrahalia and Callowfeenish (*Caladh Fínis*) in Connemara were not. The FPC had preferred the latter two, but they lacked the full local contributions. Doleen and Oranmore, on the other hand, neither of which appear to have had an outstanding case on fisheries grounds alone, were fortunate that local priests, Fr T. Flannery PP (£130 to Doleen) and Fr R. Quinn PP (£50 to Oranmore), had added to the donations of the Canadian Fund, thereby ensuring the full one-fourth contributions were forthcoming.

Staff Commander Langdon, the Treasury nominee on the FPC, confirmed that the local funding was an all-important, even the overriding, consideration. When asked for the reason why certain piers were selected for support by the FPC he replied that 'wherever money was subscribed to build a pier, or slip, we put one if possible; that is to say, if it was not too exposed.'[64] He was even more explicit when describing how the Committee proceeded: 'We had a list of distressed districts, with the wants of the fishermen set forth, and also those that wanted piers, and wherever the money was forthcoming, that is to say the one-fourth, we examined the locality, sent an engineer down, and if it was possible to build a pier there we did so.'[65] Two other piers were erected in places where there appeared to be little fisheries justification or urgent fishery need for them: Ardmore pier is only about four miles by road from Kilkieran, where a large pier with good access already existed. Glengevlagh pier is further inland than Leenaun pier, and boats would have needed to pass the latter in order to land at the former, which hardly seems warranted.

In locations like those two in Killary Harbour, the need for relief work clearly superseded fishery considerations. The people there were extremely poor, as the reported crowding onto the works at Glengevlagh testifies. Working strictly in accordance with the *Fishery Piers and Harbours Act* and the *Relief of Distress Acts*, the FPC and the OPW were, in a sense, attempting to ride two horses simultaneously, not without some success. Acting in harmony, they might have achieved that feat with even better results, but as we shall see later, relations between the two bodies were severely strained and the outcome did less good for the piers than might have been hoped for.

The Gaskell piers suffered no restrictions arising from the Acts: they were entirely charitable works carried out on an *ad hoc* basis without government support and as such they were of a far more rudimentary nature.

Some Notable Benefactors of the 1880s

William Lane Joynt

Before leaving the kindness of those who helped when times were bad, some mention should be made of some others, nearer home, whose generosity in supporting small piers and landing places has largely been overlooked or forgotten. The Irish Coastguard, and the Irish Lighthouse Service later, were quietly generous in practical and financial ways; certain religious leaders like Fr Thomas Flannery PP of Carna and Fr Charles Davis PP of Baltimore in Co. Cork stand out as champions of fishery and community development, pier building and other maritime and relief activities. Two individual benefactors deserve particular mention.

One is William Lane Joynt (1824–95) of Clareville House, Ballyvaughan. He was the agent of Baron Anally, a major landowner in north Clare at the time of the 1880 distress. Earlier on, Joynt was a fishery conservator in the Limerick district, where he took an active part in the management of the salmon fisheries and published a long paper on the topic, which he delivered in an address to the Dublin Statistical Society in 1861.[66] He was Solicitor to the Treasury in Ireland, Mayor of Limerick (1862) and Mayor of Dublin (1867). He was well familiar with the Irish coast and its fishing communities, after more than forty-three years' interaction with them.[67]

Having been a welcoming delegate in 1880 at the arrival in Cork of the US frigate *Constellation*, carrying relief food supplies from America, he subsequently toured the west coast in 1882 accompanying the Duke of Edinburgh in the latter's 'relief squadron' on board HMS *Valorous*.[68] Captain Morant, who captained the *Valorous* after the Duke's departure, mentions Lane-Joynt's personal superintendence on board of the relief goods supplied by the Dublin Mansion House Relief Fund. He was a member of the committee administering cash grants from that fund whose resources included a generous gift of £1,000, contributed mainly from Australia.[69]

Joynt allocated some of this money to small piers and landing places in the West. His procedure was to offer the sum of £50 each at about fifty coastal localities from Donegal to Cork to help with the construction of landing places.[70] In many instances, his intervention resulted in additional sums up to £100, and sometimes up to £300, being contributed locally by private benefactors. A small slip or simple quay could be made for less than £200, so his £50 could be used either to build something useful with the added help of voluntary donations, or used to leverage the official local contribution for a more substantial structure.[71] His co-operation is also acknowledged in the report of the Canadian Fund Committee. As a result of his intimate personal knowledge of the conditions of the coastal communities, he appreciated the people's needs far better than his masters in Treasury, or his peers on various committees and commissions. He recognised, for instance, that in many cases an unencumbered grant of a small sum was of more immediate local benefit than a larger contribution that carried official accountability in its wake.

Liberal in politics and liberal in his dealings, Joynt lived at 43 Merrion Square, Dublin, where he passed away in 1895. He is buried at St John's Church of Ireland, Limerick.

Sir Thomas Francis Brady

The second, possibly surprising, benefactor meriting special mention is Thomas Francis Brady, the Inspector of Fisheries. On his personal urging, many private benefactors and charitable organisations donated sums of money to be dispensed by him on numerous small works where there was little or no prospect of any public funds ever being made available. He used some to help with small slips and landing places, just as Joynt did, and there can be little doubt that many of the small facilities in Co. Galway and north Clare that do not feature in the public records (and cannot, therefore, be traced adequately or in detail here) were the result of the charitable work of these two men and others like them.

Brady also distributed part of the donated money as small loans without surety to poor fishermen for the repair of nets and boats. Singlehandedly, he made a special appeal for funds so that the fishermen of Inishbofin Island might be given interest-free loans, on their own security, in order to purchase nets and gear. Many individuals, not just in Ireland, but in England and Scotland also, responded generously with donations, which Brady applied to good causes on the island.[72]

He claimed to have administered over 2,000 private loans throughout the whole country, funded out of charitable donations placed in his hands, with no difficulty in having them repaid in good time, and very few that he had any trouble with.[73] He used some of the money to hire two fishing boats and their crews for the 1886 and 1887 fishing seasons, with a third boat in the 1887 season. These he put fishing experimentally from Aran and elsewhere (Portmagee in Co. Kerry, for example), so as to encourage local fishermen, and to demonstrate the likely profitability of fishing ventures on the west coast. It is worth recording that Irish fisheries research had its humble origin in efforts of this kind, funded by charitable donations from benefactors at home and abroad.

In retirement from 1891, he continued to devote himself to philanthropic activities, for which he had been knighted in 1886. He was especially involved in charities connected with the welfare of fishermen and mariners in general[74] and was also a staunch supporter of the Dublin Society for the Prevention of Cruelty to Animals, for which he established a subscription to provide a refuge for lost and starving dogs.[75]

After his death (aged eighty years) at home in 11 Percy Place, Dublin, in 1904, his charitable work for fishermen was continued for a number of years by his daughter, Miss A.K. Brady. Eventually, she handed over the final residue of his charitable funds (£326) to the Congested Districts Board in 1908. The Board added it to the Fishery Loan Fund already in its care.[76] That latter fund had been set up originally by the very first Fishery Commissioners in the 1820s, out of the donations generously given by the London Tavern Committee during the 1822 famine and not fully expended at the time. Adding Brady's fund to this legacy fund was a fitting end to a century-long history of charitable support of Irish fisheries and fishery piers, much of it from sources abroad, and to the life-long work of a distinguished and humane fishery administrator. Inexplicably, his name has slipped entirely from public memory and remains forgotten and unacknowledged, like the London Tavern fund, the Canadian gift, the Inishbofin fund and the very many other charitable donations made during the nineteenth century. Perhaps the immense personal and financial generosity shown by ordinary Irish people in modern-day humanitarian crises is the most fitting response to, and acknowledgement of, the kindness of strangers shown to us in earlier times.

Brady, although knighted, is not recorded in the many biographies and reference sources examined in the research for this book. His body lies in Mount St Jerome cemetery in Dublin. His friend and one time co-Inspector, John A. Blake MP, had pre-deceased him by seventeen years. Blake, too, was generous in his support of the poor, tireless in promoting the Irish fisheries, and had contributed personally to subscriptions such as that for the relief of the fishermen of Inishbofin. On his death, he bequeathed numerous sums to many worthy charitable causes. One such bequest of a more professional nature was £200 given to support the artificial cultivation of oysters, a project dear to his, and to Brady's, heart. He also left £500 to Brady, as a personal gift and token of the depth of their friendship. The bulk of the residue of his estate was left to members of his family who lived in Australia.

The Piers of the 1880–81 Famine in Co. Galway and North Clare

Smeerogue (Renvyle; Gurteen; Tully) Pier

Smeerogue, also known as Renvyle, Poulahly, Tully or Gurteenclogh, was first proposed as a site for a fishery harbour as early as 1835,[77] but nothing was ever done in the vicinity until a memorial was sent to the Fishery Inspectors in 1874 for a pier to be built there. The Inspectors approved the application without delay but in the absence of the local one-fourth contribution nothing was done for another five years. It was the intervention of the Canadian Committee in 1879 – its first venture into piers – that resurrected the proposal, not only at Smeerogue but eventually at the other twenty-eight places the Committee supported. Some local opinion holds that Smeerogue pier is not, in fact, built at the exact site originally intended, but a short distance west of it, a change allegedly occasioned by a dispute involving the Blake family of Renvyle, the owners of the land. Whatever the case may be, the structure that was built was an impressive, solid quay and pier sheltering a small beach where boats could be drawn up on the shelving intertidal sandy shore. The pier was constructed along a rocky spit that extended out at an angle, and considerable excavation was carried out alongside to clear the beach of rocks, thereby creating the sandy landing place.

The quality of the pier masonry is notable, particularly its bedding on the underlying rock. The deck is paved with stone slabs, the coping is cut limestone and overall the structure has a commanding presence. It was built five feet longer than originally designed, possibly to ensure a greater depth of water at the pier head.[78] The storm wall, about nine feet high, is massive and, for most of its length, was built later than the pier itself.[79] Estimated to cost £4,000, the whole work was completed for £3,821, with the Canadian fund giving £566, the *New York Herald* fund £189 and the Liverpool fund, £250. For all its presence, and the fact that the general site was one favoured by Nimmo, Bald and others earlier in the century, it did not prove particularly successful for the fisheries. Some fish and lobsters were landed there, but not to any great extent.[80] According to the Coastguard, the pier was not safe in bad weather and there was nowhere else to run to for shelter should the boats venture out.[81] But the main reason for its underuse may have been the absence of any convenient means of transporting fish to market when catches warranted, a problem that also affected Clifden and Cleggan before the Galway to Clifden railway was opened in 1895.[82]

Glengevlagh (Glenagimlagh) Pier

The piers at Glengevlagh (in Killary Harbour) and Bush Harbour (in Kinvara Bay) are of very similar design, each comprising a straight jetty pier in cut stone, projecting straight out from the land and supporting a central, low, cut-stone storm wall along its length. There are good cut-stone steps indented on both sides of each pier, near the head. The Glengevlagh pier is smaller (about seventy feet long and twenty-five feet wide) than that at Bush Harbour, and in contrast to the latter, the central wall extends the entire length of the pier. The deck is made of large stone slabs, now mostly overgrown with grass. Located down

FIGURE 6.3. Smeerogue (Renvyle or Tully) Pier. Left: Wharf wall of the pier. Note how the stonework is built along the underlying rock. The storm wall is a later addition as is the limestone coping. Right: View of the pier and the sandy landing beach alongside. Photos NPW.

a steep track leading off the main Leenaun to Westport road, well within the innermost part of Killary Harbour, it is unlikely ever to have been of much use either for the fisheries or for general trade, due largely to the excessive steepness of the approach track; today the pier does not seem to be frequented by any boats. Despite its apparent inconvenience for fisheries, the FPC approved it and allocated £488 for its construction. The Canadian fund gave £63 towards the cost. It is located within two miles of Leenaun pier which also received support from the Canadian Fund.

Why two locations so close to each other merited support is not clear; boats would have had to pass Leenaun pier (and Bundorragha pier on the opposite, Co. Mayo, shore of Killary harbour) to get to Glengevlagh pier, which is in shallow water and not as convenient to the fisheries as either of the other two named piers. It may have been that the dire level of poverty and misery endemic in the general locality necessitated extra relief work, notwithstanding the apparent unsuitability of the site. James Hack Tuke described Glengevlagh as 'an exceedingly poor little village' of about twenty families when he examined the place, and he was moved to give a small sum of money 'to give employment to some of the many idle men in the construction of a pier'.[83] Fifty men on average were employed daily building the pier under a contractor, T. O'Malley. According to Thomas F. Brady, a sad fate had befallen Glengevlagh by 1888: 'There used to be a fishing population at a place called Glenagimlagh [Glengevlagh], but they have all died out or nearly so'.[84]

Bush Harbour (Parkmore Pier)

At Bush Harbour, on the south coast of Galway Bay, the straight pier is about 200 feet long and forty feet wide and the central wall does not extend to the very head, so that both sides can be accessed on foot round the pier head and at its root. Bush Harbour and Glengevlagh piers are the only ones in the county with a central storm wall. Estimated at £3,130, Bush Harbour was built on contract by R.D.W. Gray for £3,113[85] (£2,160 according to a separate Return to Parliament,[86] possibly an error). Twenty-six men on average were employed daily in its construction. The Canadian fund gave £583; the Liverpool fund for the Relief of Distress gave £100 and Count de Basterot, the local landowner, £100. The free grant was £2,380. There is a plaque inserted in the central wall at the pier head with the date 1881 inscribed on it, but without acknowledgement of the contributing donors. Well located near the mouth of Kinvara Bay, this pier obviated the need to sail fully into that shallow bay to land at Kinvara and was therefore more convenient for the fisheries. Used also for landing turf from Connemara for distribution in the locality, it is still in use today for fishery and leisure purposes.

Bush Harbour is an interesting case with regard to the naming of Irish piers in the documentary records. There was a landing place at the site long before the present pier, which today is known by local residents only as Parkmore pier. The name Bush Harbour is the only name used for it in the official nineteenth century documents, but this writer has not heard that name used among modern local residents. According to Caoilte

Breathnach,[87] the pier was still known as *an Cora Mhór* or *Caladh na Sceiche* by older local residents into the 1970s. The latter Irish name translates directly as 'Bush Harbour', indicating that the official English name used in the documentary records is based on an accurate translation of the earlier traditional Irish name for the site, which has now disappeared from local usage and memory. The two older Irish names – *an Cora Mhór* ('the big spit') and *Caladh na Sceiche* ('bush harbour') – therefore reflect different, long-forgotten, topographical features of the site. Many other piers bear English names based on Irish words, and this can lead to confusion in identifying precisely the places involved. For example, another Irish word '*crompán*' means a creek, or low-lying land alongside a creek or river. It is transcribed in the official documents as 'crumpaun' or 'crampaun' and a number of piers in different places are named 'Crumpaun'. Without a place-name suffix, e.g. 'Crumpaun Carna', the generic name 'Crumpaun' alone is insufficient to identify any particular site. We will encounter similar difficulty again later with the word '*Cora*', meaning a rocky spit or a weir and '*Dóilín*', another Irish word for a creek.

The Transformation of Leenaun (Leenane)

Nimmo's 1823 works at Leenaun comprised a simple roadside quay with a single short jetty pier projecting at right angles from its eastern end, somewhat similar to the structure he built at Clonisle. (RH No. 041) He indicated that it was his practice to erect a quay wherever he had to embank a roadway skirting the sea, and Leenaun was one such instance. The pier here proved particularly useful, as it is located near where the roads to and from Westport, Cong and Clifden converge in the village. The original cost of construction in 1823 was borne by Government famine relief funds; the site received no support from the first Commissioners for Fisheries or from any charitable sources. About thirty rowboats landed fish there at the beginning, but the place was in greater use for trade and general transport.[88] It vested in the OPW in 1849 and it was listed in the schedule of piers that transferred to Galway county in 1853.[89] Like many other piers of the early period, Leenaun was largely ignored by officialdom during and immediately after the Great Famine.

By 1867, renovation was badly needed and work which was to transform the site started in July. The original jetty pier was reconstructed and raised above the level of high Spring tide for its whole length of eighty feet, and a low parapet wall was added. A new jetty pier, about 110 feet in length, was then erected at the western end of the original roadside quay. This transformed the site into a small harbour between the two jetties, with the stone-built roadside quay joining them at the roots. The new western jetty was built of random blocks laid without mortar. Battered on its inner face, and with a pronounced sea slope on the outer, it was topped with a parapet four feet wide and six feet high. The structure may have been intended primarily as a breakwater, not as a landing wharf.[90] Both jetties projected straight out from the ends of the adjoining quay, and both had a flight of steps at their heads. The steps on the new western pier were made of large slabs laid, unusually, on edge.

Still *in situ*, they are an interesting feature of the pier, marking the original head of the structure. Overall, the work carried out under the direction of James J. Boylan, an engineer with the OPW, has a decidedly rough-and-ready appearance. It had been estimated at £360 but actually cost only £310,[91] which may explain its rather crude and almost makeshift appearance. Everything was fully completed by December 1867 and it was handed over to the county.[92] The following year, a short slip and pathway were cleared at the western side of the new pier for the convenience of persons landing seaweed from small boats. This was to replace a traditional landing place at the spot, which had been obliterated by the erection of the new pier.[93]

In 1879, a memorial for the extension and renovation of the whole site was submitted to the OPW[94] and Robert Manning inspected the place in January 1880. As a result, Leenaun was selected for funding from the £45,000 granted by the amended *Relief of Distress Act 1880*, receiving a free grant of £600 and a contribution of £200 from the Canadian fund; work started immediately. On average, seventeen men were employed directly by the OPW on a daily basis[95] (twenty-three according to the Canadian Committee)[96] and all was completed by October 1881. Renovation work on the 1867 western pier comprised an extension built at a slight angle, pointing towards Nimmo's eastern pier, so that it closed the existing harbour mouth a little. The extension was done in cut-stone ashlars for the upper part that overlies, for part of its length, a layer of coarser stonework more typical of the 1867 work. A new flight of steps was added; the stones for these are set horizontally in the fashion that is more usual.

A crack appeared in the new work soon after it was completed, going right through the pier from the inner wharf wall to the outer sea slope. Manning attributed this to settlement due to the poor nature of the foundation. He monitored the situation for three months,

FIGURE 6.4. Western pier at Leenaun. Left: original steps of the 1867 section, made of blocks laid on edge. Right: Wharf face of the pier. Note the contrast between the old, 1867, section and the newer, 1881, section. Photos NPW.

noted no further deterioration and anticipated no more movement. Repairs were carried out and everything was completed by October 1881. Manning approved the work as it then stood, and the pier was handed back to the county by warrant on 15 November 1881.

Today, the western pier seems from the quality of the masonry, to comprise three distinct sections: a rough, root section with the crude steps (1867); a middle section of ashlars laid in uneven courses with one set of steps; beyond that, the jetty and its head are made of good cut-stone ashlars laid in level courses. The middle section – between the two sets of steps – appears to be the zone where the original crack occurred, and the rough, irregular nature of the stonework in this part suggests that quite substantial remedial work had been carried out to address the subsidence issue.

The masonry of the distal section with the pier head is characteristic of the good-quality work evident at nearby contemporaneous Glengevlagh and at other Canadian fund piers. Leenaun harbour is altogether most interesting, both structurally and historically, exhibiting almost 200 years of structural evolution. It does not feature in any public documents after 1881, up to the end of the Congested Districts Board's activities in 1923, just as it was overlooked by the fishery authorities after it was first constructed in 1823. Today it is little used except for leisure purposes and some aquaculture activity.

The Sad State of Doleen (Ballyconneely) Quay

Some Irish words occurring in place names can cause difficulty. One such is *Dóilín* (anglicised as 'Dooleen' or 'Doleen'), meaning a creek. No less than four piers in Connemara are known by this or a similar-sounding (Dolan, Dolan's) name: Aillebrack pier (No. 029); Ballyconneely quay (No. 032); Dolan pier (No. 036); and Dooleen pier in Lettermore south (No. 130). On careful examination of the documents and the relevant maps,[97] and by personal examination of the various sites, the Doleen quay in question here (in some documents[98] referred to as 'Doleen, Ballyconneely', and in others[99] as 'Doleen (Dolan's)') is the one now known only as Ballyconneely quay.

Although virtually a complete ruin in 2001,[100] the description given of the work in 1882,[101] and the good quality of the masonry still remaining *in situ* when examined in 2010 and 2015 (which conforms to the quality evident in other piers from the 1879–81 period) support this attribution. The estimated cost of building it in 1880 was £2,000. The Canadian fund gave £350, the Irish Land League gave £20 and Fr Thomas Flannery gave £130; the balance came from a free grant of £1,500 from the Government. Fr Flannery was Catholic curate at Ballyconneely in 1879 and 1880. His donation to the quay is evidence of his dedication to his flock in that district and a parting gift to them: he was transferred to Carna as Parish Priest in 1880 and served the remainder of his short life there. In the latter village he was a tireless supporter of the fishermen and of the community in general. Later on he acted as works supervisor on the Mweenish causeway being built by the Piers and Roads Commission, for which he was warmly thanked by the Commissioners.[102] Born in 1853, he died a very young man in 1891.[103] He is commemorated by Flannery Bridge near Kilkieran, which was named in his memory; Ballyconneely seems to have done him no such honour, nor does the present state of the quay that he generously helped to fund.

The structure is a longshore quay, about 180 feet in length, faced originally in good cut stone. It proved necessary to excavate a cut through some rocks to make a safe approach to the quay possible; all in all, the precise location was really not very convenient for boats. This fact, and the observed accumulation of sand at the quayside during the winter of 1881–82, should have alerted the engineers to an obvious problem: located behind large rock outcrops at the inner end of a narrow sandy inlet, the quay was inaccessible at most times except at high tide. A natural landing place at Doohulla, a short distance south of it, was far more suitable and accessible, but that site was overlooked at the time. Five years later, Doohulla landing place would be selected by the Piers and Roads Commission for a new breakwater,[104] suggesting that by then Doohulla was more in use by fishermen than Ballyconneely quay.

Even Staff Commander Langdon RN of the FPC was curtly dismissive of Ballyconneely: 'We built one [a quay] at Doleen. It is not a harbour but we put up a couple of hundred feet of wharfage.'[105] In light of this rather dismissive comment, one wonders why he approved the site, and why the FPC agreed to it in the first instance. Langdon was, after all, the expert placed on the Committee by Treasury in order to ensure the FPC had sufficient navigational and maritime surveying expertise at its command.[106] While he did not regard the Connemara fishermen as 'real fishermen', Langdon was personally in favour of building small piers close to villages and he may have approved Ballyconneely because of the distressed state of the local population there.

Started in November 1880, the quay was built for half its length by 1881, but was left unpaved. On average, thirty men were employed daily under the direction of the OPW.[107] It was still uncompleted in August 1882,[108] when the number of labourers had been reduced to an average of twenty-seven men in daily employment. It was finally finished, at a total

FIGURE 6.5. The sad state of Ballyconneely (Doleen) quay. Left: Half of the quay is derelict. Right: The landward end of the quay has been repaired in recent times. Photos NPW.

cost of £2,045,[109] in October of that year.[110] The total cost of the labour employed was £1,160,[111] so it had made a significant contribution to the economy of a very poor district for quite a long period. A rough estimate calculated from the figure given suggests that a man's wage was about one shilling a day, the usual sum at the time. At nearby Roundstone, work on the north pier had employed around twenty-five men on average and cost £1,178. Taken together, these two projects alone provided a much-needed cash injection in to the distressed locality.

In 1883, a further memorial was submitted for a pier at Calla (*Caladh*), located between Doohulla and Roundstone, where many of the memorialists were said to have two boats each, fishing for lobsters, cod and ling; a small rough pier was erected there some time later.

There is little or no mention of fishing activity at Doleen quay in later official documents, so the money earned by those who built it was its main beneficial effect. Located high up in Ballyconneely Bay, it is difficult to see why it should ever suffer severe storm damage; nevertheless, half of the quay (at the seaward end) has entirely collapsed and the remaining half has recently been stabilised with concrete coping and decking. The rest of the deck is entirely broken up.

When visited in 2010 and 2015, the deck of the collapsed section (nearest the open bay) was littered with large, well-shaped cut-stone blocks, lying about among heaps of rubble stone. Some show the masons' hammerings, and some bear holes for the dynamite that was used to blast them from the quarry. It looks almost as if they were brought on site dressed and ready, but never placed in the structure. Old concrete lying about indicates that the collapsed section had been capped at some stage, and that it, too, had collapsed over time. The landward end of the quay is in better shape, capped with modern concrete coping atop the wharf face. The grassy track that led to the quay in the year 2000 has been widened and surfaced with hard-core material, and some new houses have been built overlooking the quay.

The very fact that the exact site of the quay had to be established mainly from nineteenth-century maps and documents, and confirmed by the excellent quality of the remaining original stonework, speaks to the extent to which the generosity of the Canadian people, and the other donors, has been forgotten today. Its present ruinous condition and its state of virtual abandonment do nothing to honour the men who built it and the charitable intent of those who sponsored it; it shames our present generation. The Ryan Hanley engineering surveyors Colleran and Hanley said of it in the year 2000: 'Quay in ruins. Unsafe structurally. Hazardous for anyone who visits it. Demolish, grout and make safe'.[112]

Inishlacken Harbour

Inishlacken harbour is one of the most attractive small harbours in Co. Galway. It is formed by a straight pier about 120 feet long, with a breakwater jetty projecting at right angles towards it from the opposite shore, together forming a semi-enclosed harbour. Both structures are well built of roughly shaped stone, now with cement pointing. The sea face of the pier is battered and it is surmounted by a low storm wall. It was built five feet longer than originally designed[113] and there is a single set of cut-stone steps in the wharf face. The opposing breakwater is about 100 feet long, stepped at its nose. The decks of both structures are surfaced in stone.

The OPW commenced work on 21 March 1881[114] and completed everything by December at a total cost of £616 (or £603, given in a different Return),[115] slightly above the estimate of £600.[116] On average, eighteen men were employed daily. Since the total cost of the labourers was £480, the daily wage rate, calculate for 260 workdays between 21 March and the end of December (285 calendar days), averaged around two shillings per man per day. Plant and materials cost £88 and superintendence £48. The Canadian fund provided the entire one-fourth contribution of £150 and the free grant was £453. Today, the harbour is in excellent condition, a tribute to the skill of its designer (Robert Manning of the OPW) and the workmanship of the original builders, superb value at two shillings per man per day.

Ardmore (Pollnadhu) Quay and Pier

Ardmore, situated on *Iorras Aithneach* peninsula in Connemara, is located about four miles south of the pier at Kilkieran. A quay and pier were sanctioned for the place at an estimated cost of £1,400; the Canadian fund provided the £350 local contribution. The angled pier that was built is about 150 feet long, twenty-nine feet longer than originally designed.[117] It is not clear why this enlargement occurred, since the work was done by direct labour under the supervision of an engineer of the OPW. It may have been to ensure a greater depth of water at the pier head at low tide; despite the pier's relative height and mass, the whole basin it shelters dries out as far as the pier head.

Built apparently of large concrete blocks, the pier projects out from a quay about seventy feet long, made of large rocks pointed with mortar. The works were completed in July 1882 at a final cost of £1,389, just under the estimate; the free grant given was £1,034. On average, twenty-three men were employed daily. There was trouble among them at the site in course of construction and bad weather did not help either; completion of the work was delayed for a considerable period as the bollards for the wharf were held up in Clifden by severe storms, and no one was willing to risk their transport by boat around Slyne Head under the conditions.[118] Ardmore pier was transferred to the county in 1883.[119]

Oranmore Castle Pier

When the works proposed at Killala in 1880 did not go ahead, Oranmore was a fortunate beneficiary. A pier had been proposed beside the old Blake castle (which was unoccupied and roofless at the time) and the Canadian fund diverted £100 of the erstwhile Killala allocation to it. The Liverpool Relief fund and Fr Quinn, PP of Clarinbridge, both gave £50, thereby triggering the release of the £435 free grant sanctioned by Government. There had been a pier at the site reputedly since mediaeval times and the Blake family was said to have charged a toll for its use. However, in a picture dated 1792, reproduced in a recent publication,[120] there is no pier obvious at the castle site. In 1807, James Fitzgerald sought to have a pier or quay erected there to improve commercial trade, but nothing came of that.[121] Another application for a fishery pier had been rejected in 1827 by the engineer Barry Gibbons on the grounds of the site's excessive distance from known fishing banks and the shallow nature of the sea approach to it through Oranmore Bay.[122]

Edward Lambert, a poor law guardian and member of the local relief committee, complained to the OPW in 1847 that nothing had ever been done at Oranmore, although the Board had sent an engineer to survey the site and piers had been erected at many other places where they were less needed.[123] Describing Oranmore in 1849, Samuel Lewis wrote 'it is situated on an inlet of the bay; and two miles distant is New Harbour, where there is a quay with good anchorage', but he made no mention of any pier or quay existing at the village in his time.[124] The reasons why the Fishery Piers Committee favoured it above other proposals in 1880, and why the Canadian Committee considered it worthy of financial support, despite its previous failures, are discussed below.

The memorial 'to enlarge and repair' Oranmore Castle pier was submitted on 26 May 1880.[125] Fishery Inspector Brady wrote in support of it a week later, saying a pier was very urgently needed for the fishing and trading activities of the locality. It would have been

FIGURE 6.6. Ardmore pier and harbour. Photo NPW.

provided earlier, he continued, except that a Mr Meldon claimed he owned the site privately. Meldon was now prepared to give up all claim to it for the benefit of the public, and Brady urged it strongly, claiming that a suitable structure would not be very expensive.[126]

A local contractor, J.J. Brady, successfully tendered for the job at £510, an amount less than the OPW estimate. When he started work a difficulty was soon encountered, which casts some light on the pier and its history. Because the originating memorial had been drafted 'to enlarge and repair' an existing pier – although there was no evidence presented then for a pier existing at the site – strict interpretation of the law demanded that the contractor either restore the old one entirely, or build anew upon it. A minute of the OPW explains the difficulty:[127]

> [The] wharf wall was to be built on the old quay wall. When cleared of seaweed its nature is better exposed and the old work does not consist of any wall, it is merely boulders roughly built up with no backing but loose stone shingle etc. The contractor is bound to repair the wall and build on it, but no repairs will render a sound job of it, especially as there are no foundations and the foreshore slopes away from the quay to the navigable channel.

This minute signifies that there was indeed an old, rough structure at the site that had become dilapidated and completely overgrown with seaweed during long disuse. The contractor first proposed to demolish this entirely and rebuild from scratch with new foundations.

The OPW, however, needed to proceed cautiously when approving this proposal, which could be seen to breach the terms of the grant 'to enlarge and repair'. Only ten years earlier at Ballyloughane, a memorial for 'repairs to a pier' where there were only a few stones lying about in evidence of any pre-existing structure, had been refused sanction by Treasury. The Board therefore may not have cared to test Treasury in the same way again. Since the contractor had already dismantled the old structure for most of its length, Samuel Roberts and OPW chairman McKerlie signed off on a new plan and drawings. These showed a new pier said to be 'built on top of the old pier' and stating that 'the old pier wall was taken down and the new pier built over the area'.[128]

The new pier head was not founded on the old pier head, which had been completely removed. Foundations for the new head were laid one foot below the level of the strand and the new head erected on that. The drawings, still extant, include a sketch of the site by Roberts and McKerlie, showing the preliminary indicative plan and the new proposal. The new pier is positioned 'in the area of' the old pier, whereas the preliminary plan had it displaced slightly to the right.

Overall, the information accompanying the new plan and scheme gave a reasonable and understandable impression that the structure was not entirely new, when in fact it really was. The design envisaged the pier turning to the right at about eighty feet along its length, but this part was never undertaken. A very substantial, straight, cut-stone pier was erected, with slips at both sides of the pier root. It was constructed only sixteen feet wide, to correspond to the supposed width of the original old structure. J.J. Brady started work on 9 May and had completed everything on time in November, at a final cost of £575, having employed an average of fourteen men daily for the period. The iron mooring rings are recessed into niches cut out of the ashlars of the wharf face and pier head walls, constituting an attractive and unusual feature.

Another memorial was submitted by the occupiers of Oranmore village on 28 August 1883 to 'add to the pier recently built by the Board of Works'. The pier was, they claimed, too small, and the deck space too limited to permit a horse and cart to be safely turned round when bringing or removing goods. The problem was, according to Fishery Inspector Brady, that insufficiency of funds in 1880 had resulted in the 'repair' of the pier being left incomplete. To make it useful to the fishermen now, 'the repairs ought to be continued by the rebuilding of the outer part, to at least its original extent, if not some distance further.'[129] The pier had indeed been built only eighty feet long, leaving the pier head incongruously short of the navigation channel. It was also excessively narrow, to comply with the requirement of building 'on the original', a requirement that would now be regarded as unnecessarily restrictive and preventative of any enhanced development. However, the die had already been cast, the OPW was not intent on again raising the matter so soon after construction, and the pier remained as it was. In 1906, Oranmore Castle pier was in use mainly for landing seaweed and turf and there are no reports of its use for fisheries. The slip on the town side needed repair and was concreted over early in the twentieth century.

The county surveyor remarked on the principal shortcoming of this and many other piers of earlier times: when constructed, no land, or an insufficient amount, was legally acquired to ensure an adequate public right of way to them. In Oranmore's case, the approach road was covered five to six feet deep at high tide, preventing access to the pier at the very time in the tidal cycle that it was most needed. It remains so today. A long, stone mole curving to the right, easily seen in satellite images and from the shore at low tide, extends out from the pier head today; this represents the extension and pier head that were originally projected for the present structure, but never completed.

Carrowmore (Lynch's) Pier (Ballinacourty)

Carrowmore (Lynch's) pier at Ballinacourty in Inner Galway Bay received a free grant of £210, and £70 was contributed locally, probably from the landowner, John Wilson Lynch. The site lies on the east side of the small triangular peninsula on which Nimmo had built his ill-fated Ballinacourty pier between 1823 and 1826. It is marked as 'Lynch's Quay' in the first edition of the six-inch OS map, but there was no manmade quay shown there at that early date (1839). It was simply a natural landing place for boats fishing the oyster beds in the locality. The new, 1884 works comprised a small quay with a straight jetty pier projecting outwards at right angles from it. They were completed in November 1884 at a final cost of £255.[130] The space between the pier, the quay and the shore formed a small harbour where boats could be safely beached. Both the original quay and the pier are still in existence, but were greatly modified in modern times and given added protection by the construction of a massive new rock breakwater outside them.

Today the site is mainly used for inshore fishing and by oyster dredgers still working the local oyster beds. The small traditional stone enclosures, called *brácaí*, used to store oysters temporarily, prior to sale, can be seen in the intertidal zone just outside the new breakwater.

Glenina Pier, North Clare

Glenina was an important fishing centre in north Clare, being close to the fishing grounds in deep water around Black Head. Located on a steep, rocky coast with no other place of shelter in close vicinity, a harbour of refuge was badly needed there. Nimmo favoured the site, and suggested that a suitable harbour could be made for about £200.[131] Other places in the district, like Ballyvaughan and Kinvara, were not particularly suitable for fishermen, being more or less embayed in shallow water and located much further from the deep-water fishing grounds. Bindon Blood, the enterprising local landowner, made an application for a pier at Glenina in 1822 but was turned down due to a shortage of grant funds.[132] The local fishermen made a repeat application in 1848 but they, too, were unsuccessful.[133] There were said to be about 120 fishermen working in the district at the time; they fished from Black Head into Galway Bay where they came into conflict with the Claddagh men who, for an extended period, had imposed their own eccentric rules on fishing in the Bay.[134]

In 1880, the FPC agreed to fund a pier here under the *Relief of Distress (Ireland) Act* at an estimated cost of £1,200;[135] the free grant sanctioned was £900, and the Canadian Committee provided the one-fourth contribution of £300, allowing work to commence on 6 December 1880. Initially thirty-eight men, on average, were employed daily,[136] but this had reduced to twenty-nine before the work was fully completed in September 1881.[137]

An angled pier, built of limestone ashlars, with a storm wall on its west side, was erected; a slip was excavated beside the root of its east-facing wharf wall. There is a single set of cut limestone steps near the pier head. A plaque dated 1881 is inserted in the storm wall, but, just like the one at Bush Harbour pier, it makes no mention of the Canadian contribution. All the work was carried out by the OPW and completed at a final cost of £1,268 (labour £1,001; plant and materials £136; superintendence £131). Since there are about 280 work days from 6 December to the following September, average wages were about seventy-one shillings per day; if an average overall of thirty-five men were employed daily during the whole period of construction, the daily wage would have been about two shillings per person, a significant contribution to the economy of the district. Glenina pier is also known as Coolsiva pier and today it is much as it was originally constructed, but is little used for anything other than leisure activities.

Trácht Each (Traghtagh) Landing Place, Ballyhees, *Inis Oirr*

The sum sanctioned for Ballyhees (*Baile Thíos*) in 1880 was only £188; the local contribution made this up to £250, the estimated cost of the proposed works. Unlike other sites, where the provision of landing facilities involved building artificial jetties out from the land into the sea, the task at *Trácht Each* was to make a landing place by excavating inwards, forming a sheltered indentation in an otherwise rock-bound coast. A small rough landing place was excavated by blasting with dynamite alongside a rocky ledge a short distance from the village.

The work, done by OPW staff and finished in 1885, created a short, narrow inlet through the surrounding rocks, and the ledge to one side of it formed a rough wharf or quay. The following year work was taken up again by the Piers and Roads Commission (chapter 8) and a further forty-five linear feet were blasted and excavated to enlarge, extend inwards, deepen and improve the landing.[138] Again in 1887, Col. Fraser excavated a further 360 cubic yards of rock to extend the landing area and improve the quay on the rocky shelf alongside.[139] Between these two latter operations a total of £273 was expended. The final site modification would come with the building of a concrete protective wall on the western side of the artificial cut in order to break the swell of the sea at the quay face. This would be erected much later by the Congested Districts Board (chapter 11) at a total cost of £415 in 1905.[140] From the very beginning therefore, a total of £940 was spent on this small landing place. Finally, an entirely new concrete pier was built on the eastern side of the landing, along the original rocky shelf, in the late twentieth century. The site, called Traghtagh (or, in the Ryan Hanley survey, '*Trá Teach*', a misspelling of '*Trácht Each*' that changes the etymology of the place name), is numbered 185.

It was, and remains, a difficult and dangerous place for boats and people alike: rough, exposed and subject to very large swells. The Ryan Hanley survey recorded 'a general sense of danger here due to large swell ... A rough and exposed place.'[141] It is little used as a landing place today, having been replaced in usefulness by a new slip made at the village by the Congested Districts Board in 1905 and by the modern main pier of *Inis Oirr* (No. 189) built more recently.

Endnotes

1. J.A. Blake, *The History and Position of the Sea Fisheries of Ireland and how they may be made to afford Increased Food and Employment* (Waterford: J.H. McGrath, 'The Citizen' office, 1868), p. 32.
2. L.M. Cullen, *Life in Ireland* (London: Batsford, 1968).
3. 48th Annual Report of the OPW, BPP 1880 [C. 2646], p. 30.
4. 48th Annual Report of the OPW, BPP 1880 [C. 2646], pp. 24–26.
5. Inishbofin and Inishshark are usually dealt with as a single entry in the records and it is not always possible to distinguish which one is in question in every case.
6. 48th Annual Report of the OPW, BPP 1880 [C. 2646], appendix 6, p. 64.
7. 48th Annual Report of the OPW, BPP 1880 [C. 2646], p. 26.
8. Hansard, House of Commons Debates, 242, col. 84–9.

9 48th Annual Report of the OPW, BPP 1880 [C. 2646], p. 30.
10 *An Act to render valid certain proceedings taken for the Relief of Distress in Ireland, and to make further provision for such Relief; and for other purposes,* 43 Vict. C. 4.
11 Report of the Canadian Fund Committee, BPP 1881 (326).
12 M.E. Hicks-Beach to Duchess of Marlborough 13/3/1880, in Report of the Canadian Fund Committee, BPP 1881 (326).
13 Report of the Canadian Fund Committee, BPP 1881 (326), pp. 4–5.
14 Report of the Canadian Fund Committee, BPP 1881 (326), p. 5.
15 Canadian Fund Committee to Treasury 3/5/1880, in Report of the Canadian Fund Committee, BPP 1881 (326).
16 Report of the Inspectors of Irish Fisheries for 1874, BPP 1875 [C. 1176], p. 9.
17 A. Hornsby to C.F.P. Dowson, 9/2/1881, in Report of the Canadian Fund Committee, BPP 1881 (326), p. 32.
18 Report of the Canadian Fund Committee, BPP 1881 (326), p. 6.
19 Ibid., p. 8.
20 Treasury to Chief Secretary, 25/5/1880, in Report of the Canadian Fund Committee, BPP 1881 (326), pp. 9–10.
21 T.H. Burke to Canadian Fund Committee, 4/6/1880, in Report of the Canadian Fund Committee, BPP 1881 (326), p. 10.
22 Canadian Fund Committee to Chief Secretary, 7/6/1880, in Report of the Canadian Fund Committee, BPP 1881 (326), pp. 10–11.
23 Report of the Canadian Fund Committee, BPP 1881 (326), p. 11.
24 *An Act to amend the Relief of Distress (Ireland) Act 1880; and for other purposes relating thereto,* 43&44 Vict. C. 14.
25 Return to Parliament, BPP 1880 (409).
26 Return to Parliament, BPP 1881 (244).
27 49th Annual Report of the OPW, BPP 1881 [C. 2958].
28 Report of the Canadian Fund Committee, BPP 1881 (326), p. 12.
29 Ibid., pp. 13–14.
30 Ibid., pp. 33 –34.
31 Ibid., pp. 13–14.
32 Report of the Inspectors of Irish Fisheries for 1880, BPP 1881 [C. 2871], pp. 40–41.
33 Dublin City Archive, Clark Mansion House Fund Collation, Ch 1/ 21, 1/70, 1/71.
34 D.G. Morant, Report to the House of Commons on Relief of Distress on the West Coast of Ireland, BPP 1880 [C. 2671].
35 C. Langdon, in Select Committee on Harbour Accommodation, BPP 1883 (255), queries 3175–6.
36 C.W. Cole, Sketch of W. Gaskell at Loch Aconeera, NLI call no. PD 2196 TX 5 (C).
37 Gaskell, W.P. Report to the Congested Districts Board, 26 May 1892. Reprinted in Morrissey, *Verge of Want*, p. 183.
38 Minute of the Engineering Department of the OPW. NAI OPW5/3815 15.
39 Report of the Canadian Fund Committee, BPP 1881 (326), p. 11.
40 The full title of the Act is *An Act to render valid certain Proceedings taken for the Relief of Distress in Ireland, and to make Further Provision for such Relief; and for other purposes,* 43 Vict., C. 4. The short title is the *Relief of Distress (Ireland) Act 1880.*
41 43 Vict. C. 4, sections 10 & 20.
42 43&44 Vict. C. 14.
43 Report of the Canadian Fund Committee, BPP 1881 (326), p. 11.
44 49th Annual Report of the OPW, BPP 1881 [C. 2958], pp. 25–28.
45 Return to Parliament, BPP 1881(244).

46 House of Commons Debates, 22 June 1881. Hansard, 3rd series Vol 262, page column 1034–5.
47 48th to 52nd Annual Reports of the OPW, 1879–84.
48 52nd Annual Report of the OPW, BPP 1884 [C. 4068].
49 52nd Annual Report of the OPW, BPP 1884 [C. 4068], p. 26.
50 52nd Annual Report of the OPW, BPP 1884 [C. 4068], pp. 26–27.
51 50th Annual Report of the OPW, BPP 1882 [C. 3261], appendix B, p. 68.
52 Return to Parliament, BPP 1881(244).
53 Return to Parliament, BPP 1882 (203).
54 Return to Parliament, BPP 1881(244).
55 Report of the Piers and Roads Commission, BPP 1887 [C. 5214].
56 50th Annual Report of the OPW, BPP 1882 [C. 3261], appendix B, p. 66.
57 Ibid.
58 Ibid.
59 Ibid., p. 67.
60 D.G. Morant, Report to the House of Commons on Relief of Distress on the West Coast of Ireland, BPP 1880 [C. 2671].
61 Report of the Inspectors of Irish Fisheries for 1884, BPP 1884–85 [C. 4545].
62 52nd Annual Report of the OPW, BPP 1884 [C. 4068].
63 43 & 44 Vict. C. 14.
64 C. Langdon, in Report of the Select Committee on Harbour Accommodation, BPP 1883 (255), query 3116.
65 C. Langdon, in Report of the Select Committee on Harbour Accommodation, BPP 1883 (255), query 3136.
66 W.L. Joynt, *The Salmon Fishery and Fishery Laws of Ireland. A Paper read before the Dublin Statistical Society, the 21st of January 1861* (Dublin: E. Ponsonby, 1861).
67 W.L. Joynt, in Report of the Select Committee on Harbour Accommodation, BPP 1883 (255), query 2927.
68 Ibid., query 2907.
69 Ibid., query 2824.
70 Ibid.
71 Ibid., query 2850.
72 The list of donors is reproduced in Morrissey, *Verge of Want*.
73 T.F. Brady, in 2nd Report of the Royal Commission on Irish Public Works [The Allport Commission], BPP 1888 [C. 5264–I], query 19520.
74 *Charity Record* 6/10/1904, p. 13.
75 T.F. Brady, *Proposed Home or Refuge for Lost and Starving Dogs in Dublin* (Dublin: the Author, 1878).
76 27th Annual Report of the Congested Districts Board for Ireland, BPP 1920 [Cmd. 759], p. 30.
77 1st Report of the Commission of Inquiry into the Irish Fisheries 1836, p. 223 .
78 53rd Annual Report of the OPW, BPP 1884–85 [C. 4475], appendix D, p. 69.
79 50th Annual Report of the OPW, BPP 1882 [C. 3261], p. 64.
80 Report of the Inspectors of Irish Fisheries for 1889, BPP 1890 C 6058.
81 Report of the Inspectors of Irish Fisheries for 1892, BPP 1893–94 [C. 7048].
82 Report of the Inspectors of Irish Fisheries for 1887, BPP 1888 C. 5388.
83 J.H. Tuke, 'Irish Distress and its Remedies'. *The Land question; A Visit to Donegal* (London: W. Ridgeway, 1880).
84 2nd Report of the Royal Commission on Irish Public Works [The Allport Commission], BPP 1888 [C. 5264–I], query 19374, p. 621.

85 53rd Annual Report of the OPW, BPP 1884–85 [C. 4475], appendix D, p. 69.
86 Return to Parliament, BPP 1882 (203).
87 C. Breathnach, *A Word in your Ear. Folklore from the Kinvara area* (Kinvara: Comhairle Phobail Chinn Mhara, N.D. [*c.* 1990]).
88 1st Report of the Commission of Inquiry into the Irish Fisheries 1836, Appendix 19, p. 125.
89 16&17 Vict. C. 136.
90 36th Annual Report of the OPW, BPP BPP 1867–68 [4043], p. 55.
91 Return to Parliament, BPP 1884–85 (167), p. 5.
92 37th Annual Report of the OPW, BPP 1868–69 [4174], p. 15.
93 Ibid., appendix D, p. 54.
94 48th Annual Report of the OPW, BPP 1880 [C. 2646], p. 25.
95 Return to Parliament, BPP 1882 (203).
96 Report of the Canadian Fund Committee, BPP 1881 (326), p. 14.
97 59th and 60th Annual Reports of the OPW, BPP 1890–91 [C. 6480] and BPP 1892 [C. 6759]; Report of the Inspectors of Irish Fisheries for 1890, BPP 1891 [C. 6403].
98 50th Annual Report of the OPW, BPP 1882 [C. 3261].
99 59th Annual Report of the OPW, BPP 1890–91 [C. 6480].
100 Colleran, and Hanley, *Assessment of Piers and Harbours.*
101 50th Annual Report of the OPW, BPP 1882 [C. 3261], appendix B, p. 66.
102 Report of the Piers and Roads Commission, BPP 1887 [C. 5214].
103 Rev. M.H. Heaney, *Sacerdotes Tuamensis.* Unpublished digitised MS in James Hardiman Archive, NUI Galway. I am grateful to Fr Kieran Waldron for bringing this MS to my attention.
104 Report of the Piers and Roads Commission, BPP 1887 [C. 5214], p. 10.
105 Report of the Select Committee on Harbour Accommodation, BPP 1883 (255), query 3174.
106 Report of the Select Committee on Harbour Accommodation, BPP 1883 (255), queries 3134–35.
107 Return to Parliament, BPP 1881(244).
108 Return to Parliament, BPP 1882 (203).
109 Return to Parliament, BPP 1884–85 (167).
110 53rd Annual Report of the OPW, BPP 1884–85 [C. 4475], appendix D, p. 68.
111 Return to Parliament, BPP 1882 (203).
112 Colleran, and Hanley, *Assessment of Piers and Harbours.*
113 Return to Parliament, BPP 1884–85 (167).
114 Return to Parliament, BPP 1881(244).
115 Return to Parliament, BPP 1884–85 (167).
116 Return to Parliament, BPP 1882 (203).
117 53rd Annual Report of the OPW, BPP 1884–85 [C. 4475], Appendix D, p. 68.
118 50th Annual Report of the OPW, BPP 1882 [C. 3261], p. 67.
119 51st Annual Report of the OPW, BPP 1883 [C. 3649], pp. 82–3.
120 K. Davis, *Oranmore in Days of Yore* (Galway: N.D. [*c.* 2005]). (NUI Galway Hardiman Library, SCRR 941.74 Dav).
121 Memorial of James Fitzgerald, 29 November 1807. NAI OPW8/274.
122 NAI OPW/8/11.
123 Letter of Edward Lambert to the OPW, January 1847. NAI OPW8/274.
124 S. Lewis, *Topographical Dictionary.*
125 NAI OPW5/4188/14.
126 Report of T. F. Brady dated 3 June 1880, on the Memorial for Oranmore pier. NAI OPW5/4188/14.
127 OPW minute of 16 July 1881. NAI OPW5/4188/14.

128 Plan of new pier at Oranmore. NAI OPW5/4188/14.
129 Report of T.F. Brady, 30 August 1883. NAI OPW 5/4188/14.
130 53rd Annual Report of the OPW, BPP 1884–85 [C. 4475], Appendix D, p. 68.
131 Appendix 1st Report of the Commission of Inquiry into the Irish Fisheries 1836, appendix 17, p. 105.
132 NAI CSORP 1822/2237.
133 NAI OPW/8/158.
134 First Report of the Commission of Inquiry into the Irish Fisheries 1836, pp. 99ff & 105.
135 Return to Parliament, BPP 1880 (409).
136 Return to Parliament, BPP 1881 (244).
137 Return to Parliament, BPP 1882 (203).
138 Report of the Piers and Roads Commission, BPP 1887 [C. 5214].
139 Report by Col. Fraser, BPP 1889 [C. 5729].
140 Fourteenth Annual Report of the Congested Districts Board, B.P.P. 1906 [Cd. 2757].
141 Colleran, J. and D.P. Hanley, *Assessment of Piers and Harbours* (http://data-galwaycoco.opendata.arcgis.com/datasets/).

CHAPTER 7

Fisheries to the Fore: The Sea Fisheries (Ireland) Act and Fund

1883–1889

The various bodies concerned with the selection of sites ... seem to have been bewildered between exaggerated accounts of the fishing possibilities, on the one hand, and the limitations of money, on the other. They were asked nearly everywhere to build deep water harbours for large fishing boats and small slips for small fishing boats. Ideal accommodation, however, could not be provided everywhere and expenditure had to be distributed over the coast. It was stated in evidence that, as they apparently could not decide what was really required, they struck an average expenditure for each district. This average was, however, often extravagant, often niggardly, and sometimes ineffective, as at no place did it suffice to make a harbour good enough for a large boat, and the piers that were built were often quite unnecessary for small boats.

Royal Commission on Congestion in Ireland, 1908[1]

Maintaining the Piers

With the construction of new piers in suspension from 1875 to 1879, and putting aside the ongoing relief of distress operations from 1879, the OPW was free to address the maintenance of existing marine structures. It could, for example, undertake the repair and maintenance of those already existing and vested in the counties, in cases where the counties neglected that duty and the Viceroy directed them to intervene.[2] But some appropriate applicant had to raise the matter with the Viceroy in the first instance. The OPW could not initiate a memorial on its own initiative: section 6 of the 1846 Piers and Harbours Act stipulated expressly that memorials for the construction *or improvement* of any work could only come from the owners or occupiers of land on the sea coast.

In late 1879, however, as an extraordinary measure to help provide relief work, the OPW indicated to the Lord Lieutenant that repairs were needed urgently to certain of the vested piers known to be virtually derelict.[3] This caused the Viceroy to request the Inspectors of Fisheries to report to him in January 1880 on the state of all vested piers. The Inspectors immediately drew up a list of eighty-one, situated in the eight western counties between Donegal and Cork, and reported that a number had indeed been neglected and had fallen into a state of serious dilapidation.[4]

For the twenty-four they examined in Co. Galway, the news was not altogether bad: thirteen were in good order, two needed only small repairs, four were in fair condition and five were in a bad state. The OPW confirmed this after their engineer had visited the various places and drawn up estimates for the necessary repairs.[5] The five most deficient, Kilcolgan, Maumeen, Errislannan (Curhowna), Rossroe and Cashla, needed repairs totalling £2,306. The Exchequer advanced loans to the OPW to enable the works to get underway immediately: £800 for Kilcolgan, £400 for Maumeen, £406 for Errislannan (Curhowna), £100 for Rossroe, and £800 for Cashla.[6] These advances, and any excess of the actual costs over the estimates, would be charged to the county as laid down in the Act 16&17 Vict. C. 136. That meant they were neither relief funds proper nor an allocation of new voted funds; they were simply a charge on the local cess-payers to meet their obligations under the Act.

No detail was given of the exact nature of the repairs at individual places beyond the comment that 'the repairs really amounted to the reconstruction of a considerable part of the work'. Within the year, renovations were completed at all of them at final costs well above the loans advanced.[7] Kilcolgan quay, often called the Weir quay, or even Moran's quay (after the tavern close by), was reputedly built by the county before 1830 as a general-purpose landing place and it had fallen into serious disrepair since then. It was never envisaged as a pier exclusively for fisheries, but it did serve in that capacity, as well as being used to land turf and seaweed. Its renovation in 1880 involved the erection of a fine stone quay wall, almost 1,000 feet long, with a slip made of stone flags at its upstream end and a shingle slip at its seaward end. The final cost was £1,233[8] and Galway County Grand Jury made a presentment for £1,383 towards the cost in 1882.[9] This was the first recorded restoration of Kilcolgan quay, almost fifty years after its original construction. Maumeen and Cashla were piers dating from Nimmo's time. Cashla pier required £1,378 worth of repairs, the money being provided under the Public Works Loans Act 40&41 Victoria C. 27.[10]

The following year, 1881, applications for repairs were received from eight more sites in Galway: Duras, *Spidéal*, Barna, Kilronan, Killeany, Drimmeen (Errislannan), Cleggan, New Harbour (also called Ardfry or Rinville in the records). All of these, except Drimmeen, Kilronan and *Spidéal*, dated from the 1820s. The work at Duras pier in 1881 consisted mainly of clearing out the accumulation of silt from the harbour. This was a problem that had beset the place since the pier was first erected in 1823, and the District Engineer commented that, because of its situation high on the shore, the quay would always require expenditure to keep it clear of sand and debris.[11] The clearance in 1881 cost £330 which was refunded to the OPW under the 1853 Act.[12]

The harbour remains the same today as then, for the most part seriously silted and of little practical use. The *Spidéal* pier in question was not the old one built by Nimmo in 1823, but the new pier completed in 1871. The Barna pier application was directed to the Galway Grand Jury, and Manning deemed the Kilronan application, estimated to cost only £4, too trivial to bother with. The Killeany restoration cost £345 and was soon completed,[13] but the small repairs at Drimmeen would not be addressed until 1886. The repairs at *Spidéal* were started immediately and, as described in chapter 5, proved exceptionally costly.

Repeated repairs and maintenance would continue for a few more years after 1880 at other Galway locations, Inishbofin, Barna and Rossroe for example, supervised by James Perry, the County Surveyor. Continuing repairs would also be carried out at Drimmeen and *Spidéal* new pier but, in general, no more complaints of neglect by the counties were being received from 1888 onwards. All these repairs and renovations were important in ensuring that the existing structures were maintained in, or restored to, usefulness, both for the fisheries and for other purposes.

The Sea Fisheries Bill of 1883

As Chief Engineer of the OPW, Robert Manning had other responsibilities not relating to fishery piers and harbours and, by 1881, he was under considerable pressure and exceptionally busy. With fishery piers alone, he had fifty-nine cases of design, construction and repairs to deal with and had, under his superintendence, three District Engineers in charge of the several works. He complained that their time was continually being taken up with inspections in remote places, and he himself had to make two half-yearly tours of inspection in the northern and western districts involving 1,400 miles of travel.[14] Side by side with the repair and maintenance works, he and his staff were also occupied with ongoing relief of distress works, so, without any doubt, they were carrying an enormous burden. On top of all that, he had to travel to London to be examined by a committee of the House of Commons regarding the planned big harbour at Arklow – one of the 'important' works that was the cause of the suspension of fishery pier construction from 1875 to 1879.

As for building new piers and harbours separate from those under the Relief Acts, this was simply not possible without new dedicated funds being voted by Parliament. John A. Blake MP, the member for Waterford, who had been an Inspector of Irish Fisheries from 1869 to 1878, brought a new Irish Fisheries Bill[15] in to the House of Commons in February 1883, with a view to alleviating the situation. His main aim was to secure funding from an Irish source for the development of the Irish fisheries, without requiring any disbursement of funds from the Imperial Parliament – 'the expenditure of Irish money for Irish purposes', as he put it to the House.[16] His intended source of funding was the money in the hands of the Commissioners of the Irish Church Act, which had arisen from the disestablishment of the Irish Church. Significant sums from that source had already been allocated to various aspects of the relief of distress, so Blake's approach was not entirely novel or unusual.

What the Sea Fisheries Bill Proposed

As originally drawn up, the Bill envisaged radical reorganisation of the Fishery Inspectorate and of the whole administration of the fisheries. It proposed that the Lord Lieutenant should appoint and incorporate unpaid Commissioners – four persons, one for each province – who would oversee, direct and improve the deep-sea and coast fisheries, acting as 'Commissioners of Irish Sea Fisheries'. The sea fisheries were defined to include the close inshore fish and shellfish fisheries, as well as the deep-sea fisheries – in short, everything except the inland fisheries.

The proposed new Commissioners would report directly to the Viceroy and Parliament, and be entirely separate from other bodies, such as the OPW. They would be given all the powers then vested in the existing Inspectors of Irish Fisheries in all matters concerning the deep-sea and the coastal fisheries. To enable them to make advances for the construction or improvement of piers and harbours, they should receive the sum of £250,000 from the Church Temporalities Fund and apply it under terms and conditions to be determined by themselves, but subject to the Viceroy's approval. They could, the Bill proposed, fund the entire cost of construction, enlargement or improvement of any existing pier or harbour.

Crucially, the existing Piers and Harbours Acts should remain in force and 'all the powers, rights, privileges and duties vested in or exercised by the Commissioners of Public Works under said Acts or any of them, shall be vested in and may be exercised by the Commissioners of Irish Sea Fisheries under this Act.'[17] That particular section would have removed responsibility for building fishery piers entirely from the OPW, and was one that Blake and Thomas Brady had first championed sixteen years earlier at the Select Committee of 1867.[18]

Other clauses in the Bill, such as one severing the sea fisheries from the inland fisheries, were directly opposite to what Brady had espoused on the earlier occasion. That apart, its general thrust was the same as previously: it endeavoured to place all the sea fisheries responsibilities, including, critically for our purpose, the fishery piers, into the hands of an independent Fisheries Commission that would answer directly to the Viceroy. The existing situation of the Inspectors, dating from 1869, in which they functioned without the oversight of any overseeing commission or board, was quite anomalous and unacceptable to some.

The Select Committee on Harbour Accommodation

The House, however, saw Blake's Bill in a much broader context. The general matter of harbour accommodation in the whole of the UK was already exercising Parliament, which had recently established a 'Select Committee on Harbour Accommodation on the Coasts of the United Kingdom' to examine the matter.[19] Although the Irish Sea Fisheries Bill was never formally referred to it, the Select Committee took it up and subjected it to detailed examination in the light of the changes underway in the British fishing industry. It was entering the first phase of industrial expansion, driven by the extension of the railway network and the introduction of steam trawlers in place of sailing vessels, developments that necessitated radical change to fishery harbours.

Nine Irish witnesses presented a wealth of revealing evidence at the Select Committee's hearings held between May and July 1883. Since Parliament was then within a month of ending its session, and the Committee would go out of office when that occurred, the members decided to report urgently on the Irish Fisheries Bill in July. They recommended, among other things, that Parliament approve the allocation of the money from the Irish Church Fund, the Bill's main objective, but they added an admonition:[20]

> Your Committee are of opinion that a very large proportion of the money to be expended should be devoted, not to the multiplication of fishery piers, which are already numerous, but to the construction of real harbours with, where possible, a considerable depth of water at low spring tides. They also consider it of the utmost importance that, in the selection of any site for harbour works, existing, or at any rate, the probabilities of prospective, railway or tramway accommodation should be carefully taken into consideration.

A certain tension had been evident at the hearings regarding the kind of facilities – large, deep harbours or small fishery piers and quays – that were most desirable for Ireland. The Irish witnesses generally favoured both kinds, with the larger ones mostly on the East and South coasts – which were also the coasts with the most developed railway facilities – and the smaller ones in the West. The debate on this, with the resultant recommendations and admonition of the Select Committee, were early official articulations of a strategy to develop and modernise the Irish fisheries infrastructure.

The Sea Fisheries (Ireland) Act, 1883

Parliament accepted the Committee's recommendations and the Act that ensued in August 1883, *An Act to Promote the Fisheries of Ireland 1883*[21] (also known by its short title, The Sea Fisheries Ireland Act, abbreviated here as the SFI 1883 Act) had a distinctly different hue from Blake's original Bill. It enacted that up to £250,000 would indeed be made available for the Irish fisheries from sources controlling the Irish Church Funds (specifically, from the Irish Land Commission, out of the funds they raised on the security of their income under the Irish Church Act).[22] It was to be paid, not to any new Fishery Commission, but to the Commissioners for Public Works. This huge sum was entitled the 'Sea Fisheries Fund' and was to be used by the OPW for new marine works carried out under the existing Piers and Harbours Acts, modified by new provisions contained in this, SFI 1883 Act.

A new temporary Commission was set up, called the 'Fishery Piers and Harbours Commission' (hereafter the FPH Commission), comprising the existing Inspectors of Irish Fisheries (Thomas Brady, Major Hayes and Joseph Johnston, who, on his subsequent forced resignation as Inspector, was replaced by Alan Hornsby), together with a chairman who would be appointed by the Lord Lieutenant. That post fell to Blake himself in early 1884 when he lost his parliamentary seat. As it happened, his term as chairman turned out to be very brief: he resigned in 1885 on his re-election to Parliament in a by-election as the member for Carlow. He was replaced as chairman by Col. John P. Nolan, MP for County Galway.

Subject to Treasury approval, the OPW was empowered to make free grants out of the Sea Fisheries Fund of all or part of the cost of construction of any fishery pier, or make a loan at less than 3½ per cent interest, the loan being secured as provided in the earlier Piers and Harbours Acts. If necessary, and with the prior approval of the Lord Lieutenant, any grant could be for the full cost of the works envisaged. Section 5 of the Act was clear and explicit on this very important and long-sought-for alteration:

> The enactment contained in the fourth sub-section of the fourth section of the Fishery Piers and Harbours Act 1846, that the amount of any grant shall not exceed three fourths of the cost of the work, and that such grant shall be made only upon the condition that the repayment of the residue of the total actual cost of the work, with interest, shall be secured or agreed to be secured in the manner therein referred to, shall not apply to grants under this Act.[23]

At last, the fiction of a 'local' contribution from localities where no such contribution was ever likely to emanate, was removed, and local poor fishermen would no longer be penalised for the venality or alleged straitened circumstances of reluctant landowners. The limit of £7,500 in aid to any one site, in operation since 1866,[24] was also revoked.[25] However, when any work done under the new 1883 Act would be completed, the provisions of the earlier Piers and Harbours Acts regarding their maintenance and repair, and those relating to tolls and rates, were to apply. Up to £50,000 could be spent from the fund in any one year, giving it a potential five-year life-span.

The Sea Fisheries (Ireland) Fund

Even while the legislation was in preparation, Blake informed the House that there were seventy memorials for piers awaiting activation. By these he meant the residue of the memorials submitted from 1870 onwards and still unaddressed, listed by the OPW in their 48th to 51st Annual Reports[26] and by the Inspectors of Fisheries in their Annual Report for 1884.[27] The latter Report listed a greater number than the OPW, because it included every memorial received up to 1884, including those that were ineligible for whatever reason. No less than 264 had been received, which permits the demand for piers from different counties and regions to be considered in some detail.

The distribution is shown in Figure 7.1(a). The great majority (N = 223, being 84 per cent) were from counties west of a line from Baltimore, Co. Cork, to Derry; the remainder (N = 41, including Baltimore, being 16 per cent) were from east of that line. With so many memorials to hand, the FPH Commissioners were able to set to work without delay selecting suitable candidate sites from the list. They held their first meeting on 1 October 1883, the very day the Act came into effect, and within two months they had approved a preliminary group of thirty-nine sites which merited immediate surveys, plans and estimates from the OPW. All are listed in the 52nd Annual Report of the OPW for the year to April 1884.[28] Overall, nineteen were west of the Baltimore/Derry line and twenty east of it. This geographically even distribution favoured the eastern region by ignoring the 84/16 split in the distribution of memorials. The FPH Commissioners were clearly taking a more nuanced approach to their brief than the number and origin of the applications alone would have dictated. Engaging some independent engineers, the OPW completed the necessary preliminaries for thirty-two of the sites (whose locations are shown in a map in their 52nd Report), and forwarded them to the FPH Commissioners by April 1884. This enabled the remaining administrative requirements to be addressed immediately: by the end of 1884, contracts were in place for twenty-three, and thirteen were already under construction.[29] That was just the beginning.

The East–West Divide

As soon as the preliminary estimates for more piers were provided by the OPW, the FPH Commissioners formulated a final comprehensive scheme for the application of the full £250,000. They presented it to Treasury in February 1885 and it was approved surprisingly rapidly. The finalised scheme named a total of sixty-five locations throughout the whole country that the Commissioners held to be the places most in need of new fishery piers. They assigned these to two lists, published in the 53rd Annual Report of the OPW[30] and summarised here in Table 7.1.

List No. 1 was the 'top-priority' list. There were thirty-five sites in that, all of which had already been sanctioned by Treasury, or were in course of sanction. Twenty of these were located west of the Baltimore/Derry line and would require a total of £85,285; fifteen were located east of the line and would absorb a greater total of £100,337. The average allocation per site was £5,303 overall; for the western piers it was £4,264 and for the eastern piers it was over 50 per cent greater, at £6,689.

For some individual sites the estimates were remarkably large, such as £13,500 for Carrigaholt, Co. Clare, £19,300 for Ballycotton, Co. Cork, and £15,000 for Carlingford, Co. Louth. Until this Act abolished the cap on total expenditure per site, large allocations like these would not have been possible without special Acts of Parliament. Taken together, all the sites in list No. 1 would absorb an estimated £185,622, leaving a balance of £64,378 for list No. 2 sites.

List No. 2 contained thirty sites, twenty-three in the West and seven in the East, estimated to require a total expenditure above £57,300; the FPH Commissioners proposed to allocate the balance of the fund to these. The twenty-three western sites would absorb an estimated £35,500 and the seven eastern sites £21,500. The average

TABLE 7.1

Sites selected by the FPH Commissioners for the expenditure of £250,000 granted under the Sea Fisheries (Ireland) Act, 1883

List 1: First Priority				**List 2: Second Priority**			
West[1]		**East**		**West**		**East**	
Site[4]	**Est. Cost**[3]	**Site**	**Est. Cost**[3]	**Site**[4]	**Est. Cost**[3]	SITE	**Est. Cost**[3]
GALWAY	£8,000	CORK (E)	£29,650	GALWAY[2]	£2,000	CO. ANTRIM	
Cleggan	£2,000	5 sites		Collaheigue	£2,500	1 site	£1,000
Cora		DOWN	£10,650	Cashla	£300	CO. DOWN	
(Trawndaleen)	£24,300	2 sites		Kilkieran	£500	3 sites	£5,000
CLARE		DERRY	£4,000	B'dangan	£500	CO. DUBLIN	
4 sites	£26,205	1 site		Mason Isl.	£2,500	2 sites	£7,000
DONEGAL		LOUTH	£32,000	Crumpaun	£1,000	CO. WEX-	
6 sites	£2,800	2 sites		Ard West	£300	FORD	£8,500
KERRY		WATER-	£14,037	Mace	£2,500	1 site	
2 sites	£9,980	FORD		Bofin	£1,000		
MAYO		4 sites	£10,000	Gannoughs	£800		
3 sites	£12,000	WICK-		Bunowen			
SLIGO		LOW		Birterbuy Bay	£2,400		
3 sites		1 site		Bealantra			
				CORK(W)	£11,200		
				1 site			
				KERRY	£8,000		
				4 sites			
				MAYO			
				5 sites			
N (West) = 20[1]	Total est. Cost. West. £85,285	N (East) = 15	Total est. Cost. East. £100,337	N (West) = 23	Total est. Cost. West. £35,500	N (East) = 7	Total Est. Cost. East £21,500
		Total, East + West, List 1 £185,622				Total, East+ West, list 2 £57,000	
Average per pier	**West £4,264**		**East £6,689**	**Average per pier**	**West £1,543**		**East £3,071**

Source: data summarised from relevant OPW reports.
FPH Commissioners placed the sites in two separate lists: list 1 contained works of the highest priority, and list 2 the works of secondary priority. The sites in each list are separated here into western (West of a line from Baltimore Co. Cork to Derry city) and eastern sites (Baltimore and East of the line). Individual sites are named only for County Galway. For other counties only the number of sites is given. (All the sites are named in Appendix F).[1] One site in Co. Limerick, estimated at £19, omitted.[2] Road to Rossaveel, estimated at £340, omitted.[3] Grants, loans and contributions included.

allocation per site in the second list was, therefore, £1,543 in the West and £3,071 in the East. The two lists together show clearly that the FPH Commissioners were distributing the resource in a carefully considered manner and a national strategy for fishery piers, however inchoate, was beginning to emerge. This involved building a large number of small, cheaper structures in the West and a much smaller number of significantly bigger, more expensive structures in the East and South. We will return to this strategy below and again in chapter 10.

Another example of how the Commissioners exercised a judicious selection is apparent in list No. 2: fourteen of the sites in that list were in Co. Galway, a number that was inflated by the inclusion of five locations that were designated 'alternatives', that is, they would only go ahead in certain circumstances. Inishbofin, Gannoughs, Bunowen, Birterboy Bay [*sic*] and Bealantra [*sic*] would go ahead only if the proposed railway from Galway to Roundstone and Clifden was abandoned. If the railway did go ahead, the harbours of Clifden and Roundstone would be favoured, and the 'alternative' locations 'might have to yield to allow the money to be spent on the harbours near the railway terminus'.[31] This was in accord with the recommendation of the Select Committee on Harbour Accommodation, which favoured developing piers close to modern means of transport.

In Co. Down also, the three sites in list No. 2 were alternatives, one alone (to be chosen after further surveys) being judged sufficient for the needs of that county. A small number of the locations chosen in Co. Galway were entirely novel, such as Collaheigue (*Caladh Thaidhg*), Ard West (*Árd Thiar*), Crumpaun (Carna) and Mason Island, a welcome expansion from 'the usual suspects', and good news for the inhabitants of those places.

Galway accounted for only a small tally, about 6 per cent of the sites and 6 per cent of the money allocated in list No. 1 but almost 50 per cent of the sites and 25 per cent of the sums allocated in list No. 2. Apart from the 'alternatives', the OPW regarded the two lists as definitive, stating expressly in 1885 'the scheme may be considered complete and liable to little alteration.'[32] As it transpired, the Galway to Clifden railway did go ahead in 1895, almost ten years later, but notwithstanding that, all the 'alternatives' proposed for the county had already been built or were underway by then.

The western counties, especially Galway, Donegal and Mayo, had benefitted beyond others from the earlier Relief of Distress funding – eight sites in Galway, six in Mayo and six in Donegal had shared the £45,000 voted for piers in 1880, and Galway and Mayo would also benefit from piers built by the Piers and Roads Commission of 1885 to 1887, to be described in the next chapter. Because of these relief work benefits, Galway had only two sites in the top-priority list No. 1 – Cleggan and Cora (Trawndaleen) (*Inis Meáin*, Aran Islands); Mayo had three, and Donegal had six. The sites east of the Baltimore/Derry line, where British and other foreign fleets used to gather, were given the bigger allocations in order to provide larger, deeper and more sophisticated port facilities; the smaller allocations to the western sites (excluding Carrigaholt) would suffice for piers and harbours capable of hosting only small boats.

All of the Galway sites mentioned in the two SFI lists are recorded with their estimates and costs in Table 7.2. Of these, Cora on *Inis Meáin* was an unexpected beneficiary and is described below; the SFI works at Cleggan will be dealt with more fully in chapter 11.

By June 1886, fifty-seven of the sixty-five listed works had been sanctioned by Treasury. Five from list No. 1 were completed, twenty-eight were in progress and two had yet to start. From list No. 2, twenty-two had been sanctioned and twenty were underway.[33] Construction was pushed forward rapidly everywhere, because the funds were scheduled

TABLE 7.2

List of all the PIERs in County Galway funded under the *Sea Fisheries (Ireland) Act 1883.*

SITE	ORIGINAL ESTIMATE[1]	FINAL COST[2]	STRUCTURE
List 1			
Cleggan	£8,000	£8,032	Harbour improvements
Cora (Trawndaleen)	£2,000	£1,715	New pier
List 2			
Collaheigue	£2,000	£1,704	New harbour
Cashla	£2,500	£2,036	New pier
Kilkieran	£300	£773	Small improvements
Bealandangan	£500	£435	Pier and harbour improvements
Mason Isl.	£500	£590	New pier
Crumpaun }		£2455	New pier
Ard West }	£2,500	£759	New pier
Mace }		Not done	(New harbour)
Inishbofin }	£1,000	£162	Small improvements
Gannoughs }	£300	£255	Landing place
Bunowen }	£2,500	£2,465	New breakwater
Birterboy Bay }	£1,000	Not done	(New pier)
Béal an Trá }	£800	Not done	(Breakwater and slip)
Kilronan*	£3035	£3,772	Pier. Not yet completed – £868 to come
TOTAL	£26,935	£25,153	

Sites joined by brackets (Crumpaun, Ard West, Mace) (Inishbofin, Gannoughs, Bunowen) (Birterboy Bay, Béal an Trá) were 'alternatives' (see text for explanation). Three works were proposed but not carried out under the SFI Act, as indicated, and no works were proposed in north Clare.

[1] Data from 53rd Annual Report of the Commissioners for Public Works, Ireland.[2] Data from Parliamentary Return, dated 11 July 1902, on expenditure under SFI 1883 Act.

* Kilronan was a late addition (1900) to the works sanctioned under the Act – see Return to Parliament. BPP 1902 (353), p. 5.

to run out by 1888. By then, the total number completed and handed over to the counties was fifty-seven.[34]

The OPW had proceeded by employing day labour under its own engineers' supervision at more sites than usual, which reduced the delay that seeking tenders from independent contractors would have entailed: twenty-one works were done by day labour and thirty-eight under contract. In 1891, the Board was able to report that all of the works sanctioned and actually commenced under the Act – a final total of sixty-one – had been successfully completed at a final overall cost of £233,643. The savings made were spent on 'works for additional accommodation',[35] and a further small number of works would be added later to the final total. The Fishery Inspectors kept a watchful eye on developments, giving accounts of progress in their Annual Reports from 1885 to 1888.[36]

Col. J.P. Nolan Becomes Chair of the Fishery Piers and Harbours Commission

When Col. John P. Nolan replaced John A. Blake as Chairman of the FPH Commissioners in 1885, he claimed to have changed the Commission's approach in the direction of smaller piers:

> As to the deep-sea harbours, there was no money in my time voted by the Piers and Harbours Commission for deep-sea harbours. In Mr Blake's time, I believe there was some money so voted, but in my time we deliberately made up our minds that we had not enough money to spend on two or three deep-sea harbours when there was a very large fishing population scattered all over the country that wanted immediate accommodation for the actually existing class of boats, and that we were bound to spend it first of all upon that accommodation. I am now speaking of the residue of the Fund, because I believe some of the harbours decided on by the [FPH] Commission at first, when Mr Blake was chairman, might fairly be called deep-sea harbours, but in my time the money was reduced.[37]

By the time Col. Nolan was made chairman, the two SFI lists had already been completed and sanctioned and were little liable to alteration; the whole matter had already been 'done and dusted'. Most of the top-priority list No. 1 works were already underway and some even completed, so his chairmanship mainly involved overseeing the progress of the later works, and in truth, he probably had little enough influence on the overall outcome. Nolan was a Nationalist landowner from Ballinderry, Tuam, Co. Galway, who owned large tracts of land in Connemara. He was MP for Co. Galway from 1872 to 1874, and from 1875 to 1885; then he was elected MP for Galway North from 1885 to 1895, and again from 1900 to 1906. He became the first Chairman of Galway County Council when county councils replaced the Grand Juries as local authorities in 1899. His comment above regarding the 'actually existing class of boats' and their accommodation needs, and his deep local political interests may explain his heightened satisfaction that Galway (with thirteen), and Mayo (with eight), were counties that had benefitted the most from the forty sites funded in the western region under the SFI 1883 Act. However, from his comment, he does not appear to have grasped the extent to which large sailing smacks, and larger steamboats, were coming on the fishing scene, and he seems to have made no overt attempt to get even a single large, deep-water fishery port for his own county. On the other hand, north Clare did not receive even a single pier, big or small, from the Sea Fisheries Fund. Carrigaholt pier, in the south of the county, was Clare's only beneficiary and, in fact, the only large pier erected between the Shannon estuary and Donegal.

Final Outcome of the SFI Fund

Figure 7.1(a) and (b) shows the distribution by county of all the memorials in hand and the distribution of the sixty-one works actually constructed under the SFI Act; the majority (forty, being 66 per cent) were west of a line from Baltimore to Derry. However, when the sums spent in each county are considered, as shown in Figure 7.1(c), the gross amount spent on the twenty-one eastern works (£128,061) exceeded that spent on the forty western ones (£122,415). The eastern group in general received far greater amounts per site (£6,098) than the western group (£3060). Along with the voted sum of £250,000, the OPW had received other small contributions amounting to £9,864, making a grand total available of £259,864. Having spent £253,047 (including all administrative expenses) and retaining £3,594 for completions pending, there remained a balance in the Sea Fisheries Fund of £3,223. Altogether, £201,810 (78 per cent) had been awarded as free grants and £27,384 as loans repayable over twelve years. The amount given in loans, along with the £9,864 of other contributions, made up about 16 per cent of the total expenditure, a clear reduction on what the old one-fourth local contribution would have been, had it not been repealed. Of the sixty-one works completed to then, thirty-four received only the free grant (i.e. they were built entirely with SFI funds), and the remaining twenty-seven supplemented the free grant with loans from the fund, and in some cases, with voluntary contributions also. In the case of these, the combined loans and voluntary contributions made up 24 per cent of the total cost, almost the same as the old one-fourth local contributions would have provided. For these, at least, it appeared to be business as usual, with the local one-fourth contribution surreptitiously creeping back in, despite its clear repeal by the Act. Of the Co. Galway works, only one, viz. Cleggan, took up a loan (£2,000, repayable by Galway County Grand Jury), and none of the rest offered any local contributions, or took any loans, relying entirely on the free grants. Galway city had been briefly considered for a large harbour around this time, but in its case, it was for a 'convict' harbour – one that would be built with funds from a different source, by convicts who were to be housed in barracks to be built on either Mutton Island or Hare Island; this proposal came to nothing in the end.[38]

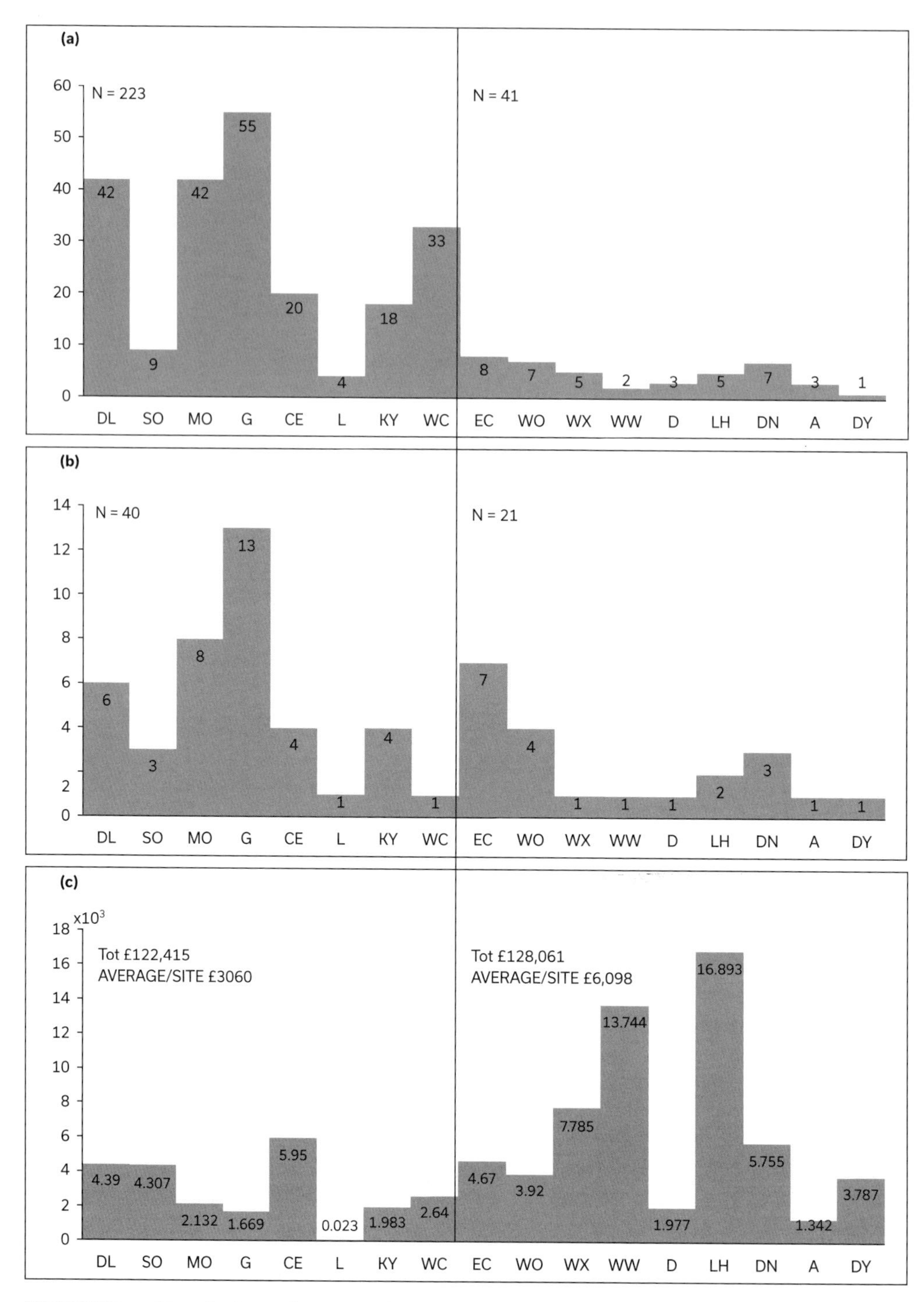

FIGURE 7.1. How the Sea Fisheries Fund was distributed in 1884. (a) The total number of memorials in hand. (b) The number of works selected for funding in each county. (c) The amount of money allocated to the sites selected in each county. Counties are indicated by initials along the horizontal axis, from Donegal (DL) on the left to Derry (DY) on the right. The division of counties into western and eastern groups by an imaginary line drawn from Baltimore Co. Cork to Derry occurs between WC (West Cork) and EC (East Cork). SO, Sligo; MO, Mayo; G, Galway; CE, Clare; L, Limerick; KY, Kerry; WO, Waterford; WX, Wexford; WW, Wicklow; D, Dublin; LH, Louth; DN, Down; A, Antrim.

When everything was finished and the FPH Commission had been dissolved, the OPW prepared a Return to Parliament illustrating the final outcome.[39] The sixty-three works carried out under the SFI 1883 Act (two extra had been added later) made a significant inroad into the number of memorials outstanding. Over the whole country, they comprised forty-one harbours, nine slips (with breakwaters where necessary), two lighthouses, three channels, deepened with associated beacons installed, six improvements to existing structures and two access roads. They are all listed in Appendix F.

Generally speaking, the various structures were built of large concrete blocks, with cut stone used for copings, bollards, steps and landings; the heartings were of rubble.[40] Thirty-three dried out at low water of Spring tide, twenty had some water at low tide and only six (about 10 per cent) had eight or more feet, the least depth considered necessary for the larger boats then coming into use.

Despite criticism of it, the SFI 1883 Act had constituted a watershed: it was the main legislation that funded the construction of piers and harbours as essential infrastructural elements of the fishing industry, rather than as public works directed to the relief of distress; it set the agenda in fishery pier building for five years; it set no limit to the total cost of any individual work; it was more liberal in its requirement regarding local financial support; and it provided a new procedure that gave the Inspectors of Fisheries, acting in their capacity as FPH Commissioners, a greater role in decision-making and in the design of piers and harbours. Because the Act did not appear to impose an unyielding, overarching Treasury-imposed policy, it resulted in the expression of a pragmatic, if still largely inchoate, piers and harbours strategy by an Irish body, and funded entirely from Irish financial resources – in short, and in its best meaning, it represented an Irish solution to an Irish problem.

In its Annual Report for 1885–86,[41] the OPW noted that 'Your Lordships [of Treasury] have not sanctioned any further expenditure under those [the Fishery Piers and Harbours] Acts', so it was clear that no further expenditure was contemplated once the Sea Fisheries Fund would run out. In a Return to Parliament dated 5 March 1885,[42] all the fishery piers and harbours built by the OPW since 1846 were listed, and the OPW reprinted it, together with a map of the various locations, as an appendix to its Annual Report for 1884–5.[43] That map illustrates dramatically the differrence in the abundance of piers and quays between the east and the west, already discussed in figures.

The Board now seemed to be drawing a line under the long phase of pier building of the previous forty years. But Tim Healy, the Irish MP who had requested the Return, was not satisfied with its completeness, so it was followed three months later by a Further Return, this time including all the sums expended on fishery piers and harbours since 1846 by the Grand Juries, the amounts of the tolls collected as reported by the harbour constables (where they existed), and the existing state of the piers and access roads, as determined by the County Surveyors.[44]

The Further Return presents little information that is new about the Galway and north Clare piers, except that no tolls were ever paid at any of them. It does not deal well with the pre-1846 structures but is otherwise useful to those interested in documenting the many episodes of repairs, modifications and renovations to a number of the piers in other counties.

The Royal Commission on Congestion in Ireland, quoted at the heading of this chapter, would later (1908) be very harsh in its criticism of the approach taken by the FPH Commissioners and the OPW, and of its outcome. Was this criticism justified? In reality, the FPH Commissioners do not appear to have been at all 'bewildered' by their task; they did succeed, largely, in tailoring their selection to the existing memorials and to the prevailing fishery needs as they saw them; they did take into account the recommendations of the Select Committee on Harbour Accommodation; and they did not simply strike an average expenditure between districts.

Nevertheless, the criticism penetrated to the core of a crucial issue: the majority of the piers built, especially in the West, were too small, or too poorly located, to be of much use for *deep-sea* fisheries; they did not have sufficient depth of water at low tide; and there were not enough large ports on the west coast to induce or even attract the bigger boats of the English, Scottish and French fleets to explore actively the deeper Atlantic waters to the west. Those fleets fished seasonally in Irish east- and south-coast waters, and they were turning increasingly to steam propulsion, which necessitated deeper and more sophisticated port facilities than were generally available in the West. Yet, the fact that the majority of the Irish piers and harbours were small was hardly surprising: the average expenditure in the whole country under the Sea Fisheries Act was around £4,000 per site, a sum far below what was necessary to provide decent harbour accommodation accessible at low Spring tides at a desirable number of sites. Indeed, if the six relatively large harbours that met the low water criteria – Carlingford and Clogher Head in Co. Louth, Greystones in Co. Wicklow, Kilmore Quay in Co. Wexford, Ballycotton in Co. Cork and Carrigaholt in Co. Clare – are excluded from the calculation, the average spent per remaining structure was just over £2,900.[45]

Average amounts spent on the individual piers built under the Relief of Distress (1880) and the Piers and Roads (1886) legislation (next chapter) were also very small, as were the piers themselves. Therefore, while a great number were built, and a national strategy regarding fishery pier sizes and locations was beginning to take shape, the decade did not result in the degree of development that might have been hoped for from a serious fisheries, especially a deep-sea fisheries, point of view. This was especially true for Galway and north Clare, whatever about the east and south coasts. This, of course, is not to suggest that the piers that were built in the West were of no value whatever, especially for purposes other than fisheries. But the question still remains as to why the SFI Fund in particular was expended as it was, on so many seemingly inadequate piers? We will discuss this apparent inadequacy further in chapter 10.

There remained a very healthy balance in the Sea Fisheries Fund in 1890,[46] made up of savings from the original £250,000, plus repayments of capital and interest coming in from some loans. Further scheduled repayments were anticipated to increase the balance to £38,223 by 1902, a sum that would be availed of for many years thereafter to fund continuing pier building and renovation.

Piers and Quays in Co. Galway Funded under the Sea Fisheries (Ireland) Act 1883

Cora (Trawndaleen) (Trá an Mhaindlín) pier, Inis Meáin, Aran Islands

Cora is sometimes called Trawndaleen, especially in the OPW reports; in some other documents (particularly in the Fishery Inspectors' reports), it is referred to as Cora, Trawndaleen. The Irish word *cora* is a generic name for a rocky ridge or spit extending into the sea,[47] and it occurs in many place names throughout the country. The use of one generic name for various disparate places can make it difficult to ascertain, from the written documents alone, the precise locations in question: there is, for instance, a pier and slip at a place called *An Cora* (Cora) near Cora Point on the east coast of *Inis Meáin*. That is now the main port of the island and is named Cora Point pier (No. 184) in the Ryan Hanley survey. It is a modern structure made of concrete, with a concrete slip alongside. Reference to the maps in the 54th Report of the OPW[48] and the Report of the Inspectors of Fisheries for 1890[49] confirms that Cora (Trawndaleen) pier is located on the north coast of *Inis Meáin*, in a bay named in the first edition of the six-inch OS map as *Trá an Mhaindlín* (Mandolin Beach).[50] How the beach got that unusual name is not known. The name Trawndaleen is an Anglicised corruption of *Trá an Mhaindlín*; the Irish form of the name is now the only legal name for the place.[51]

The first edition of the six-inch OS map shows an elongated, regular structure there; it may be a natural ledge on which landings were made long before any artificial structure was built. Today, an old causeway with a pier head/wharf at right angles to it is still extant at the site. These appear to be the structures built under the SFI 1883 Act as Cora (Trawndaleen) pier; the causeway is built along the natural ledge shown in the first edition OS map.

The memorial requesting a pier at the Trawndaleen site had been received and approved by the Fishery Inspectors before 1882 and was listed in their schedule of memorials in that year.[52] On the passing of the SFI 1883 Act, the FPH Commissioners recommended a free grant of the entire amount – estimated at £2,000 – needed to build a pier there. A contractor was engaged who agreed to do the work for £1,376, with completion scheduled for October 1885.[53] The work was largely finished that year, except for paving and coping the wharf at the head of the causeway. The OPW had difficulty in getting the contractor to complete it, so it took over the work from him and employed day labour to carry it out. Manning drew up altered specifications for the job, the deepening of the harbour, and for a short, inclined approach in 1885.[54] The work continued to struggle and completion was delayed, first to October 1886, and eventually it dragged on until June 1888, when everything was finally finished well under the original estimate.[55] Trawndaleen merited SFI funding because it would serve as the only pier on the island and would greatly improve connectivity for the local population, in addition to serving the fishermen.

At the start of the twentieth century, the CDB classified it as class 'C', saying it was 'of little importance to the fisheries' and in need of repair.[56] By April 1906, all the masonry at the west side of the outer end was cracked and there was a large split extending to sixty feet on the north side, which the Coastguard reported needed to be seen to 'as soon as possible'. Appropriate repairs were carried out in December 1906 and again in 1908. However, in truth, the wharf/pier head was not founded on rock and the whole structure was crumbling away and nearing the end of its useful life. By 1911 it had cracked again from its foundation right across the pier head.

By 1913, large stones were being washed away from it. Soon after that it was replaced by a new pier built about 150 yards south of it. That latter pier, called *An Caladh Mór* (No. 182) in the Ryan Hanley survey, became the main pier of the island until it, too, was superseded by the present pier at Cora Point (No. 184), located close to the principal village of the island. Cora Point pier is a modern construction, made entirely of concrete. The Ryan Hanley survey recorded the Trawndaleen SFI pier as follows: 'Separate jetty and causeway about 160m north of pier [No. 182]'

Collaheigue (Caladh Thaidhg) Harbour

The small harbour of Collaheigue (*Caladh Thaidhg*) is shown in the first edition of the six-inch OS map as a natural indentation of the coast, but with no manmade structures evident. Local opinion holds that the place was used as a port 'for centuries' and it sounds reasonable that such a suitable natural harbour was a regular landing place long before it was developed to its present state. According to Robinson,[57] the harbour was first developed in the 1840s by Tadhg O'Catháin, hence its name in Irish, which translates as 'Tadhg's (Tim's) landing place'. He reputedly owned a number of boats trading with Galway city.

According to a minute of the Engineering Department of the OPW, Major Gaskell had improved the place as part of the relief works he supervised in 1880.[58] This may have involved erecting a small jetty within what is now the boat slip area, shown in a sketch in that document; the structure was later demolished during the course of the OPW works in 1886. *Caladh Thaidhg*'s next appearance in the records is on list No. 2 of the FPH Commission,[59] where an estimate of £2,000 to construct a proper harbour was indicated. A grant from the SFI Fund was sanctioned in 1885 and work commenced in 1886 by day labour under the Board's engineer.

Everything was completed by 1890 at a final cost of £1,715, a saving of £285.[60] The harbour comprised a pier and slip, with a breakwater jetty on the northern side narrowing the harbour entrance and creating a sheltered basin where boats could be safely beached at the slip. Today *Caladh Thaidhg* is still an important local amenity and, despite recent renovations and a new wharf, its original form is still clearly recognisable.

Ard West Pier and Harbour (Árd Thiar)

Ard West in *Iorras Aithneach* was another place where Major Gaskell had built a small quay in 1880–81. By 1883, it had fallen into serious disrepair and was a hazard to boats entering the harbour.[61] A memorial for repairs and improvements was submitted in 1885 and Treasury sanctioned a grant of £960 (£759 according to another source)[62] to clear the harbour and repair and improve the quay. The money came from the SFI Fund and the OPW carried out the work. The old quay was completely reconstructed, extended and raised above high water for a length of 94 feet. A storm wall or protective parapet was erected along the sea face of the pier and the outer section of the quay. The inner, landward, part of the quay does not appear to have been raised: it lacks the cut-granite coping that distinguishes the raised section of the quay and the pier. The harbour basin was cleared and deepened before the whole works were transferred to the county on 20 June 1887. In 1895, further restoration was needed – the pier was said to be breaking up – but nothing was done. The local parish priest sent a plea to the Grand Jury for it to take the matter in hand; the Jurors suggested that it might be restored using the residue remaining in the SFI Fund, but once again nothing was done.

When Ard West was examined by Ryan Hanley consultants in year 2000, it was described as 'a well constructed small sturdy pier/quay, well used', so it must have been repaired eventually. In recent times, it has been extensively renovated with concrete.

Crumpaun Carna Pier and the Piers on Mweenish Island

A memorial for a pier at Crumpaun, Carna, 'for about £1,000 or whatever sum your engineer would say would build it', was submitted by Martin Lydon and Martin Mongan, occupiers of the site, on 4 September 1883.[63] The FPH Commissioners asked the OPW for a plan and specifications, which they received on 12 August 1885. The plan shows an L-shaped pier comprising a leg 300 feet in length with a short outer leg at right angles to it, about seventy-five feet long and constituting a substantial return head. The whole pier has a substantial storm wall and the site is joined to the main road by a low causeway. A coloured tracing of the plan, signed by OPW Commissioners Sankey and Le Fanu, and dated 12 August 1885, is conserved in the National Archives and the map accompanying it confirms the exact location of the pier.[64]

Few pier memorials were completely free of objections and Carna was no exception. Patrick Gorham, of Rusheenacolla near the village, held traditional seaweed rights over the rocks on which the pier was to be built and he wanted compensation for the loss of his rights. Ten pounds would meet his claim.[65] He was an old man and he sent his son to negotiate the compensation claim with the Board's representative. The claim was negotiated down to thirty shillings (one and a half pounds), a trivial sum even then, and confirmation maybe of the adage, 'Never send a boy on a man's errand'. The Board demanded and received a proper receipt for this payment before allowing the work to go ahead. The final cost of the whole project amounted to £2,455,[66] funded entirely by a free grant from the SFI Fund; there was no local contribution of any sort. In a letter from the Irish administration to Treasury, dated 8 February 1904, the pier is classified 'B', one 'of some importance to the fisheries'.[67] It still exists in its original form.

FIGURE 7.1 This small, artisanal pier was built on Mweenish Island (*Máinis*) before 1839. It was used for communication with the mainland. It does not appear to have been improved or altered since then and it is not used much today. Photo NPW.

A later CDB storehouse built at the site was converted around 1960 into a Marine Biological Station for University College Galway. Alongside that, a Shellfish Research Laboratory was built later and the whole site and pier now form part of a centre of research on fish and shellfish.

There are a number of other piers near Crumpaun Carna. Two are located on Mweenish Island. In February 1846 John Nolan submitted a memorial for a pier at *Meall Rua* on the east side of the island.[68] He claimed that a plan to build a pier there some three or four years earlier had been approved by the county surveyor but nothing had been done for lack of funds. Nolan offered £300 towards it in 1846, provided the OPW would give £500 and the county £300.[69] Barry Gibbons attempted to visit the

FIGURE 7.2. *Caladh Thaidhg* (Collaheigue) harbour today. The pier is on the left, the breakwater on the right where the entrance to the slip can also be seen. Photo NPW.

place but as it happened, he was unable to land on the island because of bad weather. (The causeway and bridge to the island were not built until 1886.) That was a fortunate misfortune, and understandably enough in the circumstance, Gibbons decided there and then that a pier would 'no doubt be a great convenience' and he recommended a grant.[70] However, it is doubtful that any substantial grant was ever made: there is a low, vernacular, pier about eighty feet long and made of loose stones there today, which does not resemble any other well-founded public pier built during the Great Famine. The existing structure was probably erected by Nolan from his own resources and using his own tenants.

The other Mweenish Island pier, at *Portach*, on the north shore, is a more significant structure and certainly a busy place today. In May 1887, the occupiers sent a memorial for the repair and extension of an existing old pier at the site, then in a useless state. A minute attached to the memorial drafts a proposed reply: there were no funds available and 'an enquiry would only raise false hopes'.[71] Because the only means of accessing the island at that time was by boat, the reply proposed was insensitive, to say the very least. Later in the century, a good straight pier was built at the *Portach* site and it was completely encased in concrete in a modern renovation. Well inside the shelter of that pier there is a small vernacular quay of loose stones with a grassy surface; this is most likely part of the original structure mentioned in the 1887 memorial.

There are many other piers and quays near Carna, which was at one time an important place for the export of seaweed, kelp and turf. None of the documents consulted make any reference to these but one can surmise from that fact that most may have been erected during the 1880s with funds provided by philanthropic donors.

Mason Island Pier

A presentment for a quay on Mason Island close to Mweenish Island had been approved in 1846 at an extraordinary presentment session of the County Grand Jury held in Clifden. For some unknown reason, it was not proceeded with immediately. William Pierce, the Clifden engineer, wrote to the OPW on 15 November 1846 requesting that it be given the go-ahead, not merely as a relief work but to promote the fisheries.[72] A collapsed jetty located outside the present angled pier is possibly the original presentment quay, but this has not been fully established.

The SFI structure, built in 1885 at a cost of £590, comprises an angled pier sheltering a small inner quay with rough wharf walls. It is far less substantial than the pier at Crumpaun Carna. Yet, it was still in use in 1911 when both the pier and the breakwater were reported to be cracked and the channel badly obstructed by stones and sand.[73] It has been renovated a number of times since then and a storm wall has been added. It is used nowadays for landing people and animals on the island, which is no longer permanently inhabited.

Two Piers Named Cashla (Casla)

There are two separate piers in Co. Galway known by the name 'Cashla pier', both located in Cashla Bay. The earlier one is that erected by Nimmo in 1823 at Costelo near Rossaveel harbour and now part of that large marine complex. Known originally as Costelo pier, it is called 'Costello or Cashla' in the 50th Report of the OPW,[74] referred to as 'Cashla Bay' in the 53rd Report[75] and simply Cashla in the 54th Report.[76] It underwent major restoration when Treasury authorised emergency loans to the OPW and extraordinary presentment sessions by Grand Juries, in order to initiate immediate relief work in 1880. Loans amounting to £1,150 were sanctioned for Cashla at that time[77] and the work was done in 1881.[78] The Grand Jury made a presentment for £1,378 to repay the loan in 1882.[79]

FIGURE 7.3. *Árd Thiar* pier in 2016. Photo NPW.

The pier was put in good order but the road to it was left undone. The Piers and Roads Commission would improve the road in 1886 at a cost of £73, reporting: 'Existing track from Rossaveel harbour to main road made passable. Length about ¾ of a mile.'[80] That was the first mention of Rossaveel harbour in the records. Nimmo's old Costelo pier was the only manmade structure at Rossaveel at the time, but even so, the whole inlet lying behind Lyon's Point (*Gob an Laighin*) was used as a place where boats would lie in safety, with ready access to deep water close by. The road to the pier was further improved later still, at a cost of £340, provided from the SFI Fund.[81] In the twentieth century, Rossaveel was developed as a major fishery centre, and Nimmo's old Costelo pier was incorporated into the modern harbour complex that exists there now. The stone structure of the old pier is still evident in parts of the quay face. The modernised pier is now used as a ferry terminal.

The second Cashla pier is the one newly built under the SFI 1883 Act, at a site in Keeraunnagark South (*Caorán na gCearc*) in the townland of Ballintleva, in the parish of

FIGURE 7.4. *Crumpán Carna* pier. Photo NPW.

Spidéal. The memorial for the pier is not extant, but a letter that accompanied it from Fr Curran, the parish priest of *Spidéal*, is conserved in the archives. He referred to it as 'a small pier at Cashla Beg coastguard station'.[82] The conflation in the records of two different piers – Nimmo's old Costelo pier and the SFI pier in Keeraunnagark – under the single name 'Cashla pier' can cause some difficulty. For example, they are confused in an appendix to the second report of the Allport Commission in 1888,[83] as pointed out in a minute of the engineering department of the OPW dated 28 March 1903: 'The old pier at Rossaveel formerly was called "Cashla pier" but it is two miles or so north of the "Cashla" pier No. 91 made under the 1883 Act.'[84]

Built in concrete with fine cut-limestone coping, the SFI pier is an angled structure, 250 feet long with a short return head. Alongside it is a stone-surfaced boat slip, ninety feet

FIGURE 7.5. The pier and slip at Cashla (Keeraunagcearc). Photo NPW.

long, well sheltered by the pier. Treasury sanctioned a free grant of £2,750 for the works, which were started in 1889 and completed in 1890, at a final cost of £2,036.[85] As if to add to the confusion, the pier is also known as *Céibh an Station* (after the now ruined coastguard station beside it), *Céibh Cheann an Bhóthair* ('the quay at the head of the road') and *Céibh Teach an tSalainn* ('salt house quay', after a ninteenth-century salt store nearby). Access to it for boats is difficult due to the island of Callowansha and other rocky obstructions in the offing, so that it never became a very important place for fishermen, although trawlers from the Claddagh were said to shelter there when rough weather did not permit an easy passage to Galway.[86] Rossaveel harbour, only two miles further up Cashla Bay, was closer to deep water, offered greater shelter and was easier and safer to access. It, rather than Cashla, would grow and expand to become the main fishing port of the county in the twentieth century. Cashla appears to be little used today.

Other SFI Sites in Galway County

In the end, only three of the thirteen Galway sites included in SFI list No. 2 (Table 7.1), viz. *Béal an Trá* (Bealantra) Birterboy Bay and Mace, failed to receive any SFI funding. A structure would be built eventually at Bealantra by the Piers and Roads Commission in 1886–87,[87] under the direction of the engineer R.F.E. Hayes. This was a granite and cement breakwater, 180 feet long, giving shelter to the natural harbour there.[88] It was lengthened and the stonework finished by Col. Fraser a year later and a slip, 170 feet long, paved in stone and faced in concrete, was added.[89] Today, the site is much as it was when Fraser finished it (Figure 7.6). It is a useful small landing place for slipping boats, although the approach road to it is narrow and inconvenient. Bealantra happens to be another pier whose original official name has fallen from memory. It is not mentioned in Tim Robinson's map and gazetteer,[90] and is not known to him or to others in the locality.[91] From the description of the work given in Fraser's and the Piers and Roads reports, it is certain that the site is the place now known only as Ervallagh, located between Roundstone and Gorteen Bay.

As regards Birterboy Bay (Bertraghboy Bay), it is not clear what exact site was intended there, but a solid granite pier, sixty-two feet long and twenty-five feet wide, with cement coping, was erected at Cashel, and a granite protective wall, 144 feet in length, was built at Ailnacally, both places in Bertraghboy Bay, by the Piers and Roads Commission in 1886–87. Ailnacally is most probably the one recorded as 'Birterboy Bay' in the SFI list.

FIGURE 7.6. Ervallagh (Bealantra) breakwater and slip in 2009. Photo NPW.

Mace Harbour was built some years later (1898) by the Congested Districts Board and will be discussed in a later chapter. Therefore, all of the Galway piers mentioned in the SFI lists were built eventually, although not all with SFI funds.

Endnotes

1 Final Report of the Royal Commission on Congestion in Ireland, BPP 1908 [Cd. 4097], p. 27.
2 16&17 Vict. C. 136, section 11.
3 48th Annual Report of the OPW, BPP 1880 [C. 2646], pp. 30 and 34.
4 Report of the Inspectors of Irish Fisheries for 1879, BPP 1880 [C. 2627], pp. 9–10.
5 49th Annual Report of the OPW, BPP 1881 [C. 2958], appendix B, pp. 68–69.
6 Ibid., appendix A6, pp. 52–3.
7 50th Annual Report of the OPW, BPP 1882 [C. 3261], pp. 33–34.
8 Return to Parliament, BPP 1884/5 (167).
9 Ibid., (266).
10 Ibid.
11 51st Annual Report of the OPW, BPP 1883[C. 3649], appendix B, p. 64.
12 Ibid., p. 48.
13 50th Annual Report of the OPW, BPP 1882 [C. 3261], p. 68.
14 Ibid., appendix B, p. 68.
15 Bill 31 of 1833, BPP 1883 [31].
16 House of Commons debates, 20 June 1883. Hansard, 3rd series, Vol. 280, page column 1047–1099.
17 BPP 1883 [31], Bill 31, section 12.
18 Select Committee on the Sea Coast Fisheries (Ireland) Bill, BPP 1867 (443).
19 The Select Committee on Harbour Accommodation on the Coasts of the United Kingdom was set up on 13 March 1883 and reported on 13 July 1883.
20 1st Report of the Select Committee on Harbour Accommodation, BPP 1883 (255), p. 5.
21 *The Sea Fisheries (Ireland) Act 1883,* 46 & 47 Vict. C. 26.
22 *Irish Church Act* 1869, as amended by the *Irish Church Amendment Act* 1881.
23 *The Sea Fisheries (Ireland) Act 1883,* 46 & 47 Vict. C. 26, section 5.
24 The limit had been raised to £7,500 by the Act 29 & 30 Vict. C. 45.
25 *The Sea Fisheries (Ireland) Act 1883,* 46&47 Vict. C. 26, section 5.
26 Annual Reports of the OPW: 48th for 1879–80, BPP 1880 [C. 2646]; 49th for 1880–81, BPP 1881 [C. 2958]; 50th for 1881–82, BPP 1882 [C. 3261]; 51st for 1882–83, BPP 1883 [C. 3649].
27 Report of the Inspectors of Irish Fisheries for 1884, BPP 1884–85 [C. 4545], pp. 5–8.
28 52nd Annual Report of the OPW, BPP 1884 [C. 4068], p. 28.
29 53rd Annual Report of the OPW, BPP 1884–85 [C. 4475], p. 24.
30 Ibid., pp. 25–26.
31 Ibid., p. 26.
32 Ibid, p. 24.
33 54th Annual Report of the OPW, BPP 1886 [C. 4774].
34 57th Annual Report of the OPW, BPP 1889 [C. 5726].
35 59th Annual Report of the OPW, BPP 1890–91 [C. 6480], p. 26.
36 Report of the Inspectors of Irish Fisheries for 1885, BPP 1886 [C. 4809]; for 1886, BPP 1887 [C. 5035]; for 1887, BPP 1888 [C. 5388]; for 1888, BPP 1889 [C. 5777].
37 J.P. Nolan, in Royal Commission on Irish Public Works [The Allport Commission], Second Report, BPP 1888 [C. 5264-I], query 9977, p. 194.

38 Select Committee on Harbour Accommodation, BPP 1884 (290), queries 2103, 2194–2198, 2224–2232.
39 Return to Parliament, BPP 1890 (276).
40 Ibid.
41 54th Annual Report of the OPW, BPP 1886 [C. 4774], p. 25.
42 Return to Parliament, BPP 1884–85 (167).
43 53rd Annual Report of the OPW, BPP 1884–85 [C. 4475], appendix D, pp. 66–71 and chart of Ireland.
44 Return to Parliament, BPP 1884–1885 (266).
45 Return to Parliament, BPP 1890 (276).
46 Return to Parliament, BPP 1890 (276), page 3.
47 N. O'Donaill, *Foclóir Gaeilge-Béarla [Irish-English Dictionary]* (Dublin: The Stationery Office, 1977).
48 54th Annual Report of the OPW, BPP 1886 [C. 4774].
49 Report of the Inspectors of Irish Fisheries for 1890, BPP 1891 [C. 6403].
50 OS map, six-inch series, 1839, Galway, sheet 120.
51 Placenames (*Ceantair Ghaeltachta*) Order, 2011, Government of Ireland Statutory Instrument 599/2011.
52 Report of the Inspectors of Irish Fisheries for 1882, BPP 1883 [C. 3605], p. 7.
53 Report of the Inspectors of Irish Fisheries for 1884, BPP 1884–85 [C. 4545], p. 10.
54 NAI OPW5 2907/85.
55 Report of the Inspectors of Irish Fisheries for 1888, BPP 1889 [C. 5777], p. 18.
56 NAI OPW5/1524/14
57 Robinson, *Connemara Part 1*, p. 121.
58 NAI OPW/5/3815/15.
59 53rd Annual Report of the OPW, BPP 1884–85 [C. 4475].
60 Return to Parliament, BPP 1902 (355).
61 OPW minute, 6 September 1897 on Ard West. NAI OPW/5/3800/15.
62 Return to Parliament, BPP 1902 (355).
63 NAI OPW8/59.
64 NAI OPW5/3803/15.
65 Letter from Patrick Gorham, 22 August 1885 seeking compensation for seaweed rights. NAI OPW/5/3803/15.
66 Return to Parliament, BPP 1902 (353).
67 Letter to Treasury, 8 February 1904, reg. no. 2018-04. NAI OPW/5/3803/15.
68 NAI OPW8/265.
69 NAI OPW8/248.
70 Report of Barry Gibbons dated 9 May 1846, NAI OPW8/248.
71 Memorial dated 9 May 1887, and Minute regarding pier at Portach, Mweenish Island. NAI OPW8/281.
72 W. Pierce to OPW, 15 November 1846. NAI OPW 8/244.
73 Report of the coastguard, 10 March 1911. NAI OPW/5/3801/15.
74 50th Annual Report of the OPW, BPP 1882 [C. 3261], p. 27.
75 53rd Annual Report of the OPW, BPP 1884–85 [C. 4475].
76 54th Annual Report of the OPW, BPP 1886 [C. 4774].
77 49th Annual Report of the OPW, BPP 1881 [C. 2958], pp. 52–53; 50th Annual Report of the OPW, BPP 1882 [C. 3261], p. 50.
78 50th Annual Report of the OPW, BPP 1882 [C. 3261], p. 27.
79 Return to Parliament, BPP 1884–85 (266), p. 16.
80 Report of the Piers and Roads Commission, BPP 1887 [C. 5214], p. 8.
81 Royal Commission on Irish Public Works [The Allport Commission], Second Report, BPP 1888 [C. 5264-I], list No. 5, handed in by General Sankey.

82 Letter from Fr Curran PP dated Spiddle, 10 August 1883, accompanying a memorial. NAI OPW8/69.
83 Royal Commission on Irish Public Works [The Allport Commission], Second Report, BPP 1888 [C. 5264-I], lists Nos 1 and 5, handed in by General Sankey.
84 OPW minute dated 28 March 1903. NAI OPW/5/4186/14.
85 Return to Parliament, BPP 1890 (276); Return to Parliament, BPP 1902 (355).
86 H. Ree in Royal Commission on Irish Public Works [The Allport Commission], Second Report, BPP 1888 [C. 5264-I], appendix, p. 725.
87 Report of the Piers and Roads Commission, BPP 1887 [C. 5214].
88 Ibid., p. 9.
89 Col. T. Fraser, Report to the Lord Lieutenant, BPP 1889 [C. 5729], p. 9.
90 Robinson, *Connemara Part 1*.
91 T. Robinson, personal communication, 2015.

CHAPTER 8

More Distress and Another Colonel: The Piers and Roads Commission

Great harbours have not been suggested, because it is believed it would be the height of folly to throw large sums of money into the sea in the hope, unsupported by evidence, that then and there fish must come out of it.

Col. T. Fraser 1889[1]

The Relief of Distressed Unions Act 1883

The construction of sixty-one piers under the Sea Fisheries (Ireland) 1883 Act and Fund did not obviate the need to start or to sustain relief works to assist the destitute. While the threat of major famine abated as the 1880s progressed, it never entirely went away, flaring up spasmodically in various Poor Law Unions in the West. In April 1883, the Government had to introduce legislation, the *Relief of Distressed Unions Act*, providing £50,000 for grants towards public works to address the crisis in the worst affected areas.[2] This Act was only a temporary, one-year crisis measure scheduled to terminate on 31 March 1884.[3] By that date only £10,400 of the sum provided had been spent. In order that the unused balance could be applied to the original purpose, the Chief Secretary was obliged to bring forward another Relief Bill in 1886, reviving those sections of the 1883 Act that had expired.[4]

The new Bill was limited to Poor Law Unions named in a schedule, all of which were in counties Galway and Mayo, where distress was most acute due to failure of the potato crop, poor markets for cattle, and a shortage of credit.[5] As originally drawn up, the Bill dealt only with outdoor relief (consisting of free food and fuel given out from the Poor House to persons who congregated outside), and with general-purpose grants to the Guardians of the specified Unions. The Bill was generally welcomed with enthusiasm by the Irish members at Westminster (prominent among them were Col. J.P. Nolan of Galway, John Dillon of Mayo and T.M. Healy of Derry). During its passage through the House, they requested that additional specific provision be made for the construction of piers and slips. These were particularly desirable, they claimed, because the scheduled Unions were ones that included many of the islands, where distress recurred most frequently and with the greatest degree of severity. Even in the best of times, piers were essential there. John Dillon went as far as to propose that funds be made available directly to Fishery Inspector Brady and James Hack Tuke, the renowned Quaker humanitarian, to be spent on piers and slips at sites of their own choosing. In response, the Chief Secretary claimed that the urgency of getting relief measures in place precluded adding piers and slips to the draft Bill as Members requested and he said that an entirely new Bill would be needed for that purpose.[6]

The *Relief of Destitute Poor in Ireland Act, 1886,* and the Piers and Roads Commission

As it transpired, the Act that emerged from that Parliamentary discussion was passed on 10 May 1886, as *An Act to make Temporary Provision for the better Relief of the Destitute Poor in Ireland* (short title, *The Poor Relief (Ireland) Act 1886*) and comprised two distinct parts.[7] Part 1 concerned grants to the amount of £20,000 for the relief of distress and the provision of outdoor relief by the Local Government Board for Ireland. Part 2 was entirely new and, in light of the Chief Secretary's earlier comment, completely unanticipated: it granted a further £20,000 to be spent in the scheduled Unions – Belmullet, Clifden, Galway, Oughterard, Westport and Swinford – for making and improving roads, bridges, piers, quays, slips and landing places, under the direction of a new body to be set up and called the Piers and Roads (P&R) Commission.

The members of this, a Chairman and two others, would be appointed by the Lord Lieutenant and would hold office during his pleasure. By a warrant dated 11 May 1886, the day after the Act passed, the Viceroy appointed Christopher T. Redington as Chairman. Col. Thomas Fraser RE and Mr Pierce Mahony, Nationalist MP for Meath North, were appointed the two ordinary members. The warrant of appointment was a document which, considered in the light of what was happening in the background between the OPW and the Fisheries Inspectorate (described in chapter 9), was most astutely drafted. The new P&R Commissioners were to make an enquiry as to where the £20,000 granted by the Act for piers and roads could be spent to best advantage. They were to make the fullest use possible of all existing establishments of the public service, whose chief officers were *required* [emphasis added here] by the warrant to be helpful and to give assistance to the fullest extent that their other duties permitted.

Once the new Commissioners' initial enquiry was concluded, they were to report back to the Viceroy on the locations selected, together with 'the nature, character and dimensions [i.e. the designs and plans] of the piers and landing places to be erected, and the probable cost of each of them' for his approval. This instruction put responsibility for the precise locations, the exact plans, the specifications and estimates in every case on one single body, the P&R Commission. Finally, the warrant charged the Commissioners, none of whom had any previous connection with either the OPW or the Fishery Inspectorate, to start, progress, and complete the works in the most expedient manner. In this way, the warrant removed, formally and entirely, but very deftly, both the OPW and the Inspectors of Fisheries from any stated role in the selection, design, funding or construction of the marine works to be done under the *Relief of the Destitute Poor Act*.

The P&R Commissioners, who may have been aware of the possible reason and the import of this instruction, engaged the County Surveyors of Galway and Mayo, rather than the Engineers of the OPW, to supervise and carry out the works, although they did later acknowledge the assistance the OPW gave in clerical and accountancy matters.

The Piers and Roads Commissioners Set to Work

The Commissioners proceeded immediately to Galway – a short enough journey for Redington, whose seat was at Kilcornan in Clarinbridge – where they made arrangements to meet the County Surveyors, James Perry of Galway and Peter Cowan of Mayo. Accompanied by them, the Commissioners made a tour of inspection along the coast and islands from Galway city to Broadhaven in Co. Mayo. Notices were issued to various influential persons advising them of the impending visit, and inviting written suggestions for works that would be most appropriate for their neighbourhoods. On receipt of replies, the Commissioners held meetings in different centres to consider and discuss the proposals. After that, they visited the exact sites recommended, in order to examine at first hand their general suitability. This task was not to be a particularly easy one, as they recorded: 'The investigation was unavoidably protracted, as the most out-of-the-way districts had to be examined. Some of them could only be approached on foot, and the islands were constantly inaccessible, owing to the rough seas.'[8]

To meet the requirement of relieving distress in an equitable fashion, they decided to apportion their funds among the various districts in accordance with the extent of poverty proven to exist in each, information that was available in tables of the value of land per head of population supplied by the Local Government Board. Choosing between the 240 different projects that were submitted for their consideration was a more difficult task, even though only 125 were at all practicable. How were they to balance the need for a new turf road – opening a fresh bog – against a new pier or landing place elsewhere? Or the value of a causeway between two islands, against a new mainland road to market? They perceived that the most remote and inaccessible districts, and those least able to draw attention to their plight, had hitherto received the least support, so to these they decided to give special attention. As soon as the Viceroy approved their selection, they set about getting the support of the landowners and the tenants concerned. All the landowners gave permission freely for works on their property, and helped in other practical ways (it would cost them little or nothing); tenants' claims were numerous and sometimes more difficult to deal with, but eventually a list of eighty-six projects was fully approved so that work could start.

The operations were grouped into ten districts, and a number of engineers were employed to take charge of them (Table 8.1).[9] Where suitable candidates were available locally, they

TABLE 8.1

Engineers engaged by the Piers and Roads Commissioners and the works they directed in 1886 to 1887.

Engineer's name	Marine works done	Other works done
Henry Abbott	Crusheen wharf; *Sruthán* pier; Tooreen harbour	Tooreen road; Muckinagh road; Rossaveel road
G. Armstrong	Slackport slip; Fahy slip; Turbot slip; Inishturk slip; Aughrus Beg breakwater; Atalia pass; Leitir pier; Glassilaun pier	Fahy road
John Harris	Killeany pier; Kilmurvey landing place; Ballyhees landing place	
R.F.E. Hayes	Cashel pier; Ailnacally pier; Ervallagh breakwater; Slackport landing place; Doohulla breakwater	Inishnee bridge
R.H. Head	Carrigaluggaun Pass causeway; Rankin's Pass	Screebe road; Furnace bridge
T.F. Plummer	Garafin Harbour; Rossmuck Harbour	Inverbeg and Invermore bridges; Keelsalia road and bridge; Aconeera bridge; Mweenish causeway

Source: data from the Report of the Piers and Roads Commission.

were engaged as foremen; otherwise competent men had to be brought from elsewhere. As for labourers, they were generally available on site in abundance, except when seasonal agricultural employment took them away to their holdings, which retarded the completion of some works. Gangs of men were changed regularly to ensure that the relief work was shared by as many families as possible; this had a detrimental consequence in that men were rotated off the work just as they were becoming more skilled at it, and therefore more useful. Labourers' wages were one shilling and four pence a day in the summer months and one shilling on the shorter winter days.[10] This was much less than the wages estimated earlier for the 1879–80 relief works using Manning's data (chapter 6), further cautioning that such data needs to be treated with circumspection. Nevertheless, it provides solid evidence that wages of one shilling a day on relief works were not unknown at the time.

What the Piers and Roads Commission Achieved

Between May 1886 and May 1887, the P&R Commission successfully completed twenty-three small harbours, four small piers, three quays, eleven slips, seven rock clearances in harbours, five causeways to islands, twenty-four miles of road and eleven bridges in counties Galway and Mayo, at a total cost of £21,129. They are indicated in Appendix G.

By any measure, it was a remarkable achievement. Thirty-four of the marine works were in Co. Galway: twenty-one piers and quays are listed in Table 8.2.[11] The average expenditure per project in Co. Galway was about £234, distributed relatively evenly between roads and bridges (twenty-one works, total £4,553, average £219) and marine works (thirty works, total £8,345, average £278). Up to seventeen of the marine works were entirely new structures,

TABLE 8.2

Marine Works carried out by the Piers and Roads Commission in Co. Galway, 1886–7

Sites and Structures	RH. No.	Works Done
Killeany harbour	178	Outer entrance channel widened to 162 feet
Kilmurvey landing	174/75	Cement breakwater, 45 feet long
Ballyhees *(Traght Each)*	185	Excavation extended 35 feet; landing placedeepened
Forramoyle slip	144X	Slip made, 241 feet, with wall and sea pavement, 3 to 6 feet high in granite
Bunnahowna Harbour	048/048A	Small harbour made, dry at low tide
Crusheen wharf	135	Existing wharf extended 45 feet. River diverted for scouring
Sruthán pier	133	Existing harbour partly rebuilt in concrete, extra landing places and road made
Tooreen Harbour	127	Small harbour made with breakwater, 210 feet
Slate Harbour, Lettermore	124	Pier and wharf, 117 feet, added to existing landing place
Lettercallow harbour	099/098	Breakwater and wharf 70 feet long added to extend existing small harbour
Sruthán Bhuí wharf	083?	Rocks cleared and wharf made to landing place
Teernea harbour *(Glas na nUan)*	120	Pier raised and extended. New breakwater added
Trawbaun *(Trá Bán)* harbour	118	Sheltered creek given entrance piers and improved
Glentrasna harbour	102	Wharf, 90 feet long, made in dry granite
Drinagh pier, Inishbarra Isl.	172	Pier extended 85 feet into deeper water
Garafin harbour	088	Existing harbour strengthened
Rossmuck harbour	089	Existing pier extended and widened. Harbour cleared
Cashel pier	047	Dry granite pier built, 62 X 25 feet. Cement coping
Ailnacally pier	040A	Small protective wall made in granite, 144 feet long
Béal an Trá pier (Ervallagh)	037	Breakwater, 180 feet long, made in granite and cement. Entrance cleared
Doohulla harbour	033	Breakwater 230 feet long made in stone and cement, and breakwater 160 feet long made in dry stone. Entrance cleared

Source: data from the Report of the Piers and Roads Commission.

erected at sites not previously recorded in public documents, or at natural landing places where no manmade structure existed previously. The latter included Forramoyle slip (No. 144X), Kilmurvey breakwater (No. 174), Slate Harbour (Lettermore) (No. 124), Glentrasna harbour (No. 102), Cashel pier (No. 047) and Aillenacally pier (No. 040A).

The Final Report of the Piers and Roads Commission

The P&R Commission closed its last project on 3 May 1887, exactly one year from its inception in May 1886. The amount it had spent exceeded the Government grant by £1,129, but various small contributions were made by Mr Berridge of the Ballinahinch Estate, Mr Tuke, the philanthropist, the Mansion House Fund, the Rev. B. McDermot and two of the Commissioners, Mr Redington and Col. Fraser.[12] Captain Alexander Boxer of the Commission for Irish Lights helped by generously offering to build some small slips from his own resources. The Grand Jury, on the other hand, proved unwilling to make any contribution.[13] The Royal Navy, the Irish Coastguard and the OPW also gave assistance in kind to the Commission, which meant that more of the grant money went directly to relief work than might otherwise have been the case.

All in all, the P&R Commission had been very successful in what it achieved in a quiet, efficient manner. In presenting their Report in August 1887, the Commissioners appended,

FIGURE 8.1. The breakwater and small harbour at Doohulla (RH 033) in 2014. Photo NPW.

in a separate memorandum, a schedule of twenty-seven sites where they felt additional works were advisable, and which would require only a small amount of extra funds, should they be sanctioned.[14] They classified these additions into 'extremely desirable', 'very desirable' and 'desirable' categories, perhaps little expecting that any but the first category, if even that, would ever be sanctioned. The total extra cost was estimated at £6,409. Fifteen of the twenty-seven extra works were for marine sites in Co. Galway.

Col. Fraser appended to the Report a separate memorandum under his own signature, explaining how the Commissioners had gone about their business, and recommending procedures that might be useful in the future to obviate some of the obstacles they had encountered. He was blunt in his account of the difficulties they had faced:[15]

> Many of the works were in such wild places that it was difficult for the engineers to find anywhere to live ... Storehouses had to be built, and watchmen paid for keeping stores, and in spite of these precautions, large numbers of cement bags, and in some cases other stores, disappeared ... It was on account of them that the Commissioners were obliged themselves to contribute to avoid an excess of the vote [the grant]. The cost of getting stores to these islands and other out-of-the-way places was considerable, and the delays and uncertainty of communication added to the expense. For the latter reasons the cost of paying the labourers amounted to about 2% of the cost of the works.

From his last comment, and the table of receipts and expenditure given in the main Report,[16] we know that about £435 in total was spent on labourers' wages. At an average wage of one shilling and two pence per day (in summer, one shilling and four pence, in winter one shilling per day),[17] the work must have taken about 7,461 man-days to complete. This equates to about twenty-five men working for a total of 300 days (fifty weeks, six days per week), hardly a massive contribution to relief of the general distress, but important nonetheless. For comparison, the two consultant County Surveyors received £140 each, and the eleven resident engineers working under them were paid a total of £1,844, an average of just under £168 each, spread over one year. The benefit of the P&R Commission may not have been huge in monetary terms, but the nature, scale, usefulness and locations of the works it carried out were of greater long-term value at a local level than many works of much greater cost.

The Public Works and Industries (Ireland) Special Grant

All in all, the P&R Commission had been an efficient and well-managed operation, and one suspects that Fraser, in particular, was one good reason for that. The Government, too, must have been very impressed with the results: within two months of the Final Report, a special grant of £6,500 was voted by Parliament (the 'Public Works and Industries (Ireland) Special Grant') to permit all the extra works, in all of the 'desirability' categories, to be undertaken straight away. This meant that, on average, £240 could be spent on extra work at each of the twenty-seven sites, almost as much as had already been expended on them by the P&R Commission! The Commission having already come to an end by then, the Lord Lieutenant appointed Col. Fraser, alone, to administer the special grant. His brief was 'to start and superintend the new expenditure for the completion or improvement of certain of the works carried out in Mayo and Galway by the Piers and Roads Commission in 1886–7'.[18]

Since the special grant had not become known until late in 1887, Fraser could do little immediately. Winter was approaching, and, to the great relief of all, an abundant potato harvest had reduced the degree and the extent of distress. He therefore confined his immediate efforts to revisiting the various sites and consulting the County Surveyor(s), whom he engaged subsequently to carry out his instructions regarding what was to be done. They were to receive 5 per cent of the cost of the works for the engineering and accounting services they would provide.

TABLE 8.3

Works directed by Col. Thomas Fraser RE in Co. Galway, 1887–89, under the Public Works and Industries (Ireland) Special Grant.

Sites and structures	RH No.	Works done
Killeany harbour	178	Entrance channel widened, rocks blasted, beacons erected
Kilmurvey landing	174/5	Pier lengthened 15 feet. Rock landing place improved
Ballyhees landing (*Trácht Each*)	185	Rock landing place extended further inwards
Forramoyle slip	144X	Damaged portion replaced in solid concrete work
Ballinahown harbour	140	Entrance widened. Pier ends rebuilt in concrete and strengthened
Teeranea harbour *(Glas na nUan)*	120	Breakwater raised above high water for full 240 feet in dry stonework
Drinagh pier, Inishbarra Island	172	Original design completed with 300 cubic yards of dry granite work
Ailnacally pier	040A	Pier completed to its full height
Béal an Trá pier (Ervallagh)	037	Breakwater extended to full 200 feet. Inner slip of 170 feet faced and paved in concrete
Inishshark Island.	164B	Breakwater and handrail extended
Doohulla harbour	033	Ladder steps added
Aughrusbeg wall	016	Parapet of breakwater completed. Slip added
Glassillaun harbour	005	Breakwater raised throughout and inner rocks blasted away

Source: data from Col. Fraser's Report to the Lord Lieutenant.

Operations commenced in May 1888 and were virtually finished by the end of the year. Fraser was meticulous in ensuring that the full twenty-seven sites were attended to in an even-handed way, although he admitted to having given a little more attention to the Aran Islands than the late Commissioners had proposed. In fact, he spent £3,170 in Co. Galway, of which £720 (over 22 per cent) was spent on three Aran Islands sites, Killeany, Kilmurvey and *Trácht Each.* The P&R Commission had already spent £886 on these, out of a total of £12,928 spent in Galway (about 7 per cent), so that, in total, the Aran sites received £1,605, 10 per cent of all the money spent in the county.

Fraser also undertook some additional new small works at four sites, including the repair of Forramoyle slip with concrete, and the addition of steps at Doohulla harbour, while still saving over £500 of the grant. The largest and most expensive project he undertook was the construction of five miles of new road, with small bridges and culverts, across the neck of the *Iorras Aithneach* peninsula in Connemara. This route was part of the coast road begun by Alexander Nimmo over sixty years earlier and left uncompleted since then. Fraser inspected the finished works in January 1889, found all to his satisfaction, including the parts completed earlier in 1886–87, and made his Report to the Viceroy and Parliament in May 1889, complete with a sketch map of all the locations serviced.[19] A list of the thirteen marine works that Col. Fraser completed in Co. Galway is given in Table 8.3.

Col. Fraser's Final Report

Fraser's Report commenced in a thorough and clear fashion, showing that the additional works had been well managed and efficiently progressed. Already they were proving useful to the fisheries and beneficial to the people by facilitating the distribution of turf and seaweed, the movements of cattle, and generally improving communication.

However, he went on to make an extended commentary on the state and prospects of the population of the West of Ireland and on the nature of the Irish character. When confining his comments to engineering prospects and technical developments, he displayed good knowledge and perspicacity. For example, he predicted the inevitable development of hydro-electricity ('water power will, doubtless, be converted into electricity for the profitable working of machinery')[20]; the pressing need to manage the turf resource ('the rocky beds of the turf in many places near the sea have been scalped and will be useless for generations ... [persons of influence] must persuade them to husband their turf resources')[21]; and the absolute requirement for fish-curing stations and good roads to market, if the fishing industry was ever to develop successfully ('Fish curing would appear to be a special necessity here ... Great harbours have not been suggested, because it is believed it would be the height of folly to throw large sums of money into the sea in the hope, unsupported by evidence, that then and there fish must come out of it')[22].

As for his sociological opinions, these might be regarded as more controversial today, and represent an early example of what is now called social Darwinism: 'In India we have extended the survival of the "fittest" to that of the whole, with the result that we have to organise beforehand for inevitable famines. I fear that it may not always be unnecessary to do the latter in these parts of Ireland, where, however, the poorest are, happily, often among the hardiest of the population.'[23]

Although an avowed believer in a *laissez-faire* philosophy, he did not allow this to hinder his work:

> Money, subject to a labour test, was brought to the doors of those who could work, and whose needs were supposed to be the greatest; but, apart from the momentary relief of distress, I believe the system to be demoralizing and nationally injurious. The expectation is encouraged that, if only things become bad enough, Government can and will provide a *deus ex machina* in all cases of difficulty. This belief saps the self-reliance, and dulls the foresight of those who, unhappily, can only live, even in favourable times, by energetic self-help.

He continued his Final Report, published in 1889, with an unusual comment: 'The works of 1888 were in completion of those of the Piers and Roads Commission and the experiment of an executive commission has thus been tried, on a very small scale it is true. The competent can judge how far the money's worth has been obtained in addition to the relief of distress.'[24]

One wonders what exactly he meant by this 'experiment'; he cannot have been unaware of the interdepartmental discord affecting the pier works being carried out at the very same time under the SFI 1883 Act. Whatever he meant, he was well satisfied that he himself, and his P&R Commission colleagues, had acted with efficiency, speed and success, and had proved that an executive commission, focussed exclusively on problems in the poorer western counties – soon to become more widely known as the 'congested districts' – could be of measurable advantage. Arthur Balfour, Chief Secretary for Ireland at the time, was surely listening intently.

Why Was the Piers and Roads Commission So Successful?

The works done by the P&R Commission were indeed numerous (over eighty separate constructions and twenty-five miles of roads), widespread, useful and beneficial to the local communities in the distressed Poor Law Unions. This can be attributed partly to the words and the thrust of the Act under which they were funded, and partly to the ability and determination of the personnel involved.

The *Relief of Distress Acts 1880* and the *SFI 1883 Act* were, under statute, strictly to be 'read as one with' the *Piers and Harbours Act 1846*.[25] On the other hand, the *Relief of the Destitute Poor Act, 1886* was not proscribed in that manner. Because of that, the P&R Commissioners were freed from certain burdens of the 1846 Act: proposals for piers were not restricted to coastal proprietors, but could be advanced by the Commissioners themselves, or in consultation with others; they were not limited to considering memorials already in hand; there was no requirement for a one-fourth local contribution; none of the excessive administrative requirements for preliminary surveys, plans, estimates, advertisements, etc., called for in the 1846 Act, were necessary; final approval of works proposed was by the Lord Lieutenant in Dublin, not by Treasury in London; and engineering supervision was provided by the relevant County Surveyors along with other engineers engaged by the Commissioners themselves.

Freed from such administrative shackles, the Commissioners were able to address issues of their own choosing, at their own pace and in their own way. They focussed their approach on small works of a utilitarian nature that had a predominantly local rather than a wider impact. They could assess local needs at first hand because they visited the localities in person and consulted the local interested parties in the company of the County Surveyors before making their recommendations. Nor were they restricted to marine works or to fishery considerations alone: the Act made absolutely no reference to *fishing or fisheries* in connection with any of the duties or powers of the Commissioners; their piers and quays could be, and in most cases were, made for any locally relevant purpose, entirely independent of any fishery usage.

Many of the roads they made were specifically declared to be for the movement of turf or the delivery of seaweed inland, and many of the piers were built at precise sites chosen by the inhabitants themselves. They could provide access roads from bogs to piers, build causeways between islands, build bridges, make rock clearances, erect piers or quays for the turf and seaweed trades, and so on; their works could be new, or they could be modifications and repairs to existing structures. Utility rather than conformity with previous schemes or statutes was the overarching tenor of their approach. Even the requirement to relieve immediate distress was met with prudence, and not permitted to swamp the overall effort: the shorter winter days paid lower wages than the longer summer days and the total spent on labourers did not exceed 2 per cent of the overall costs. This seeming niggardly approach was balanced by the instruction to the engineers to employ only one able-bodied man from each family at a time, and to change the gangs each fortnight when there was an excess of labourers, so that the small benefit was dispersed among families as widely as possible, even at the cost of some inefficiency. They gave special consideration to the islands, so that their relatively greater isolation was duly acknowledged and addressed. And most importantly, the P&R Commission was completely independent of the Fishery Inspectors and the OPW who were otherwise engaged, not only with the SFI piers being built contemporaneously, but with their own interdepartmental machinations, which we will discuss in the next chapter.

A Good Outcome for Galway in a Distressful Decade

Combining the outcome of the SFI 1883 Act with that of the *Relief of Distress Acts, 1880*, the Commissioners of Public Works reported that a total of ninety-eight piers and quays had been completed nationally between 1880 and 1890, at a cost of £307,475.[26] Co. Galway had benefitted with thirty-three valuable new works. When the various repair and renovation works (which were funded separately out of the Consolidated Fund for Ireland) are added,[27] together with the Piers and Roads Commission structures (the latter were not included in the OPW Reports, because they were directed by the County Surveyors and did not involve the OPW engineers), Co. Galway can be seen to have gained no less than eighty-six new or renovated structures during the decade 1879 to 1889, considerably increasing the stock of manmade landing facilities in the area.

The decade had seen a veritable blitz of pier building and restoration in Galway and the West in general, undertaken predominantly as relief works. (Only fifteen of the Galway works were done under the SFI 1883 Act; the rest were carried out under the different relief of distress Acts of 1880 and 1886.) North Clare benefitted only minimally in that great decade: Glenina pier was the only structure erected on that coast under the *Relief of Distress Act 1880* and nothing was done there under the SFI 1883 Act or by the P&R Commissioners. While the famine of 1879–80 had not proved as long-lasting as first feared, destitution and distress continued episodically through the decade, symptomatic of the difficulties peculiar to the West, and a local manifestation of the great depression that afflicted even industrialised Britain during that decade.[28] The small works executed under the named relief Acts and those built by the charities undoubtedly helped to ameliorate conditions in some of the most disadvantaged places. They certainly contributed greatly to the abundance of small piers and landing places for which Co. Galway is notable. Although their beneficial effect on the fisheries was less than overwhelming – as Fraser emphasised, the success of the fisheries needed more than just piers and harbours – the works facilitated communication, improved the harvesting and distribution of turf and seaweed and helped to support a very fragile cash economy in the west of the county by providing money for labour and goods. Arguably, the P&R Commission's construction of proper, raised causeways between the islands was the first real step in the opening-up and development of Connemara.

More importantly, along with other developments, the Piers and Roads 'experiment' was quietly influential in the crystallisation of policy that lead to the setting-up of the Congested Districts Board in 1891, whose marine works will be discussed in a later chapter. In a way, the Piers and Roads Commission was a model for the nature and operation of the Congested Districts Board later, at least in the Board's early years. It is unlikely that Board would ever have been set up in the manner it was, or have taken the role that it did, had the 'experiment' of the P&R Commission not been undertaken and proved so successful.

Some Co. Galway Works of the Piers and Roads Commission and Col. Fraser, 1886–1888

Forramoyle Slip

Forramoyle slip is located at a place on the north shore of Galway Bay named in the 1839 edition of the six-inch OS map as *Calahanafamosa* (= *Caladh na feamainne*? = 'seaweed

FIGURE 8.2. Forramoyle slip (*Céibhín na gCurach*). The low, granite wall is evident in the centre; to the left of it is the slipway, now largely filled with loose stones. To the right of the wall the sea slope is somewhat broken up but still evident. Photo NPW.

quay') in the townland of Forramoyle East, just west of Barna village. The old name suggests that the site may have been a landing place for seaweed long before 1839. The work carried out in 1886 consisted of clearing a sandy landing strip, about nine feet wide, and building a low wall of granite blocks, two or three courses high, along its western side. The wall, about 240 feet long and sloping in height down shore from about three feet to one foot, was backed with a low, manmade sea pavement of granite blocks held in mortar, and extended up the shore to high water mark.

When Col. Fraser returned to the site a year later, he found the upper shore section of the wall was damaged and he repaired it with concrete.[29] Today there is still evidence of the concrete section at high tide level, very much broken up; some granite blocks are also still held in cement near this section. Structurally the site is otherwise largely as Fraser must have left it. The slip is known locally as *Céibhín na gCurach* ('little quay of the currachs').[30] Some of the stones forming the low wall and the sea pavement have been dislodged and driven by the sea into the cleared landing place. It would take very little effort to restore the site to its original state, as a reminder not only of the P&R Commission's work, but of the traditional maritime activities that prevailed historically at the site. It is not included in the Ryan Hanley survey of 2000–01 and is given the number 144X in this study.

A turf road, about 770 yards long, named Derryloney Road, was made by the P&R Commissioners from Forramoyle slip into the nearby bog so that the quay site was rendered valuable for a range of purposes. The section of that road from the shoreline to the main Galway/*Spidéal* (R336) road is now reduced to a narrow track, largely overgrown, whereas the section from the R336 northwards into the bog is now a tarred, macadamised suburban residential road.

According to the available records, Forramoyle slip was not transferred to the county until 30 January 1899.[31] The place was strongly favoured in 1918 for the proposed

FIGURE 8.3. The 'breakwater' erected at Aughrusbeg seen from Illaunageeragh island. The sand-blown upper section of the original inlet is above the structure. The lower section of the original inlet is below the breakwater, now part of a lobster-holding facility. Note that the parapet of the dam, built by Fraser in 1888, differs from the rest of the structure built by the P&R commission in 1886. Photo NPW.

development of a major transatlantic port for Galway, a small indicative plan of which is shown in the proposal published at that time.[32] That scheme was the fourth such proposal for a major commercial transatlantic port in Galway Bay but, like the others before it, nothing ever came of it.

Clerical Intervention at Aughrusbeg

A very peculiar circumstance arose at one site, Aughrusbeg (No. 016), where twenty-six men were employed. Their task was to make a 'breakwater' crossing the channel thereby joining the mainland to a small islet called *Illaunageeragh* ('Sheep Island').[33] The breakwater was really a large dam that divided the channel into north and south sections, each of which could then be used as separate small harbours. The new structure, a dam wall about eight feet thick, thirty feet long and twenty-five feet high, was pierced by two tunnels at ground level so that water could pass from one section to the other depending on the ebb and flow of the tide. The men worked under the direction of a resident engineer, G. Armstrong. The site supervisor, William Geraghty, was seconded to the place by John Perry, the County Surveyor, to act as foreman.

On 28 June 1886, Armstrong sent a note to the Commissioners that the men would not work under Geraghty, because their priest ordered them not to – and not to work at all, unless a Clifden man recommended by the priest, Fr Linskey PP, was put in charge of the work.[34] The men informed Armstrong that they were, in fact, anxious to work and had nothing at all against Geraghty, but they did not like to act contrary to the instructions given by the priest. A minute written on the margin of this archived note states 'Col Fraser has, I understand, settled this matter on the spot on June 30th and the men promised to go to work under Geraghty'. In this, Fraser had showed himself to be an excellent practical manager, as he proved to be in other, later operations for the Commission.

The intervention of Fr Linskey was not typical of the response of other members of the clergy: in their Report, the Commissioners warmly thanked the many Catholic clergy whose intimate knowledge of the condition and needs of the people had 'rendered their cooperation of the greatest value'.[35] Fr W. Conway of Carraroe and Fr T. Flannery of Carna even acted as supervisors of construction work in their respective parishes. Fr Flannery, for example, directed the building of Mweenish Island causeway, achieving excellent results. Foremen singled out by the Commissioners for special praise included Mr P. Toole of Lettermore Island, Mr Coneys of Aughrusbeg (who replaced Geraghty when the latter had to return to Perry's work) and Mr Nixon of Lettermogera.[36]

The tunnels in the base of Aughrusbeg wall were largely closed up later and today the place is occupied by a lobster-holding operation, taking up the whole southern section of the original channel. The section north of the dam has become completely filled with blown and drifted sand; the closure of the tunnels converted that part of the hitherto open channel into a blind *cul de sac* where sand accumulated, forming a raised sandy beach. There were fifteen risers in the original stone steps built into the north face and these are now completely buried in sand, which reaches up to the deck level of the dam walkway.

Some years later, in 1892, Mr Robert Ruttledge-Fair would survey the site for the Congested Districts Board. He considered that the northern section had been too shallow and too narrow from the very beginning and he recommended that this should be remedied; quite how, he did not say, and by then it was far too late for any useful or effective remedy. A slip, made of large flat slabs, lies partly buried in sand high up on the shoreline of the northern section, and the remains of a winch for drawing up boats are located at the top of it. The exact location of the slip suggests that the sand had already filled the upper section in the two years after the dam was first built. The slip was made by Fraser in 1888, at the time that he added a parapet to the breakwater. The section south of the dam is still open to the sea, but partitioned into lobster-holding pens, and it still fills and empties with the tide.

Clais na nUan (Teeranea) Harbour and Caladh Thaidhg (Collaheigue) Harbour

Completely new marine structures could be a boon to any coastal area. Extensions or modifications to pre-existing structures could also alter a site in a fundamental way, which could add significant benefit. For example, a pier already existed at Teeranea (*Clais na nUan*) on Gorumna Island before the P&R Commission arrived. No manmade structure is shown there in the first edition of the six-inch OS map (Nimmo's 1823 Maumeen pier, a little distance to the north of it, is shown), but an indent of the shoreline shown on the map suggests that *Clais na nUan* (its name in Irish translates as 'the foamy channel') was suitable at the time for use as a natural landing place. A petition for a pier 'at Greatman's Bay' had been refused by the first Commission for Irish Fisheries in the 1820s. That was probably for Clais na nUan and the refusal may have resulted because the pier at Maumeen nearby was already underway. The first manmade quay and pier at *Clais na nUan* must have been erected there some time after 1839. A possible date of construction in the 1840s is suggested by reference to another harbour in the locality, *Caladh Thaidhg* (Collaheigue) near Carraroe, reputedly built in the 1840s.[37] However, Major Gaskell of the CDB, claimed that a pier at *Clais na nUan* was first built only in 1880,[38] probably under his own supervision during that summer, using funds from the Duchess of Marlborough Relief Committee.[39] He had built a small pier at *Caladh Thaidhg* that same year under the Committee's relief scheme.[40] *Céibh Clais na nUan* and *Céibh Caladh Thaidhg* lie across from each other on opposite shores of Greatman's Bay, and boat crossings between them would have been a great convenience to people on both sides.

Whatever their true date of origin, the quay and pier at *Clais na nUan* were 'improved, raised and strengthened, and a breakwater thrown up outside' by the P&R Commission in 1886.[41] The core of the original quay and pier (now almost completely encased in

FIGURE 8.4. The breakwater erected by the Piers and Roads Commission at *Clais na nUan* in 1886 and subsequently raised above high water by Col. Fraser in 1887. Photo NPW.

modern concrete) are, therefore, older than the Commission's renovated surfaces of them. The 1886 breakwater was an entirely new construction, 240 feet long, made of rubble stone and built so low it was normally inundated at high water. Two years later, Col. Fraser would raise it fully above high tide level, greatly improving its appearance and effectiveness.[42] It extends outwards from the shore opposite towards the pier, converting the site into a distinct harbour. It is the juxtaposition of the new breakwater and the original pier that made the place really useful and gave it the attractive rustic character it had at the time of the Ryan Hanley survey in the year 2001. Today the harbour has been considerably changed with heavy rock armour and modern concrete works so as to be hardly recognisable.

Obviously, it is patently misleading to give a single date of origin to a complex and changing site like this. All harbours evolve over time, expanding as they become more important and more specialised for trade, communication, fishing or leisure activities and decaying as they go out of use. A similar situation – a new pier erected alongside an existing small jetty, converting a simple jetty site into a small harbour – occurred also at Leenaun in Killary Harbour, as we saw in chapter 5, and at Roundstone, as we shall see in chapter 11. Happily, at the latter two places, modernisation has not yet been as severe as at *Clais na nUan*.

Other Works of the Piers and Roads Commission

Ten works of the P&R Commission were extensions, modifications or improvements to sites where artificial harbours or landing places were already known to exist, such as Garafin harbour in Rosmuc (No. 088), *Sruthán* (Sheffaun) pier (No. 133) and Crusheen wharf (No. 135), the latter two located in Cashla Bay. In these and other places, the fact that the Commissioners referred to them as 'already existing' is useful confirmation that some landing facilities already existed there before 1886.

Naturally enough, the Commission's works of improvement were small, and they varied in cost from £20 for Glantrasna ('Small landing place in dry granite work, ninety

FIGURE 8.5. Garafin harbour at low tide. Photo NPW.

feet long. Sub-contract') to £556 for *Leiter* pier in Ballinakill Harbour ('This work is a concrete breakwater, 170 feet long, forming a shelter for boats during westerly gales; there is also a slip-landing. The hamlets here have no road communication with the neighbourhood').[43] Unusually for the Commission, no access road was made to *Leiter* pier, despite the absence of roads in the neighbourhood being pointedly noted in their report. The pier itself is a narrow, stepped breakwater, rather than a true pier with a wharf wall; the slip alongside is also narrow and surfaced in stone. There is still no road to the site today. Some other works of the Commission will be touched upon in later chapters.

Endnotes

1 Col. T. Fraser., Report to the Lord Lieutenant, BPP 1889 [C. 5729].
2 46 & 47 Vict. C. 24.
3 Act 46 & 47 Vict. C. 24, section 1.
4 Bill 155 of 1886.
5 Chief Secretary John Morley, House of Commons debates, 1/4/1886, Hansard, vol. 304, third series, cc 566ff.
6 Chief Secretary John Morley, House of Commons debates, 1/4/1886, Hansard, vol. 304, third series, cc 577.
7 49 Vict. C. 17.
8 Report of the Piers and Roads Commission, BPP 1887 [C. 5214], p. 5.
9 Report of the Piers and Roads Commission, BPP 1887 [C. 5214], appendix 4, p. 17.
10 Report of the Piers and Roads Commission, BPP 1887 [C. 5214], p. 6.
11 Report of the Piers and Roads Commission, BPP 1887 [C. 5214], pp. 8–11.
12 Report of the Piers and Roads Commission, p. BPP 1887 [C. 5214], 11.
13 A. Boxer, in Report of the Select Committee on Harbour Accommodation, BPP 1883 (255), query 2101,
14 Report of the Piers and Roads Commission, BPP 1887 [C. 5214], appendix 1, p. 13.
15 Report of the Piers and Roads Commission, BPP 1887 [C. 5214], appendix 2, p. 14 .
16 Report of the Piers and Roads Commission, BPP 1887 [C. 5214], p. 11.
17 Report of the Piers and Roads Commission, BPP 1887 [C. 5214], p .6.
18 Col. T. Fraser, Report to the Lord Lieutenant, BPP 1889 [C. 5729], p. 3.
19 Col. T. Fraser, Report to the Lord Lieutenant, BPP 1889 [C. 5729].
20 Ibid., p. 6.
21 Ibid., p. 5.
22 Ibid., p. 7.
23 Ibid., p. 7.
24 Ibid. pp. 4–5.
25 *Sea Fisheries (Ireland) Act 1883*, 46&47 Vict. C. 26, section 5; *Relief of Distress Act 1880*, 43 & 44 Vict. C. 14.
26 59th Annual Report of the OPW, BPP 1890–91 [C. 6480], p. 26.
27 50th Annual Report of the OPW, BPP 1882 [C. 3261], p. 33 and p. 68.
28 Great depression in UK.
29 Col. T. Fraser, Report to the Lord Lieutenant, BPP 1889 [C. 5729].
30 Seamus Hickey of Barna, personal communication, 2015

31 Archive, Galway County Library. GC/CSO/2/58.
32 J.P. Griffith and J.W. Griffith, *Western Harbours of Ireland. An Enquiry into the Suitability of various Bays and Estuaries on the West Coast of Ireland for the formation of a Commercial Port and Naval Base, March 1918.* Privately printed. (available in NUI Galway, Hardiman Library, SPCOL 387.109415Gri.).
33 Report of the Piers and Roads Commission, BPP 1887 [C. 5214].
34 NAI OPW/8/15.
35 Report of the Piers and Roads Commission, BPP 1887 [C. 5214], p. 13.
36 Report of the Piers and Roads Commission, BPP 1887 [C. 5214], p. 15.
37 Robinson T., *Connemara, Part 1*, p. 121.
38 W.P. Gaskell, Report to the Congested Districts Board, 26 May 1892. Reprinted in Morrissey, *Verge of Want,* p. 183.
39 G.D. Morant, Report on the Relief of Distress on the West Coast of Ireland, BPP 1880 [C. 2671].
40 Minute of the Engineering Department of the OPW. NAI OPW/5/3815/15.
41 Report of the Piers and Roads Commission, BPP 1887 [C. 5214].
42 Col. T. Fraser Report to the Lord Lieutenant, BPP 1889 [C. 5729], p. 9.
43 Report of the Piers and Roads Commission, BPP 1887 [c. 5214], p. 10.

CHAPTER 9

Retrospective on Troubled Waters: An Awkward and Delicate Position

We cannot shut our eyes to the fact ... that there has been often serious personal friction between the two departments such as ought not to exist between different servants of the Crown ... and furthermore that the Irish Executive ought to have taken care that the two bodies acted together, and that no works were carried out without the full approval of the professional advisers of Government in matters of engineering. The result has been that sites have been selected, estimates made, and the amount of money to be expended at various spots settled, without any competent professional advice, and without any proper certainty that the sums spent would produce any valuable result.

Report of the Royal Commission on Irish Public Works, 1888[1]

A Restricted Flow of New Blood

From the Fisheries Act of 1842 onward, more than twelve statutes concerning Irish fisheries and fishery piers had been enacted by Parliament. Aspects of the administration of these laws had been investigated by at least six distinct committees: the Select Committee on the Sea Fisheries of the UK in 1863, the Select Committee on the Sea Fisheries (Ireland) Bill in 1867, the Lansdowne Enquiry of 1872, the Departmental Committee on Irish Public Works (the Crichton Committee) in 1877, the Select Committee on Harbour Accommodation in the UK in 1883 and the Royal Commission on Irish Public Works (the Allport Commission, see below) in 1887. The great decade of pier building from 1879 to 1889 alone saw the operation of three separate Commissions on fishery piers: the Fishery Piers Commission (FPC) of 1879–80, the Fishery Piers and Harbours (FPH) Commission of 1883–89 and the Piers and Roads (P&R) Commission of 1886–7.

Notwithstanding all the laws and their administration, and the huge increase in the number of fishery piers and quays, the numbers of boats and fishermen showed a severe decline from the Great Famine onwards. The provision of piers, therefore, had not fostered growth of the Irish fishing industry in any real way. No one was sure to what extent, if at all, this disappointing outcome should be laid at the door of the various Commissions and Select Committees, the Inspectors of Fisheries, the OPW, the Grand Juries or the County Surveyors. The decade of the 1880s is therefore an appropriate period from which to look back and review what had been done and to discuss the persons and events that had dominated the fishery piers scene since the 1842 Act.

A relatively small number of persons had been involved during that long period. Looking back from 1877, Col. John Graham McKerlie (1814–1900), who had joined the Board of the OPW as a Commissioner in 1855, was still First Commissioner and Chairman, a post he had taken up in 1864.[2] William Richard Le Fanu (1816–94), was the Second Commissioner; he had joined the OPW in 1863. Samuel Ussher Roberts (1821–1900) was Assistant Commissioner, having joined as an engineer in 1841 and (after a period as County Surveyor of Galway) had been appointed Assistant Commissioner following the recommendation of the Lansdowne Enquiry in 1872.[3]

Almost incredibly, Sir Richard Griffith (1784–1878), who had entered public service with the Bogs Commission in 1809 and was in his 93rd year in 1877, was still the third, albeit Honorary, Commissioner for Public Works, although he had not darkened the door of the office since retiring as Chairman in 1864. Edward Hornsby (*c.*1810–90), who had entered public service in 1837, was still Secretary of the Board, a post he held from 1849. The engineer Robert Manning (1816–97), who designed and directed the pier works after 1870, was Chief Engineer from 1874 on. In 1877, he was a relatively young man of sixty-one years of age, whose first work with the OPW dated from 1846 when he spent ten of his early professional years working mainly on drainage issues.[4]

On the fisheries side, Thomas F. Brady (1823–1904), John A. Blake (1826–87) and Major Joseph Hayes (? –1889), were the Inspectors of Fisheries. Brady had joined the OPW in 1846, around the same time as Manning, and was attached to its fishery division from 1850 until he became Secretary to the Special Commissioners for Fisheries from 1864 to 1868. Appointed an Inspector of Fisheries in 1869, he had served continuously in fisheries administration since 1850, so he already had twenty-seven years of practical experience in the field. John A. Blake was both a writer on fisheries matters and a politician.[5] He was MP for Waterford city from 1857 to 1869, Inspector of Fisheries with Brady and Hayes from 1869 to 1878, MP again from 1880 to 1884, and yet again from 1886 to his early death in 1887. Joseph Hayes was already eight years in office as an Inspector of Fisheries by 1877.

Given the restricted flow of 'new blood' into both services that these facts indicate, it was almost inevitable that some of the personnel would have grown comfortable with – and some maybe resentful of and antagonistic to – practices that had become customary over the thirty-five-year interval since 1842. It is appropriate, therefore, that any retrospective should include the publicly espoused views and opinions of these officials, before we come to meet the new *dramatis personae* and the new institutions (such as the Congested Districts Board and the Department of Agriculture and Technical Instruction) that would come on the scene in and after the 1890s.

Over-Strict Legal Interpretation

The absence of a clearly defined policy, aggravated by anomalies in the early legislation, contributed significantly to the complexity and the uncertainty associated with the various pier-building schemes since 1842. When giving the Commissioners for Public Works the role of Commissioners for Fisheries, the 1842 Fisheries Act had expressly severed fishery piers from their new statutory fishery role. Nevertheless, the Commissioners, in their 'normal' capacity as Commissioners for Public Works, continued to have responsibility for the piers (both fishery and commercial), which had been given to them by the 1831 Act.[6] But neither Act had imposed any positive obligation on them to do anything constructive about piers. Nor had the 1842 Act voted any financial resources for fisheries or for fishery piers.

The Piers and Harbours Acts of 1846 and 1847 were the first Acts of the modern legislation that provided money for grants in aid of fishery piers. The latter Acts also made the Board responsible for the design and construction of the structures that would be grant-aided, and laid down in great detail the procedure to be followed in allocating and administering the relevant funds. However, the OPW could not initiate, on its own behalf, any proposal for any pier. That power was reserved in the 1846 Act exclusively to 'any person residing in or being proprietor or occupier of land in any district or place adjacent to the sea coast of Ireland.'[7]

In this way, and in the words of Inspector of Fisheries James Redmond Barry, the 1846 Act made the Board 'an engine', rather than the driver, of pier development.[8] In routine practice, the Board of Public Works consulted the Inspectors of Fisheries (who, from 1842 to 1869, were members of its own staff) only with regard to the general proximity to known fishing grounds of any location proposed for a fishery pier. That apart, the Inspectors had no formal input into, or any authority to comment on, the precise location, structure, design, cost or advisability of any proposed pier; strict interpretation of the exclusion clause of the 1842 Act, quoted earlier (chapter 3), proscribed them from having any say whatsoever. Indeed, the Inspectors seemed to give very little advice of any sort to the Board regarding the sea fisheries or the piers; whatever information the Board did receive came mainly from the Coastguard.[9] Barry admitted that, as a Fishery Inspector, he made 'only a kind of desultory visitation' to the sea fisheries; that most of his time was occupied with inland fisheries; and that there were no standing orders that he should carry out periodical visits to the fisheries, or to the fishery ports, at any time or season.[10] 'To be candid,' he ventured, 'I have at all times considered that our Department was a species of delusion.'[11] He went on to affirm: 'I think it is rather unfortunate that there existed any

TABLE 9.1

Number of boats, men and boys engaged in the coastal fisheries from 1846 to 1876. Note the decline in both over the period.

Year	Total boats	Men and boys	Year	Total boats	Men and boys
1846	19,883	113,073	**1861**	11,845	48,624
1848	19,652	81,717	**1862**	11,590	50,220
1849	18,100	71,505	**1863**	11,375	48,601
1850	15,247	68,380	**1864**	9,300	40,946
1851	14,756	64,612	**1865**	9,455	40,802
1852	11,789	58,863	**1866**	9,444	40,663
1853	12,381	49,208	**1867**	9,332	38,444
1854	11,079	49,227	**1868**	9,184	39,339
1855	11,351	47,854	**1870**		
			1871	9,099	38,650
1856	11,069	48,774	**1872**	7,914	31,311
1857	12,758	53,673	**1873**	7,181	29,307
1858	11,823	52,101	**1874**	7,246	26,924
1859	11,881	50,115	**1875**	5,919	23,108
1860	13,483	55,630			

Source: data from Report of the Inspectors of Irish Fisheries on the Sea and Inland Fisheries of Ireland for 1875, p. 4. BPP 1876 [C. 1467].

Department at all ... even the very Coastguard, who are our executive Department, in their reports and returns, in answer to queries, invariably complained of the want of any species of encouragement'.[12]

It was a sad state of affairs for the fishery piers and the fisheries in general, and no wonder that so little was done once the first wave of relief works in the Great Famine had ended. The post-famine period was, generally speaking, a lean one for fishery pier construction and, indeed, for fishermen, the numbers of men and boats continuing to decline year by year from 1846 to 1875, as recorded by the Fishery Inspectors (Table 9.1).[13]

William Joshua Ffennell, who had joined the OPW as an Inspector of Fisheries in 1845, left for England in 1860 to become a Commissioner for Fisheries there, and Thomas F. Brady was 'appointed specially as Acting Inspector' in his place.[14] Up to then, Brady was clerk to the Inspecting Commissioners (the posts to which both Ffennell and Barry had been elevated). Secretary of the Board, Edward Hornsby, seemed to be particularly zealous in emphasising that Brady had never been made an Inspector, let alone an Inspecting Commissioner, within the OPW. Nor was Brady ever subsequently made a full Inspecting

Commissioner, despite his long service under Barry, a failure that cannot have been entirely fulfilling for him. Before and during his tenure of office as Acting Inspector from 1860, the OPW was in serious decline [15] which, along with his own lack of formal career advancement, can only have exacerbated his general dissatisfaction.

In 1863, he left the OPW to take up the post of Secretary to the newly appointed Special Commissioners for Irish Fisheries. That body was one set up to regulate the inland, principally the salmon, fisheries for a few years. It ended abruptly within three years, damaged irretrievably by internal dissent between its members, the worst negative consequences of which Brady was fortunate enough to avoid. At the same time, he was also Secretary to the Special Commissioners for Inland Fisheries in England,[16] but that did not impact noticeably on his Irish activities.

When John A. Blake's first Sea Coast Fisheries (Ireland) Bill[17] came before a Select Committee of Parliament in 1867, Brady's evidence enthusiastically supported the proposal in it to have the fisheries, and the fishery piers, removed entirely from the OPW's control. His general remarks exhibited a profound antipathy to the involvement of the OPW in fisheries, and also hinted at some personal animosity towards J. Redmond Barry, his erstwhile superior. When the Sea Fisheries (Ireland) Act 1869[18] transferred responsibility for all the fisheries, both sea and inland, to three new officers termed 'Inspectors of Irish Fisheries' operating under the direction of the Lord Lieutenant, it was very much what Blake and Brady wanted, and what they had worked assiduously towards in the Select Committee that had examined the Bill.

Although the Act was not as complete a victory as they might have wished, it was sufficient for that day, especially since both men were appointed by the Lord Lieutenant as two of the new Inspectors of Irish Fisheries. The third post went to a Major Hayes. Responsibility for the fishery piers and harbours, however, remained vested in the OPW, an outcome not greatly to the new Inspectors' liking. However, they were given a tentative foothold in the matter when charged by section 18 of the Act to 'report for the information of the Commissioners of Public Works in Ireland in regard to the necessity for, and the advantage to be derived by the fisheries from any work which may be proposed to be carried out under the provisions of the [Piers and Harbour] Acts'.[19]

Rumblings of Discord

Although the 1869 Act gave the Inspectors this independent, statutory, small footing in the matter of the fishery piers, the wording of the section (section 18, quoted already in chapter 5) was not at all satisfactory, especially when subjected to the very strict and illiberal interpretation that characterised decisions of the OPW in the 1870s.[20] It was the over-strict legal interpretation of the exclusion clause twenty years earlier that first caused the Board to exclude the Fishery Inspectors – its own staff – from any meaningful input into discussions or decisions regarding the fishery piers. Under the 1869 Act, the new procedure became one where, on receipt of a memorial for a pier from a coastal applicant,

The remains of the quay at Furbo erected by Andrew Blake with tools supplied by Alexander Nimmo in 1823. Photo NPW.

Maumeen pier (RH 121), Garmna Island, *Ceanntar na nOileán*, in 2009. Photo NPW.

The 'paved road' through the intertidal shore at *Béal an Daingin* is still clearly evident when the seaweed is pushed aside. Photo NPW.

The sea slope at *Spidéal* pier today. Photo NPW.

Photo taken from the upper boat quay in Clifden at low tide. Note the half-tide barrier of 1869 on the left, with the navigation beacon erected at its near end in 1895. The ship quay is shown on the right, and note also the angle between the boat quay (foreground) and the ship quay. Photo NPW.

Forramoyle slip (*Céibhín na gCurach*) (RH 144X2). The low, granite wall is evident in the centre; to the left of it is the slipway, now largely filled with loose stones. To the right of the wall the sea slope is somewhat broken up but still evident.

A small roadside quay (right-hand side) and disengaged pier/breakwater (left-hand side) at Rusheen Maumeen (RH 101), Garmna Island, *Ceanntar na nOileán*. Photo NPW.

The old pier and harbour at *Sruthán Bhuí*, Kilbrickan (RH 083), in 2014. Photo NPW.

The breakwater at Glasillaun (RH 005). Note the stepped nose and the central section, never completed. Photo NPW.

The pier at Glengevlagh (RH 001) in Killary Harbour. Note the central wall extending the full length of the pier. Photo NPW.

the OPW informed the Lord Lieutenant, who then directed the Inspectors of Fisheries to hold an enquiry in the relevant locality, and to report whether the pier proposed was one that would be important to the fisheries. On receipt of a positive recommendation from the Inspectors, the OPW would carry out a full site survey and draw up a suitable plan, design and estimate. It would then forward these, with its own recommendation regarding funding, to Treasury for sanction. When sanction was received, work could commence, either by contractors or directly by the OPW. Since the Inspectors were an independent body under the 1869 Act, their initial recommendation was all-important (the statute specifically required it) and, once that recommendation was positive, the OPW did not query it at all. When asked whether, in the event the Board of Public Works itself thought any pier would not be of much use to the fisheries, the Board would still recommend it to Treasury. William Le Fanu was unequivocal:

> We do, we have to do so. Under former Acts we made the whole investigation, and if we thought that a pier was applied for that would do no good to the fisheries we would not recommend it. But when that power is taken away from us [by the 1869 Act] if they [the Inspectors] recommend a pier we don't go into the matter ... We are, by the Act, relieved from that altogether.[21]

When the boot was on the other foot, as it were, for example, when the OPW was considering the most suitable design or size of a pier, the Fishery Inspectors were not invited to comment on anything the OPW proposed. As Le Fanu asserted: 'We never ask them [the Fishery Inspectors] anything connected with works. We never do; it would be entirely an engineering matter'.[22] Brady confirmed that this strict interdepartmental division existed: 'The Board of Works have their engineers and they carry them [pier works] out according to the plans approved by the engineers. The Inspectors of Fisheries have nothing to do with the approval of the plans or in fact with the plans at all, or with the carrying out of the works.'[23]

The Crichton Committee Report

The Crichton Committee was set up in 1877 to examine the constitution and duties of the OPW, which was widely considered to have been dysfunctional for many years. When Crichton reported in 1878, he gave the Board a reasonable bill of health, at least with respect to the fishery piers.[24] However, noting the longevity and long service of members of the Board and its servants, the Committee made some hard proposals regarding them. It recommended that Griffith be allowed to relinquish his Honorary Commissionership, thereby facilitating Samuel Ussher Roberts's promotion to full Commissioner status, and restoring the Board to its statutory complement of three fully active members. Griffith, in fact, died later that year soon after Roberts's promotion. The Committee further proposed that the Secretary of the Board, Edward Hornsby, should retire on pension; he would do so, eventually, five years later, in 1883. More controversially, it recommended that the Board chairman, Col. J.G. McKerlie, should be offered the opportunity to retire on a good pension, with plaudits for his long and productive service on the Board.[25] Viscount Crichton did not support this last recommendation of his own Committee, which he felt did an injustice to McKerlie's laudable tenure of office.[26] McKerlie stayed on until 1884, retiring then with both dignity and a knighthood. Independently of these changes, Blake resigned from his Fishery Inspector post in late 1879 to return to politics and was elected to Parliament in 1880 as the Member for Waterford County. He was replaced as Inspector by Joseph Johnston. The changing of the old guard was getting underway at long last and in earnest.

Further Discord under the Relief of Distress Act 1880

From the time that the fisheries were removed from the OPW in 1869, clear lines of demarcation, nascent within the OPW since about 1845, increased between the separate departments, driven by their strict interpretation of their distinct roles and responsibilities. It was a situation unintended and unforeseen in the legislation, but one that reduced relations between the Fishery Inspectors and the OPW to a formal, barely civil level. Dividing responsibility between two separate departments in the uneven way that the law seemed to demand proved to be counter-productive. Professional advice and input from a fishery expert (on the proximity and accessibility of proposed pier sites to the fishing grounds and to fish markets), a nautical person (on the range, strength and direction of the tides, currents and winds) and a marine engineer (on the physical and topographical features of differing sites and the kinds of structures that were most appropriate to them) were all essential for a proper decision on any pier application. Neither the OPW nor the Inspectors of Fisheries, each on its own, had this full suite of skills, but together they had a good many – if only they could co-operate amicably.

Treasury must have been aware early on of a certain unsatisfactory tension when it determined that a balanced Fishery Piers Committee (FPC), comprising a member of the OPW (Le Fanu), a Fishery Inspector (Brady) and a neutral member of the Local Government Board (Henry Robinson) was needed to select the piers to be aided under the *Relief of Distress Act* of 1880.[27] Treasury even went one step further, insisting that a 'nautical man' recommended by the Admiralty, namely Staff Commander Charles Langdon of the Royal Navy, be added to the FPC, to complete the range of skills deemed requisite for the success of its operation.[28] Langdon looked after the nautical aspects and selected the precise sites: 'I selected the spot, as a rule, trying to put it as near as possible in the place where the fisherman wanted it. Then the OPW engineer surveyed the ground and from that, made a plan and an estimate. If the plan could be implemented within the money subscribed or allotted by the Board, it was done.'[29] Although an improvement, this compromise arrangement had rankled with Brady. There was still, he complained, no independent review of the designs drawn up by the OPW, and no independent monitoring of the progress of the works.

An Awkward and Delicate Position

Brady did not have too long to wait for an appropriate occasion to air his dissatisfaction in public. His friend and erstwhile Fishery Inspector colleague, John A. Blake, had been re-elected to Parliament in 1880 and he, Blake, along with others, brought in the Sea Fisheries (Ireland) Bill in February 1883 (chapter 7). As drafted, the Bill proposed, *inter alia*:[30]

> The plans and specifications for all [piers and harbours] works to be undertaken shall be first approved by the Commissioners under this Act, and be submitted for the approval of the Lord Lieutenant before such works are undertaken ... All statutes now in force for such works to be executed or for their maintenance shall remain in full force and effect ... and all the powers, rights, privileges, and duties vested in or exercised by the Commissioners of Public Works under said Acts shall be vested in and may be exercised by the Commissioners of Irish Fisheries under this Act.

If passed by Parliament without alteration, that would have meant the removal entirely of the OPW from the business of fishery piers, to be replaced by the new Fishery Commissioners proposed in the Bill. When this issue was taken up by the Select Committee on Harbour Accommodation in 1883, it created the opportunity that Brady grasped.[31] In the course of his extensive evidence he alluded to the plans and designs of piers: 'I am always fearful about those designs. I think we always want some marine engineer to be consulted with regard to those harbours. I think that is a most important thing'.[32] Did he not realise that this comment could be construed as criticism of the engineers of the OPW with regard to marine constructions? When pressed further to say whether it was regrettable that more care was not taken to see that the money spent on piers was profitably expended, he waded in even deeper:

> I agree all together with you on that; but it is a question which the committee will see places me in rather an awkward and delicate position in offering an opinion with regard to another department, the Board of Works. It might look as if I were making some unpleasant commentaries upon either their engineer or upon the Board of Works itself, which I would rather be saved from if possible.[33]

That, naturally enough, introduced a more personal tone to the proceedings. Robert Manning, the Board's Engineer who was responsible entirely for the plans and estimates,[34] responded calmly enough that he had no problem with Brady expressing his opinion, but he himself could only say that:[35]

> no design of the Board of Works has failed, and I would be very glad if Mr Brady would put forward a series of specific complaints; do not let him mind my feelings at all, but let the complaints be direct and specific, and not made by innuendo which has not been the case in this instance.

Manning was concerned that his own competence or dedication was being impugned, and that the Bill, if passed as it stood, would result in his replacement:

> I can well see from the examination before the committee, that there will be a marine engineer instead of the engineer of the Board of Works. I can only say, personally, that my duties have been very arduous, and they have been altogether unpaid for; and although, personally, I would be delighted to be relieved of them, the only thing that I would avoid would be running away from a duty that is cast upon me, which I shall never do, or being sent away because I was incompetent, and that I would not like.[36]

Blake brought these fears out into the open during the questioning, asking him what had evinced this negative disposition? Manning instanced Brady's evidence, and when Blake explained that the appointment of a marine engineer, or a navigating naval officer, might be helpful to him, Manning replied 'I do not think so, because on numerous occasions I cannot fail to find that Mr Brady has not the slightest confidence in me or in anything that I do.'[37] To pour oil on clearly very troubled waters, Blake averred that he and the other Irish members of the Select Committee had no other belief but that Manning was a most competent and most zealous, but too much overworked, officer. Manning's reply was waspish, to say the least:

> You can compensate me for overwork by giving me, not overpay, but pay, for you do not pay me at all ... I have done all the work that has been put upon me, and I am bound to say, without any want of modesty, that I have done it well, and no person has been able to say that I have not.[38]

In fairness, Blake responded graciously and unequivocally that he would be happy to have Manning administer the whole £250,000 envisaged in the Bill, and Manning, for his part, was satisfied that he could carry out this huge expenditure 'although subject to great responsibility and anxiety, and occasionally to being urged to perform impossibilities'.[39]

Later in his evidence, Manning seemed to impugn Commander Langdon's reputation, saying that Langdon, 'imagined what Admiralty officers often do, that we landsmen know nothing at all',[40] whereas he, Manning, claimed to know more about the force of waves than a nautical man generally did. If he needed information on the direction and the height of the tide, or the wind, he would go to a nautical man and ask him, but if he could find it out more accurately himself he would not go to him. To be fair, Manning was a most competent engineer of very high repute, especially regarding hydrology: he was the eponymous author of Manning's Formula for Open Channel Flow, a formula used throughout the world to this day for the calculation of fluid flow in open channels.[41]

Discord Escalates under the Sea Fisheries (Ireland) Act of 1883

As it turned out, the Act that eventuated – the *Sea Fisheries (Ireland) Act, 1883*, abbreviated here as the SFI 1883 Act – pitted the various protagonists against each other once again, mainly due to obscurities in its wording. Blake had triumphed in successfully getting £250,000 allocated to the Irish fishery piers and harbours by the Act. While that outcome had never really been in serious doubt, its confirmation in legislation provided the single, biggest financial contribution to Irish fishery piers and harbours that they had ever received. He had failed, however, with regard to its administration: for this, the Act allocated the funds to the Commissioners of Public Works, to be applied by them for the purposes of the Piers and Harbours Acts.

Under the SFI 1883 Act, the Lord Lieutenant set up the Fishery Piers and Harbours (FPH) Commission, whose duties were laid out in Section 8:

> It shall be the duty of the Fishery Piers and Harbours Commission to give such assistance to the Commissioners of Public Works as the Inspectors of Irish Fisheries have hitherto given in the execution of the Fishery Piers and Harbours Acts; and to confer with the Commissioners of Public Works relative to the works proposed from time to time to be executed out of the Sea Fisheries Fund; and generally to aid in carrying this Act into effect, in such manner as the Lord Lieutenant may from time to time direct. [42]

Quite what 'the assistance they have hitherto given' actually meant was not, as a legislative provision, at all precise or obvious. It most certainly did not help in resolving the issue of co-operation that clearly existed at the time. The FPH Commissioners, comprising the existing Inspectors of Fisheries (Brady, Hayes and Johnston), with Blake as Chairman, were the very persons who were most involved in the dispute with the OPW. The requirement 'to confer with', and 'to aid' the OPW was, presumably, meant to encourage and facilitate greater co-operation between them.

However, it failed to acknowledge or appreciate one fundamental incompatibility underlying the discord: the OPW was a department answering directly to Treasury in Whitehall – in essence, as Manning and Joynt pointed out separately, the OPW was a Treasury Board[43] – but the FPH Commission, and the Inspectors of Fisheries who comprised it, came directly under the Lord Lieutenant, in effect the Irish Executive. It was almost inevitable that this 'dual responsibility' arrangement would complicate, and very possibly hinder, free and frank communication between the separate departments, as subsequently proved to be the case.

Section 8 did, however, go on to remove from the OPW a little of its exclusive control over the plans and specifications, by giving the new FPH Commissioners a formal role in assessing them in all applications: 'The Commissioners of Public Works, before forwarding to Treasury the report they are required to forward under S10 of the 1846 Act, shall furnish to the Fishery Piers and Harbours Commission a copy of the plans and specifications and the Fishery Piers & Harbours Commissioners may make such observations for Treasury as they see fit.' This was a significant intercalation of the FPH Commission into the area of competence and responsibility of the OPW that Brady seems to have overlooked in his evidence to the Allport Commission:

> The Piers and Harbours Commission was only appointed by Act of Parliament in 1883 for allocating the expenditure of the £250,000 that was granted and we were made *ex officio* members of that Commission along with Mr Blake, the Member of Parliament. The Board of Works were given all the powers for building these harbours; we were given none. We had nothing to do but to recommend the sites. I felt originally the great difficulties we should have to encounter when we first commenced to work upon that Commission without independent engineering advice. [44]

Certainly, the provision appeared on the face of it to imply that the FPH Commissioners would have some meaningful influence on, or input into, the plans, precisely as Brady had always wanted. In fact, they immediately sought to employ their own independent engineer to help them with their assessment of the plans. The Viceroy, however, would not concede that such expertise was necessary, except in very special cases.[45] Brady felt this refusal constrained the FPH Commissioners excessively, since they were not themselves engineers, and without independent engineering advice they could not properly evaluate the plans and estimates that the OPW sent them. Indeed, they complained that they lacked the power even to question the precise location for any pier: the 1883 Act, they pointed out, referred them back to the 1846 Piers and Harbours Act, which expressly laid down that applications could come only from coastal residents or landowners.[46]

While the Commissioners could, and did, hold public meetings in the relevant localities from which memorials were received, they had no power, on a strict interpretation of the statutes, to change the locality, or the precise site within the locality, or to choose sites they thought would be better. So they reverted merely to reporting to the OPW whether or not, from a fisheries point of view, a pier was necessary in any locality in question – and they usually found that it was. Further co-operation between the departments then seems to have declined very quickly to a very low level. As for the duty to confer with the Board, the FPH Commissioners never met the Commissioners for Public Works face-to-face but 'conferred with them on the piers and harbours question wholly by correspondence'.[47] Even that formal contact was soon to sink into absurdity, as Brady admitted:

> We really got so little [information from the OPW] that on one occasion I said, I remember, 'There is no use our asking any more questions about the plans, we cannot get any information"; and then we threw the responsibility on the Board of Works and their engineer. We actually commenced to approve of the plan before submitting it to the Lord Lieutenant, so that that plan would be put down and not be departed from. We wrote 'approved' on it, and the chairman signed it; and when we went on a little

> further, and could not get the information from the Board that we considered necessary, we then abandoned approving the plans, and threw the responsibility on the Board of Works engineers themselves, and from that out we never approved of a plan.[48]

For his part, Manning took to accepting the initial reports of the FPH Commissioners at face value, without ever even reading them – not even, he said, as a matter of curiosity![49] With this complete breakdown in communication and patent lack of meaningful co-operation, it is no wonder that a body entirely independent of both the OPW and the Fishery Inspectorate – the Piers and Roads Commission – was set up to administer the funds of the 1886 *Relief of the Destitute Poor Act.* Yet, despite all the discord, the number of structures (sixty-one) that were completed under the SFI Act is impressive, and more money was spent on them over a three-year period than had been expended over the previous three decades. Whether they were useful, properly sited, and of the right size, or whether they did any real good or not, was an entirely different matter: depending on the perspective of the commentator, quantity does seem to have replaced quality.

The Allport Royal Commission

Needless to say, the persistence of the strained relationship did not go unnoticed when the operations of the OPW came under scrutiny in 1886 by the Royal Commission on Irish Public Works, known also as the Allport Commission.[50] That Commission comprised the civil engineers John Wolfe Barry (1836–1918) and James Abernethy (1814–96); and Joseph Todhunter Pim (1841–1925), the only Irish member, a Quaker, Liberal businessman with interests in socioeconomics. Sir James Allport (1811–92), an English railway magnate, was Chairman. Wolfe Barry was one of the foremost English engineers of his day. Coincidentally, he had trained under Sir John Hawkshaw, who himself had trained as an engineering assistant in Alexander Nimmo's Liverpool office sixty years earlier. The Royal Commission examined the issues of drainage and inland navigation, the organisation and management of the Irish railway system and, in its second report, it addressed the question of fishery piers, one of the principal objects of its terms of inquiry:

> [To inquire into] the extent the harbour accommodation on the coast of Ireland, either completed or in course of construction, meets the requirements of vessels suited for deep sea fishing; and whether that industry can be promoted by the construction of new harbours, the improvement of existing natural or artificial shelter, the provision of better means of communication with markets, or in any other manner which may appear desirable.[51]

Having reprised the evidence of the various witnesses regarding the existing and historical state of the fisheries, and the *modus operandi* of the Inspectors of Fisheries and the OPW, the Commission was uncompromising in its analysis, which is worth quoting at length:

> We cannot shut our eyes to the fact, disclosed by this and other circumstances, that there has been often serious personal friction between the two departments such as ought not to exist between different servants of the Crown ... We must say that we fail to see why, with some personal forbearance, the Inspectors and the Board of Works should not have co-operated with the view to securing the best possible expenditure of the available funds. We can, to some extent, understand the action of Government in refusing to engage a special engineer to assist the Commission under the Act of 1883 ... Seeing that the works were to be carried out by the Board of Works, which is the engineering department of Government in Ireland, we think the contention that the Fishery Board should have a special engineer attached to it is untenable, so long as the two departments could have been expected or compelled to work in harmony; and it is remarkable that the Fishery Piers and Harbours Commission did not avail themselves of the permission to apply for professional assistance in special cases. At the same time we must record our opinion that the Board of Works should have insisted that the works that it would have to carry out were properly considered from the point of view of locality and general design, and to have formally objected to undertake any one which did not satisfy this condition *ab initio*; and furthermore that the Irish Executive ought to have taken care that the two bodies acted together, and that no works were carried out without the full approval of the professional advisers of Government in matters of engineering. The result has been that sites have been selected, estimates made, and the amount of money to be expended at various spots settled, without any competent professional advice, and without any proper certainty that the sums spent would produce any valuable result.[52]

It was indeed a frank, even scathing, account of what had transpired in the previous decade, from which no party escaped. Of the various enquiries into the Board of Works – the Lansdowne, the Crichton and the Allport – this was the first that was entirely free of serving politicians (although Joseph Pim had stood unsuccessfully for election in 1886), and therefore was more willing and able to name the Irish executive among those it indicted.

The Allport Commission regarded the piers built under the SFI 1883 Act with less than whole-hearted enthusiasm, considering them to be generally too small, and quite unsuitable for the larger boats that were needed if the Irish fishing industry was ever to take on a deep-sea aspect. The erection of small piers had, in its view, been greatly overdone. These findings, together with its criticism of the discord between the departments, caused Allport to recommend that the Inspectors of Fisheries be replaced by a new fishery administration, placed fully within the OPW. From its inception in 1869, the independent Fishery Inspectorate was unusual in never having been subject to, or having to operate under, any controlling Board. It had come directly under the Irish executive and was located in its own offices, first in 12 Ely Place, Dublin, and later in Dublin Castle.

Allport's recommended dismissal of the Inspectorate was indeed a harsh judgement on its work over many years, carried out under conditions (admittedly partly of its own making) that were, at very best, trying. Blake died suddenly in 1887 before the Allport Commission reported, leaving Brady and his Inspector colleagues isolated in facing the harsh criticism. But in recommending in their place the appointment of new 'sub-inspectors' working under 'a single head inspector ... who should be a man practically acquainted with marine matters, as a sailor, as a fisherman or as both', and in further declaring that 'such an organisation ... would be better calculated to do useful work, and to satisfy popular ideas, such as those expressed by Mr Blake',[53] the Commission was damning the Inspectors in earnest.

Existing Piers Too Small

With regard to the size of piers, the Allport Commission was clear and unequivocal in its opinion that 'the object to be aimed at is the provision of harbours with an adequate *minimum* [*sic*] depth at low water Spring tides i.e., where possible, of not less than twelve feet'.[54] No one could deny that large, deep harbours were essential if a deep-sea fishing industry were ever to develop and thrive, but the west coast was in a different situation to the east and south coasts regarding its potential for deep-sea fisheries. The difference in the number and cost of fishery piers constructed in the western and the eastern regions, documented in Figure 7.1(b) and (c), is solid evidence that the FPH Commission had sought to tailor its allocations to the different kinds of fishing activity on the different coasts. The reality, that there were no *proven* deep-sea fish stocks off the west coast on which to base a viable fishery, was something the Allport Commission expressly accepted.[55] That deficiency had nothing to do with the Inspectors and it should never have been laid at their door, which it seemed to have been. Allport went on to point out that no complete, systematic knowledge of the Irish fishing grounds existed, yet the Commission refused, paradoxically, to recommend a scientific investigation or survey of the coast – something that was repeatedly urged by the Inspectors of Fisheries over many years[56] – simply because, it alleged, no other European country had performed one. This negative attitude led to a laconic riposte by the Earl of Howth (William St Lawrence, MP for Galway Borough from 1868 until he succeeded to the earldom in 1874) in the House of Lords:

> [A survey] was essential for the sake of Irish fishermen who were seeking a means of livelihood; in the interests of English fishermen who were on the look-out for new fishing grounds; in the interests of the British public who wanted larger and cheaper supplies of fish; and in the interests of the British taxpayer who was anxious to see more fish caught in return for the money that had been expended upon the Irish fisheries.[57]

Having reviewed the state of the Irish fisheries and fishery piers, Allport went on to recommend that a sum of £400,000 be made available to develop Irish deep-sea fisheries, even in the acknowledged absence of any evidence of accessible stocks in the deep Atlantic off the west coast. And it linked the recommended new expenditure to the fishery piers:

> Deep sea fishing, in our judgement, as a national Irish industry in connection with the British and foreign markets, can only be efficiently and profitably carried on by the construction of harbours with a suitable depth accessible for fishing craft at low tides, and it is most desirable that any future expenditure should be mainly devoted to the construction of such harbours, together with such wharves and other appliances as may be necessary, within naturally sheltered areas.[58]

While this recommendation might have greatly enhanced the development and expansion of the Irish fishing fleet and Ireland's fisheries infrastructure, it represented exactly the kind of political error that the Commission itself had accused previous advisory bodies of falling into: it proposed the expenditure of public money without any certainty that the sums spent would produce any valuable result, particularly in the West. It was, as Col. Fraser had put it, 'throwing large sums of money into the sea in the hope, unsupported by evidence, that then and there fish must come out of it'.

Where Did It All End?

At whose door, then, should responsibility properly be laid for the alleged general failure since 1842 to provide suitable and adequate piers, located judiciously on the Irish coast, and properly serviced and maintained? The Government certainly had provided most of the funds for the existing stock of piers and quays; both the OPW and the Inspectors of Fisheries had a major input into them; the Viceroy had an important hand in approving them; Treasury sanctioned them; the various Commissions that oversaw their selection and progress were accountable for them also, even if that accountability lapsed when the Commissions themselves went out of office; landowners had failed broadly to provide funds for them; the maritime communities using them treated them with less than proper care; and the Grand Juries had responsibility for their maintenance, although they were reluctant to look after them, and generally failed to do so. The fishery piers were, in a sense, everyone's business, and therefore no one's. Some might contend that half a century of legislation had helped and hindered in equal measure, so that many piers were neither as suitable, nor as well located, nor as adaptable to the changing needs of the fishing industry as they should have been, and might otherwise have been. It is little wonder that they largely failed to stimulate new or more sophisticated fishing activity, especially in the West. The discord between the OPW and the FPH Commission (effectively the Inspectors of Fisheries), which legislative uncertainties and ambiguities had partly helped to create and sustain, undoubtedly meant that the opportunity presented by the SFI 1883 Act and Fund had, to an extent, been squandered.

On the other hand, for Co. Galway at least, famine and distress, rather than any clear fisheries policy, formed the backdrop to most of the piers that were built, and they were not entirely inadequate or unsuitable for the local inshore fishing that was then being pursued, and for other indigenous maritime occupations being carried on there. Without them the coastal communities would have been even more deprived than they actually were, as we will see in the next chapter.

As the century progressed, Ireland had slowly evolved its own pragmatic strategy for fishery piers, resulting in a large number of small, inexpensive structures in the West and fewer, more sophisticated piers and harbours in the East and South. In this way, it had addressed the needs of the indigenous inshore fisheries while, at the same time, had responded as best it could to developments in the British and foreign deep-sea fishing fleets that worked out of Irish ports. Underlying this approach was a dilemma that the Earl of Howth put succinctly in the House of Lords:

> [T]here were in Ireland, he was sorry to say, in fishery matters as in politics, two programmes: the popular programme was that the fishery grounds should not be made subservient to the interests of the coast population; while the other view was that the fishery grounds were not to be made subservient to supplying the demands of England with fish.[59]

He believed it was 'pretty clear' that the Allport Commissioners exceeded their powers in making an enquiry into the Irish fishery administration, and he hoped that the sweeping measure of removing Sir Thomas Brady, who was a gentleman of great experience and ability, and two other commissioners, which had been suggested by the Royal Commission, would not be adopted by Her Majesty's Government.

The Old Guard Changes as a New Century Approaches

The work of the Piers and Roads Commission had come to an end in 1887; that of Colonel Fraser ended in 1888. In 1889, the Public Works Loans Act finally brought to an end the FPH Commission which, along with the OPW, had administered the piers programme of the SFI Act.[60] Responsibility for new fishery piers reverted to the Commissioners for Public Works alone, who, from then on, would have, in their own words, 'the guidance of the Irish government in the selection of localities to be benefitted and in determining the kind of work to be executed with the moneys still available'.[61] That money referred to the residue of the Sea Fisheries Fund, since no new money was voted for fishery piers.

The Inspectors of Fisheries also reverted to their normal duties under the 1869 Fisheries Act, administering the fishery laws, issuing and enforcing bye laws, and gathering statistics. But that did not imply simply a return to the situation as it had been previously. Blake's sudden death in 1887 left the Inspectorate without its greatest Parliamentary advocate. Inspector Joseph Hayes died at the end of 1889, and Sir Thomas Brady retired under the age rule, with pronounced reluctance, in 1891. In January 1890, William Spotswood Green (1847–1919), late alpine explorer, late minister of religion, was appointed Fisheries Inspector in place of Joseph Hayes. Green was fresh from a survey of the fishing ports and fisheries of the south and west coasts, carried out under the auspices of the Royal

Dublin Society.[62] He had followed this with a visit to the USA and Canada, where he was instrumental in fostering a new market for Irish cured mackerel.

At the political level, Arthur Balfour, appointed Chief Secretary for Ireland in 1887, undertook a tour of the western counties in autumn 1890, to see for himself the state of affairs there, which a local Government official, William Micks, (1851–1928), and the philanthropist James Hack Tuke (1819–96) had reported to be precarious. What Balfour saw and experienced greatly enhanced his grasp of the economic dichotomy that existed in Ireland, between the impoverished West and the more prosperous East and South, something to which Tuke had been drawing attention for almost fifty years.

The situation of the piers – with many small, cheap structures, mainly relief works, in the West and fewer but larger, more sophisticated, structures supporting viable commercial fisheries in the South and East – was a microcosm of the economic imbalance of the country at large, at least as Tuke perceived it. Spurred on by Tuke and bolstered by his own observations on tour, Balfour moved to address the regional disparity by setting up a new, long-term, developmental agency aimed specifically at improving the poorer western region. That was the Congested Districts Board of Ireland (CDB), which was established by part 2 of the Land Purchase (Ireland) Act, passed in 1891.[63]

Endnotes

1 Royal Commission on Irish Public Works [The Allport Commission], Second Report, BPP 1888 [C. 5264 – I], p. 11.
2 J.G. McKerlie, in Select Committee on Sea Coast Fisheries (Ireland) Bill, BPP 1867 (443), queries 289–293.
3 Report of the Committee appointed to enquire into the Board of Works, Ireland [The Crichton Committee], Minutes of Evidence, BPP 1878 [C. 2060], queries 32–33, p. 3.
4 Report of the Committee appointed to enquire into the Board of Works, Ireland [The Crichton Committee], Minutes of Evidence, BPP 1878 [C. 2060], queries 624–25, p. 27.
5 See, for example, J.A. Blake, *The History and Position of the Sea Fisheries.*
6 *An Act for the Extension and Promotion of Public Works in Ireland*, 1 & 2 W IV C. 33.
7 9 Vict. C. 3, section 6.
8 J.R. Barry, in Select Committee on Sea Coast Fisheries (Ireland) Bill, BPP 1867 (443), query 3983.
9 J.G. McKerlie, in Select Committee on Sea Coast Fisheries (Ireland) Bill, BPP 1867 (443), query 357.
10 J.R. Barry, in Minutes of the Commission of Enquiry into the Sea Fisheries of the UK, part 2, BPP 1866 [3596-I], queries 37015–24.
11 J.R. Barry, in Minutes of the Commission of Enquiry into the Sea Fisheries of the UK, part 2, BPP 1866 [3596-I], query 37050.
12 J.R. Barry, in Minutes of the Commission of Enquiry into the Sea Fisheries of the UK, part 2, BPP 1866 [3596-I], queries 37057–58.
13 Report of the Inspectors of Irish Fisheries on the Sea and Inland Fisheries of Ireland for 1875, BPP 1876 [C. 1467], p. 4.
14 E. Hornsby, in Minutes of the Commission of Enquiry into the Sea Fisheries of the UK, part 2, BPP 1866 [3596-I], query 35291.
15 Griffiths, *Board of Works.*
16 T.F. Brady, in Select Committee on Sea Coast Fisheries (Ireland) Bill, BPP 1867 (443), query 5281.

17 A Bill to amend the law of Ireland as to Sea Coast Fisheries, BPP 1867 (50).
18 32 & 33 Vict. C. 92.
19 32 & 33 Vict. C. 92, section 18.
20 Report of the Committee appointed to enquire into the Board of Works, Ireland [The Crichton Committee], BPP 1878 [C. 2060], para. 281, p. lxi.
21 W. Le Fanu, in Report of the Committee appointed to enquire into the Board of Works, Ireland [The Crichton Committee], BPP 1878 [C. 2060], queries 1758–59.
22 W. Le Fanu, in Report of the Committee appointed to enquire into the Board of Works, Ireland [The Crichton Committee], BPP 1878 [C. 2060], query 1783.
23 T.F. Brady, in Select Committee on Harbour Accommodation, BPP 1883 (255), query 1791.
24 Report of the Committee appointed to enquire into the Board of Works, Ireland [The Crichton Committee], BPP 1878 [C. 2060].
25 Report of the Committee appointed to enquire into the Board of Works, Ireland [The Crichton Committee], BPP 1878 [C. 2060], para. 324, p. lxix.
26 Report of the Committee appointed to enquire into the Board of Works, Ireland [The Crichton Committee], BPP 1878 [C. 2060], minority addendum by Viscount Crichton, p. lxix.
27 House of Commons Debates, 22 June 1881, Hansard vol. 262, col. 1034–35.
28 Select Committee on Harbour Accommodation, BPP 1883 (255), queries 3049 & 3134–5.
29 Cmdr C. Langdon, in Select Committee on Harbour Accommodation, BPP 1883 (255), query 3141.
30 BPP 1883 Bill 31, sections 11 & 12.
31 Select Committee on Harbour Accommodation, BPP 1883 (255).
32 T.F. Brady, in Select Committee on Harbour Accommodation, BPP 1883 (255), query 2022, p. 145.
33 T.F. Brady, in Select Committee on Harbour Accommodation, BPP 1883 (255), query 2025.
34 R. Manning, in Select Committee on Harbour Accommodation, BPP 1883 (255), queries 4227–28.
35 R. Manning, in Select Committee on Harbour Accommodation, BPP 1883 (255), query 4090.
36 R. Manning, in Select Committee on Harbour Accommodation, BPP 1883 (255), query 4105.
37 R. Manning, in Select Committee on Harbour Accommodation, BPP 1883 (255), query 4179.
38 R. Manning, in Select Committee on Harbour Accommodation, BPP 1883 (255), queries 4180–82.
39 R. Manning, in Select Committee on Harbour Accommodation, BPP 1883 (255), query 4253.
40 R. Manning, in Select Committee on Harbour Accommodation, BPP 1883 (255), query 4191.
41 J. Robins, *Custom House People* (Dublin: Institute of Public Administration, 1993).
42 *Sea Fisheries (Ireland) Act* 1883, 46 & 47 Vict. C. 26, section 8.
43 R. Manning, in Select Committee on Harbour Accommodation, BPP 1883 (255), query 4226.
44 T.F. Brady, in Royal Commission on Irish Public Works [The Allport Commission], Second Report, BPP 1888 [C. 5264-I], query 19474, p. 11.
45 T.F. Brady, in Royal Commission on Irish Public Works [The Allport Commission], Second Report, BPP 1888 [C. 526-I], query 19478, p. 11.
46 *Piers and Harbours Act*, 1846, 9 &10 Vict. C. 3, section 6.
47 T.F. Brady, in Royal Commission on Irish Public Works [The Allport Commission], Second Report, BPP 1888 [C. 5264-I], query 19503, p. 11.
48 T.F. Brady, in Royal Commission on Irish Public Works [The Allport Commission], Second Report, BPP 1888 [C. 5264-I], query 19492, p. 11.
49 R. Manning, in Select Committee on Harbour Accommodation, BPP 1883 (255), queries 4215–18.
50 Royal Commission on Irish Public Works [The Allport Commission], Second Report. BPP 1888 [C. 5264-I].
51 Royal Commission on Irish Public Works [The Allport Commission], Second Report, BPP 1888 [C. 5264-I], p. 3.

52 Royal Commission on Irish Public Works [The Allport Commission], Second Report, BPP 1888 [C. 5264-I], p. 11.

53 Royal Commission on Irish Public Works [The Allport Commission], Second Report, BPP 1888 [C. 5264-I], p. 17.

54 Royal Commission on Irish Public Works [The Allport Commission], Second Report, BPP 1888 [C. 5264-I], recommendation 29, p. 18.

55 Royal Commission on Irish Public Works [The Allport Commission], Second Report, BPP 1888 [C. 5264-I], para. 7, p. 4 and para. 25, p. 16. The Commission's view was based on information given in queries 9870, 10353–56 and elsewhere in the evidence.

56 Report of the Inspectors of Irish Fisheries for 1874, BPP 1875 [C. 1176]; Report of the Inspectors of Irish Fisheries for 1882, BPP 1883 [C. 3605], p. 3.

57 House of Lords Debates, 22 June 1888. Hansard vol. 327, col. 963.

58 Royal Commission on Irish Public Works [The Allport Commission], Second Report, BPP 1888 [C. 5264-I], p. 13.

59 House of Lords Debates, 22 June 1888. Hansard Vol. 327, col. 964.

60 *Public Works Loans Act,* 52 & 53 Vict. C. 71, section 4.

61 67th Annual Report of the OPW, BPP 1899 [C. 9465], p. 18.

62 Report of the Inspectors of Irish Fisheries on the Sea and Inland Fisheries of Ireland for 1891, BPP 1892 [C. 6682], survey of the Fishing Grounds on the West Coast undertaken conjointly by Government and the Royal Dublin Society, report of W.S. Green, pp. 43–59.

63 54 & 55 Vict. C. 48.

CHAPTER 10

The Abundance of Small Piers and Quays: Humble Works, indeed, for Humble People

The great bulk of the population of Connemara live upon the coast. Upon the sea coast they live by the land, by kelp making and by fishing. These are the three sources of livelihood.

Fr Thomas Flannery PP, 1885

Ireland Compared with Britain

Serious as it was, the lack of harmony at interdepartmental level was not the only problem that affected the development of Irish fishery piers from the Famine onwards. Legislation that failed to formulate any clear, overall, national policy also contributed in a way that was little different from the situation in England, as Jarvis pointed out regarding the fishery piers of that country:[1]

> At no stage since the early eighteenth century have the questions underlying fishing port provision been addressed at a national level, despite fairly regular Government enquiries into the fishing industry in general. These questions begin with: what were these ports wanted for? An answer would have helped formulate a policy of port construction designed to meet the need. Instead there was an amazing jumble of reasons for investing in fishing ports.
>
> Some investments were local initiatives to stimulate the local economy, others were exercises in predatory pricing by railway companies, and others represent uses being found for obsolete trading docks. Some were built or modernised to make work for building labourers, others to make work for convicts, yet others to support non-viable fishing communities ... Fishing ports, large and small alike, were invariably impossible to fund on dues income. We can hardly be surprised, therefore, that they have often been poorly or under-engineered in the interests of cheapness, out of date, inadequately maintained, and almost never big enough.

Notwithstanding these strictures, many English fishing ports succeeded in growing significantly in size and sophistication over the second half of the nineteenth century in response to changes in the fisheries and the fishing fleets in that period. Sailing smacks used for trawling increased in overall length from around forty-five feet in 1850 to around seventy-four feet in 1890 and increased in draft from a few feet to around ten feet, displacing about forty to fifty tons.[2] These bigger, twin-masted, decked boats could travel much further afield and trawl, or long-line, in poorer weather conditions, but they also required larger and deeper harbour accommodation both for themselves and for the associated fast cutters that transported their catches to market.

By 1890, the spread of even bigger steam-powered trawlers and motor fishing vessels necessitated yet more advanced port facilities, like lifting gear, coal and oil bunkering, packing and curing sheds, and telegraph and railway services.

Jarvis's comments about English ports can be applied broadly to the Irish situation also. Up to 1885, Ireland was not unique in lacking a coherent national policy for fishery ports, and their development and evolution took essentially the same path here as in England, although not at the same pace, or to the same extent. Like England, many of Ireland's fishery piers were built to make work for unemployed destitute people, to support unviable or unproven fisheries, and to stimulate local curing initiatives. The obsolete postal harbours of Dunmore and Howth were transformed into fishery harbours, and it was even suggested that Galway should be developed as a fishery port using convict labour.[3]

However, in contrast to England, Ireland suffered a chronic shortage of investment capital, and for that reason was more dependent on State assistance for development, making its response to modernisation much slower and less pronounced. Still, while the overall number of Irish fishermen and fishing boats continued to decline nationally, the number and the size of first-class boats in the indigenous Irish fleet increased, albeit from a very low base. These bigger boats worked predominantly from east coast ports, within easy reach of the rich Irish Sea fishing grounds and within convenient communication distance of the huge and virtually insatiable English markets. Irish east coast fishermen did not need to venture far west in search of either fish or markets; they, together with the British and foreign fleets of large vessels, ventured only as far as the south coast of Ireland to work the seasonal mackerel and herring fisheries. Kilkeel, Ardglass, Clogher Head, Howth and Arklow in the east; and Ballycotton, Kinsale and Baltimore in the south, were the main ports that had developed during the century to accommodate these fleets.

The French fleet at Kinsale in 1880, for example, consisted of 160 vessels of forty-three to ninety tons, with crews of sixteen to twenty-four in each. All but three had steam winches to haul their nets and two had auxiliary screw propellers; 130 were from Boulogne, twenty-eight from Fecamp and two from Dieppe.[4] The competitive advantage they enjoyed over the local small sailing boats needs little imagination. Their presence was one reason why William Le Fanu of the OPW could state confidently that 'Howth, Ardglass, Arklow and Kinsale are more important than all the rest put together, as fishery piers.'[5] The fishery Inspectors concurred with that broad view and even hoped that Westminster would agree:[6]

> The question is not simply an Irish one. It should, in our opinion, be regarded as imperial [a matter affecting the UK as a whole, warranting UK exchequer funding]; and we submit we are warranted in forming this opinion from the very large proportion of fishing vessels from England, Scotland and the Isle of Man making use of it [Ardglass] as compared with Irish vessels ... and memorials were forwarded to the authorities, not only from Ardglass but also from large numbers of Scotch [*sic*] fishermen.

This comment was in regard to improvements proposed for Ardglass, but extended by implication to the other east and south coast ports. Larger, deeper ports like these made up a relatively small *proportion* of all the Irish fishery ports in 1890; this is explained more by the very large number of small piers and quays, rather than by any major deficiency in the number of larger ones. The main question was not, therefore, why there were *so few large* fishery piers in Ireland (almost all located along the east and south coasts where the fisheries and markets were) but why there were *so many small* ones in western counties, especially in Galway and north Clare?

The Abundance of Small Piers in Co. Galway

A clear distinction between the western region, west of a line from Baltimore to Derry City, and the region east of that line, had already become apparent in the number of memorials for piers submitted between 1870 and 1884, as shown earlier (Figure 7.1a).[7] Even allowing for differences in length of coastline between them, the greater number of memorials from the West is striking: of the total 264 memorials, no fewer than 84 per cent came from the West and they were mainly for small structures.

The FPH Commissioners balanced this pronounced demand 'pull' for small piers in the West with their own developmental 'push' for larger ones in the East and South. The fact that most of the western demand was for small piers is hardly surprising, for three main reasons. Firstly, certain restrictions in the 1846 and 1847 *Fishery Piers and Harbours Acts* prevented an optimal use of the funds granted, which had been, in truth, relatively generous. For example, the Acts were never intended to promote anything other than small structures, as expressly stated in section 4 of the 1846 Act. The limit of £5,000 placed on individual grants, raised to £7,500 by a later Act,[8] reflected this. It cannot be denied, however, that during the Great Famine, £5,000 was sufficient to build a substantial structure, albeit one that dried out at low water, and that some of the piers of the Great Famine – Tarrea, Kilkieran, Bunowen and Claddagh – are among the most substantial ones ever built in Co. Galway. In gross, the amount actually given by Government in free grants in the thirty years from 1846 to 1877 (about £125,289) was a large sum, at least at first sight.[9] But it equates only to spending about £4,176 annually, spread over the whole country – less than the cost of one useful pier – or an average of £278 annually for each of the fifteen main maritime counties. Even when the £50,000 provided as loans in addition to the free grants are added, the total expended was less than £6,000 a year nationally, or an average of £390 annually for each maritime county. Had the total sum expended been spent only on the six western counties of Donegal, Sligo, Mayo, Galway, Clare and Kerry, it would have represented an average of just £1,000 *per annum* for each county.

Such minimal funding would have needed to be allocated very judiciously to a significantly smaller number of sites than the number actually applied for, if it were ever to be of any real benefit. The detrimental effect of this strict limitation of funds was not lost on the Allport Royal Commission in 1887:[10]

> The deliberate policy of the Legislature down to ten years ago was to limit the expenditure at any one point to a sum far too small to provide, in ordinary cases, works useful at low water; and during the last ten years the administrative procedure, though freed from this legal restriction, has mainly been conducted along the old lines. As a natural consequence of these facts, only three out of the special fishery works are deepwater harbours, in the sense of having a maximum depth of 12 feet at low water spring tides.

Secondly, the small size of the piers erected in the West is further explained by the nature and purpose of the funds allocated to them: eight Acts that provided funds for fishery piers from 1824 to 1890 were enacted primarily to provide relief work in response to famine and only one, the *Sea Fisheries (Ireland) 1883* Act, was enacted specifically to provide piers exclusively for fishery development. As we might have expected, Galway benefitted more than most from the various Relief Acts, being one of the largest counties and one where famines were most frequent, most severe and most destructive. Building small piers at numerous sites had the effect of spreading the relief funds over a greater geographic area than building one or two large piers in selected districts would have achieved. The exigency of the times dictated this and other dispersive stratagems, such as the rotation of men off the work after short intervals and the employment of only one man from each household in a district.

Nevertheless, under the Sea Fisheries 1883 Act, which took no account of poverty, famine or the relief of distress, only two large piers – Carrigaholt in south Co. Clare and Cleggan in Galway – were built in the West. Figure 7.1 (b) and (c) confirm that allocations from the Sea Fisheries Fund to the western piers were significantly less than those made to the east coast piers. As a result, in Co. Galway the majority were small structures, built with an eye to the existing pattern of localised inshore fishing, as this was perceived by the FPH Commissioners, rather than in a serious effort to stimulate more sophisticated, deep-sea fishing.

Thirdly, the contentious one-fourth local contribution exacerbated the funding situation in a number of ways. As we saw, the rule had influenced the sites chosen for the early fishery piers by aligning their exact locations with the personal demands of the local contributors, rather than with the requirements of the fisheries or the fishermen.[11] Later, when the maximum permitted free grant was increased to £7,500, the absolute amount of the one-fourth contribution increased *pro rata* beyond the means of many willing local contributors. For example, to draw down the maximum grant of £7,500, the local contribution needed to be £2,500 (one fourth of £10,000), a significant sum at the best of times. If only £1,000 were contributed locally, the free grant could not exceed £3,000, limiting the total available to £4,000 at most. Because of considerations like these, the total amount spent on individual piers was severely restricted in practice, leading to an excess of smaller, cheaper structures. From the OPW's viewpoint too, there were considerable disadvantages associated with the one-fourth rule. Once the amount available as a local contribution was known, the OPW would compute the maximum grant allowable for the particular site and this, in turn, determined the nature and the size of the structure that the Board's engineer would design. When asked, for example, whether he thought the local contribution was a good provision generally, the Board's engineer Robert Manning replied: 'I think that it is not a good provision generally. If I was bound to do the best I could for the fishery, I would keep the three fourths, and not be troubled with what the one fourth carries with it.'[12] Pressed as to whether the one-fourth rule had diminished a great deal the efficiency of the piers that had been built, he went on: 'it certainly does diminish the efficiency, because there is scarcely a pier that I have designed, where, if left to myself, I would not have spent more money upon it and made it more efficient.' Contractors, too, learned to manipulate the tender process: knowing the amount of the local contribution they could compute the total cost estimate the OPW would arrive at, and pitch their tenders accordingly.

However, it was at the human level that the iniquity of the universal application of the local contribution was most pernicious. We have seen how the boatmen of Kinvara Bay had only 'the freight of their boat' to offer towards the pier at Tarrea. In Carraroe south, on Lettermore Island, the fishermen signed a memorial for a pier at *Doilín* (Doleen) in July 1846, at the height of the Great Famine. They offered £150 in labour as the local contribution because 'their poverty and destitution does not permit them to contribute otherwise.'[13]

What, one wonders, was the advantage in a local contribution, if it meant that the destitute had to forego some of the small wages they would earn from the resultant 'relief' works? How did that alleviate distress, or reduce destitution? Later, in a slightly different instance in 1883, the fishermen at Derrynea sent a memorial to the OPW for funds to finish the small harbour at that place in accordance with previously approved specifications. At an earlier date, they had seen a newspaper advertisement indicating that money was available from the Mansion House Relief Fund for small slips and harbours, up to a maximum of £60 for any one work, subject to a matching local contribution of the same amount. They had received an allocation of £50 with which they undertook the work, charging only one shilling per man per day, whereas the going rate for labourers on other works in the locality was one shilling and sixpence. They claimed never to have received the final £10, so they had left the work unfinished and were now seeking a supplementary grant to finish it off.[14] These applicants from Derrynea were tenants of Charles Cottingham of Oughterard, who does not appear to have heard the plea of his poor tenants for payment of the small sum, a mere pittance, which they felt was owing to them.

Cases like these were well known to the OPW. All parties, the OPW, the Inspectors of Fisheries, contractors and local contributors, were against the rigid application of the local contribution in all cases and circumstances, but no one, least of all the OPW, seemed to be willing to take on Treasury regarding the matter. Perhaps the principal justification for the local levy was to induce recalcitrant landowners to meet their responsibility in assisting with local development. However justified or prudent in political economic theory this might be, for any Government to enforce it strictly and rigidly without reference to exceptionally adverse circumstances was perverse to say the very least. Amazingly, it was the Crichton Committee – a Departmental Committee appointed directly by Treasury – that criticised the timidity of the OPW in situations where the local contribution was patently oppressive and unjust:[15]

> Although we fully appreciate the desire of the Board to carry out as closely as possible any directions they may receive from those who exercise a control of them, yet a timid acquiescence without remonstrance in instructions which may have a harsh and injurious effect ... cannot be justified. There may be circumstances which render it the duty of a subordinate department firmly to represent to the Treasury that a strict compliance with a rule will be inexpedient, if not impracticable.

Not only did the Board not remonstrate vigorously about the rule to Treasury, it was known, on occasion, to apply it almost with a vengeance: if a pier were deemed to be exclusively for fishery purposes, the OPW recommended the maximum free grant of three-fourths; if it had some other, commercial, usefulness, the Board recommended a reduced grant of only two-thirds.[16] Even when the *Sea Fisheries Act* of 1883 removed the local contribution, the OPW still ensured that 16 per cent of the total expended came from loans and contributions.

In the light of this, it is little wonder that the Inspectors of Fisheries treated all memorials for piers as being exclusively for fisheries purposes, or that some applicants may have framed their memorials with an undue emphasis on fishing, when in truth they had other uses in mind. In such instances, greater co-operation and harmony between the Inspectors and the OPW might have proven more conducive to their mutual benefit, and been of more value to the fisheries: if applications for fishery piers that openly admitted to having uses for general trade could have received the three-fourths grant, surely they would most likely have been designed and built with increased accommodation for larger trading boats in mind and, in consequence, would have been more suitable and attractive for large deep-sea fishing boats as well?

The 1846 Piers and Harbours Act, it must be acknowledged, permitted the one-fourth contributions to be advanced as loans, and that certainly helped in some cases. However, local proprietors were usually reluctant to put a charge on their own property as surety for such loans and, where the county or the district was the memorialist, the Grand Jurors were often reluctant to provide surety for them. Where individuals agreed to pay their contributions in cash, the OPW had to insist on receiving the full amount before starting the work because, in its experience, it rarely got it once the work was completed.[17]

Finally, and perhaps most importantly, in accounting for the small size of the western piers, we should recognise that the maritime communities of Co. Galway and north Clare, no matter how poor, were never given to the active pursuit of offshore deep-sea fishing. Ciara Breathnach, in her study of the Congested Districts Board, remarked that 'By 1890 the art of fishing was practically extinct along the west coast.'[18] In fact, there is no convincing evidence that the fishermen of the region were ever, at any time, committed deep-sea fishermen, regularly working far from shore, or ever engaged to any appreciable extent in deep-sea fishing as a full-time activity, notwithstanding historical anecdotes of occasional, allegedly stupendous, catches. The ever-astute Col. Fraser of the Piers and Roads Commission had deduced this well:[19]

> On the whole, however, my impression is that there is less fish on these [west] coasts than on the south and east coasts of Ireland, and perhaps less than there used to be, although when one comes to analyse the traditions of vast takes of fish, it generally turns out that the event only happened once in a lifetime on the occasion of some shoal getting grounded in one of the many landlocked creeks.

According to the Coastguard, fishing boats on the west coast never went out as far as forty miles, and they generally worked within eight miles of the land.[20] They rarely, if ever, stayed out at night, or went further than a distance from which they could run for shelter at short notice. This does not mean that they did not fish at all, or that there were no competent and seasonally active fishermen in the West. There were, but they were never *full-time deep-sea fishermen*, which was the main focus of attention of the Select Committees and Commissions that investigated the Irish fisheries from about 1860 onwards. Le Fanu's answer to a query at the Crichton Committee in 1878 is indicative of the prevailing attitude:[21]

> Query 1754. But is it not the case that in many places, although there cannot be said to be an actual fishing season, yet the poor people of the neighbourhood go out to fish, to obtain fish for their own food and for sale in the neighbourhood, during the fine weather?
>
> Reply. In a great many cases; and these are the cases that were of no importance as encouraging the fisheries – the sea fisheries – which was the object of the [1869] Act. It is a great boon to these people to have a harbour to get into, but in the proper

acceptation of the term I don't look upon these as an important class of fishermen as fishermen. They don't go out except on a summer's evening and they only get fish for their own houses, and not for trade. They are not, as fishermen, a very important class, I think; but it is a great boon to them, no doubt, to have a safe harbour and many of these have been built almost for that class, though no large or well-found fishing boats come, or could come, to them. It is of great importance to these poor people to have shelter for their boats.

Indeed it was: their boats were essential for the trade in turf, the harvesting and distribution of seaweed, the transport of animals from place to place, the delivery of household goods and the general communication and intercourse of people living along a highly indented coast lacking in roads but abounding in small islands. James Perry, the Galway County Surveyor, acknowledged this when he reported in 1886 that 'the Connemara piers are of more consequence for the turf traffic, seaweed traffic and as shelters and landing places for hookers bringing flour and other goods from Galway to various points on the coast, than for fishery purposes.'[22]

He went on to draw an interesting distinction between the north and south shores of Galway Bay: 'There is a great traffic (comparatively) in turf and seaweed, and I should say there is more fish brought in along the whole Connemara coast from Oranmore to Duras [*sic*] than along the Connemara coast from say, Spiddal [*Spidéal*] to Leenane.' Blake, Brady and the other Inspectors of Fisheries, men personally familiar with conditions in the West and generally sympathetic to the coastal communities, knew the people's real needs and they encouraged the building of small piers to meet those diverse needs. In his booklet on the Irish sea fisheries, Blake wrote that he 'believed it would be so much better for the interests of the country if most of the fishing would be carried on by the coast population', and that he 'would much prefer, even at some sacrifice of supply, that the trade should remain in the hands of humble fishermen'.[23] It was his opinion that unless small piers were built by Government to encourage these humble fishermen, they would not be built at all.[24]

Many of the piers and quays that were built may seem nowadays to have been sited incongruously far up shallow bays and inlets, rather than close to the open headlands and the fishing grounds, but their locations were eminently sensible at the time: in a region lacking roads, and deficient in wheeled carriages and carts, it was better to locate piers near homes and settlements, so as to reduce the labour involved in transporting heavy loads overland on the backs of men and women. The Inspectors appreciated the adventitious nature of fishing for such communities: it was never a sole occupation, but was ancillary to a range of land-based and maritime activities that made up the fabric of peoples' lives.

William Lane Joynt was clear at the Select Committee on Harbour Accommodation that 'at present the fishery is in so backward a state that you want very humble works for humble people, and little things of that sort are a great deal better than having heavy works and no boats to come there.'[25] 'Heavy works' were piers and quays with minimum water depths in excess of eight to ten feet at low water of Spring tide, conditions required only by large boats engaged in the deep-sea fisheries.

Traditional Occupations on the Galway and North Clare Coasts

So, what work of a marine nature did the people of the Galway and north Clare coasts do, if they did not engage in deep-sea fishing? Fr Thomas Flannery, PP of Carna, answered that question succinctly in 1885. The coastal people of Connemara, he said, had three sources of livelihood: 'upon the sea coast they live by the land, by kelp making and by fishing.'[26] The land gave them turf and potatoes, and the sea gave them manure, kelp, fish and shellfish.

The Turf Piers and Turf Trade

Throughout Connemara, turf was cut, dried, transported and sold to buyers elsewhere. Turf is absent from the Aran Islands and north Clare, so the opportunity existed for a vibrant trade across and through Galway Bay, into the city and out to the islands. It was a trade that was more securely in the hands of the men themselves than the kelp or fishing activities: they had full control of production, distribution and sale. All members of a family would help in the cutting, footing, stacking and carrying of turf to the boats that were largely owned by the turf cutters themselves. As Fr Flannery pointed out, every family had 'a boat, or two, or three, or even four, but there was only one cart for about twenty families'.[27]

The long inlets of the Connemara coast made it possible to bring the boats in very close to the bogs. The closer the boat, the shorter the distance the turf burden had to be carried. Before the piers were built, the boats were beached near high tide mark at the nearest suitable place, levelled for loading and steadied with small piles of stones placed under the gunnels. They were loaded while the tide was ebbing and filling. As late as 2008, the *Irish Cruising Club Sailing Directions* cautioned sailors, as they were coming alongside at *Sruthán* quay, to beware of the hazard presented by these small 'stone pillars'.[28] At the flood, the turf boats would lift and float, and could then start to make passage to their desired destinations.

While it would be misleading to say that the many piers built in Co. Galway and north Clare were specialised for specific purposes, it is nevertheless natural that over time, certain activities came to predominate at different sites. This is most clearly evident with the turf trade. About twelve places evolved to become the main shipping piers for the distribution – the export – of turf from Connemara. These were: Derrynea (No. 136), Clynagh (No. 135), Cashla (Costelo) (No. 137X1) and *Sruthán* (No. 133) in Cashla Bay; *Caladh Thaidhg* (No. 129), Tooreen (No. 127), Lettercallow (No. 099), Maumeen (No. 121) and *Clais na nUan* (Tiernea) (No. 120) in Greatman's Bay; and *Rosdubh* (No. 082),

FIGURE 10.1. The sea face of the pier at Tooreen, Carraroe (*Tuairín, an Cheathrú Rua Thuaidh*). Built by the Piers and Roads Commissioners in 1886 under the direction of the engineer Henry Abbott, Tooreen Harbour was a busy place for the export of turf. It remains busy today for landing seaweed and for inshore fishing. Its surface and inner face have been extensively repaired with concrete.

Sruthán Bhuí (No. 083) and Garafin (No. 088) in Kilkieran Bay. Generally speaking, these piers and quays are structurally more substantial than many of the others in those bays. Their decks are raised well above high tide level, a feature essential for the deposition and temporary storage of dry turf prior to shipment; turf that became sodden with sea water was a useless commodity. All these turf stations are located well within the bays and easily accessible from the local bogs where the turf was drawn. The P&R Commissioners showed themselves sensitive to the need for good access to such landing places: they made a number of roads, like Tooreen road and Muckinagh road, specifically to ease access from the bogs to the piers, and they improved loading conditions at the piers. Derrynea, Tooreen and *Sruthán* benefitted particularly.

Once loaded, the boats, of all sizes from large hookers (*báid móir*) to smaller *púcáin* and *gleoiteóga*, departed the piers and distributed their turf cargoes to different destinations depending on their size and sailing capacity. The large hookers made the long haul to

Galway city, to north Clare and to the Aran Islands; the smaller boats generally serviced the local islands and places around Connemara and along the north shore of Galway Bay. On arrival, the boatmen would seek out a buyer, negotiate a price, and the turf would be offloaded, simply by tossing the sods onto the quay, to be taken onwards in the buyer's horse or donkey cart.

At Kilmurvey on *Inis Mór*, the turf was originally offloaded directly onto a natural rocky platform until the P&R Commission built a forty-five-foot breakwater and reshaped and refaced the landing to facilitate the turf trade. Once again it was that Commission's acute awareness of the nature of the existing use of the site that resulted in this beneficial modification.

When necessary, the boatmen slept on the boat until a sale was finally completed. When it was, the deal was often sealed with a drink in a local pub, part of the social intercourse that cemented relationships between the communities of the two sides of the bay and the islands.[29] The extent of the turf trade is evidenced by the large number of old, so-called 'turf quays' recorded on the south shore of Galway Bay from Oranmore to Ballyvaughan.

FIGURE 10.2. The quay at Bell Harbour (Bealaclugga) (CS 14), north Clare, as it is today, much restored in recent times. Photo NPW.

They were predominantly turf and seaweed landing places, not significant fishery stations, although some, mainly those facing the open sea like Glenina and Bush Harbour, were important in the fisheries also. Well into the mid-twentieth century, turf boats plied their trade to Maree, Prospect (No. 149), Corranroo (CS 002), Kinvara (No. 160), New Quay (No. 190), Bell Harbour (CS 014), Bealaclugga, (CS 014), Ballyvaughan (CS 029) and other quays along that coast. If a return cargo could be acquired, like dry goods, or timber, or young livestock, or shop goods for the home (when trading with Galway city), it was a decided bonus. Otherwise the empty boats would be ballasted with limestone rocks, which would be jettisoned on arrival back in Connemara, possibly to be used later to make lime or to build a house. Many boats delivering turf or seaweed engaged in such ballasting, sometimes even taking stones from the very fabric of the piers.

Well into the twentieth century, there were complaints of the structural damage that particular activity caused. For instance, a letter from the OPW to Galway County Council in 1905 concerning behaviour at *Clais na nUan* pier was only one of many that complained of 'the removal of the pavement of the pier by the boatmen for ballast'.[30] The Coastguard blamed boat users for similar damage at *Sruthán*: 'It is the people for whose benefit the pier was constructed who are causing the damage by tearing up the pier for ballast'.[31] Even the engineer of the OPW wrote a minute in exasperation: 'This is a common trick of the turf boatmen in these parts, rather than put a little shingle in their boats for ballast which would be too troublesome, they prefer to destroy the structure which was erected at the public cost for their benefit.'[32] At many places, turf, turf-dust and mud accumulated on and alongside the piers, which led to complaints of serious obstruction. Such occurrences led to continual demands on the County Councils to repair, maintain and manage the affected structures.

Trading with the Aran Islands sometimes took the form of bartering, turf being exchanged for dried cod and ling.[33] Some of the dried fish brought back was kept for home use and some was sold to cadgers who carried it inland for sale through the countryside, or it was sold to the resident shopkeepers who sent it to Galway for resale. Trading in turf was a steady and reliable occupation, one that was more attractive to the men of Connemara than deep-sea fishing ever was: it was a small certainty, whereas deep-sea fishing was a big chance.[34] As W.S. Green put it: [35]

> The turf-carrying is a constant source of income to these people. For generations they have lived by it, and such a population would not exist in South Connemara if it were not for the turf-carrying. There is a constant demand for turf in the north of Clare, Galway and in the Aran Islands, where there is no turf. That part of Connemara is almost a mine, and those men are the miners and that is the real reason of the present large population in that district.

Although it was reliable, sustainable within reason and in the hands of the people themselves, the turf business was a poor one, and its economy precarious, as described by Major Gaskell in 1892.[36] Turf was carried in baskets from the bog to the nearest road, chiefly on the backs of women and girls, an arduous task when the bog was located distant from a useful road. There the baskets were loaded onto carts pulled by donkeys, or occasionally by horses, and taken to the pier. When the pier was reasonably close, or the family very poor, the baskets were carried all the way to the boats. Each loaded basket weighed about one hundredweight; twenty made one 'creel' or cartload. Seven full creels made a boatload of about six to seven tons in weight. Delivered at the quayside, the turf was unloaded, stacked if necessary and eventually sold to a boatman for about 1¼ pence per basket, about fourteen to fifteen shillings a boatload. From that amount, the cost of cartage, if the cart were not owned by the turf cutter and he had had to engage a carrier, was deducted. Depending on the distance travelled, this could amount to three to five shillings per boatload, 'beside the driver's food and an armful of green fodder for the horse, if any were growing', leaving less than ten shillings net for the turf cutter. To cut a boatload of turf required about 100 hours work by one man, and the turf needed to be 'footed', stacked and carried away by other members of the family. This meant that the fourteen shillings paid to the family for a boatload was a little over one penny per hour's work, say one shilling per day, a poor business indeed: relief work, if one could get it, paid up to two shillings (twenty-four pence) a day at a maximum.

The boatman loaded his cargo and took it to wherever he knew there would be buyers. For large hookers this was usually in Co. Clare, in Aran or in Galway city; for smaller *gleoiteoga* or *púcáin* it could be anywhere along the local coastline. Each boatman would hope to sell the cargo for double what he had paid, about twenty-eight shillings in a normal year.[37] He might be fortunate enough to make two such journeys a week but that depended on many factors beyond his control, like tides, weather and distances. Delayed at his destination, he would need to feed himself, pay his crewman and maybe meet wharfage costs, so that a prolonged delay could absorb much of his profit. Gaskell estimated that possession of a hooker could be worth £30 net *per annum* to the owner. That was a useful addition to the family income, although one that was contingent, requiring sustained and dangerous passages.

Kelp Making

Kelp making was hard, wet, dirty and poorly paid work. For all that, it lasted as an occupation in Connemara and north Clare for most of the eighteenth and all of the nineteenth century. First, suitable kelp weed (*Laminaria* sp.) had to be gathered at low tide, or else harvested by boat from sub-tidal rocky outcrops and small islands, using a special long, homemade implement called a *croisín*. This was a long pole with a short cross-piece attached near one end. The pole was plunged through the water, cross-piece down, into the kelp bed and twisted so that the weed became entangled in it. The pole was then pulled up sharply, freeing the tangle of weed from the rock holding it and lifted into the boat. Driftweed (old, cast *Laminaria* fronds) cast up on the shore (sometimes called '*bualadh isteach*', meaning

FIGURE 10.3. *Climíní* (seaweed rafts) beached beside *Cé na dTracht* pier (RH 125A) in Connemara in 2015. The pier is on the left; the bridge causeway is on the right. Photo NPW.

'material that drifted in') generally was not much good for kelp making, unless it was still fresh and of the best kind (*L. hyperborea = cloustoni*), when it was called May-weed or Autumn-weed, depending on the season. At worst, this could be used, along with fucoid species, as fertilizer, applied directly on to the land, and was therefore not without value when carried inland for sale in the raw state. The weed used to make kelp had to be brought ashore, either in boats or towed in as floating seaweed rafts called *climíní*, and taken on to land to be dried.

The boats used were the smaller kind, easily manoeuvrable in close inshore rocky places where the weed grew in abundance. Sheltered slips and gently sloping, sandy landing places beside grassy drying areas, rather than big piers and quays, were the most suitable facilities for these boats and for *climíní*. They are the kind of landing places that are most common along the Galway shores, but especially on the coasts and the islands of Connemara. Small islands like *Inis Múscraí* off Carna and *an tOileán Mór* near Rosmuc were well-known centres of kelp making, but the occupation was generally widespread throughout Connemara and the Aran Islands, which accounts for the very large number of small landing places and slips that mark the region.

On landing, the wet weed was spread out on the ground or on the stone walls to dry in the air, after which it was stacked and then burned in a kiln built close-by until the ashes liquefied and eventually hardened into slabs of kelp. Kelp kilns, makeshift and ephemeral structures at the best of times, once dotted all of these places but today there are very few early examples still in existence. As Tim Robinson found, they have 'long been brushed aside by winter storms and have left scarcely a trace on the shores of today'.

The solidified slabs of kelp were brought to the agent, who determined the quality and offered a price for it. Early in the nineteenth century, kelp fetched up to £15 a ton. Towards century's end, good kelp fetched only from about £5 to £7 a ton, often very much less. In eight days two men could gather twenty-four tons of raw, wet weed, which weighed about five tons when dried, enough to make one ton of kelp.[38] Delivery to the agent could take two days, since the agents' stores were not everywhere convenient, and transport by boat was slow and transport by land was poor to non-existent. From the sale of a ton of kelp (the work of two or three men) each man might expect to make £2 to £3, about five shillings for every day spent on the work. But the weather did not always allow the work to be completed without interruption: weed could take at least one week of good weather to dry (a rare enough occurrence in the west of Ireland, even in the kelp-drying season of May to October), and more than a day to burn. Kelp ash was a valuable commodity in the chemical industry and had been exported from Irleand to the Continent since the eighteenth century, and to chemical works in Great Britain from early in the nineteenth century.

In some instances, access to the raw weed had to be paid for. Mrs Blake, of that family's Renvyle estate, demanded, for example, a 'royalty' of one-third of the kelp made on her land by her tenants. She later agreed with the kelp agent to accept *in lieu* a fixed sum of twelve shillings and sixpence per ton of kelp produced and sold. She justified her 'royalty' by saying the weed was a product of her land, and in addition she had allowed the tenants to dry the raw weed free of charge on her property which, she alleged, reduced the area she could let to others for grazing. If the tenants gathered inferior weed (for use as manure) from certain sea-bound rocks owned by her, she charged them one shilling a year, even though the weed was used to improve the land that was owned by herself! She offered the use of some boats to the men in order to gather better kelp weed from below low tide – no doubt at another charge – but they wisely declined the offer.[39]

In Carna, where about 1,500 tons of kelp was produced annually, Martin Mongan was the tenant of some islands where the best kelp weed in Connemara was reputed to grow.[40] He supplied boats and access to the weed for the kelp makers, and gave them other assistance, mainly of a supervisory nature. For these favours, he demanded one-third of the men's earnings from the kelp, which was sold in his name. He claimed that one of his 'employees', with the help of two sons, had made '£30 in two or three months', of which £10 was Mongan's share; the three workers shared the remaining £20. Mongan was once the kelp agent for a Mr Beit of Guernsey, but he gave that up and began to supply kelp to Mr

Hazell, the agent of a Glasgow company, who had a store near Cashel in Kilkieran Bay. The storehouse is still in existence alongside a small pier (Doonreaghan 2, No. 044B) that was used exclusively for the reception of kelp made elsewhere in Connemara. When various 'royalties' such as those mentioned and the cost of transport were deducted from the selling price, the men rarely made anything like five shillings a day for all their labour. Still, Fr Michael McHugh of Kilkieran estimated that the people of the whole peninsula of *Iorras Aithneach* made more than £3,000 from kelp every year.[41]

Allegations of carelessness in burning the weed, poor quality of raw material, adulteration with stones and gravel, obscure tests of quality and other unacceptable practices were rife in the kelp business. To ensure the quality of his kelp, Mongan sent 'trustworthy men' around to see if black weed (fucoid species, unsuitable for good kelp) was being put into the kiln.[42] Christopher O'Connor, a county councillor and local Connemara resident, called such men 'spies'.[43] The kelp agent, too, got information from his own spies, who sought out kelp makers with cocks of black weed stacked alongside their cock of good red kelp weed; for these producers, the price Hazell would offer at sale would be much less than usual, without ever bothering to test the quality. When used, the quality test was invariably done behind closed doors by the agent himself, and the sellers were not given any information on how it was carried out, or any opportunity to corroborate or contradict it.[44] Many producers suspected that the test was no more than a charade. In some instances, when the men arrived with their kelp, it was rejected as inferior quality or they were offered a very low price, 'take it or leave it'. Sometimes, kelp rejected from one man was brought back later by a neighbour, and was purchased. There was no appealing the decision of the agent. J.M. Synge described how he once saw 'a party of old men sitting nearly in tears on a ton of rejected kelp that had cost them weeks of hard work, while, for all one knew, it had very possibly been refused on account of some grudge or caprice of the buyer.'[45] As there was only one buyer in each district a monopoly prevailed, and prices could drop to £1 a ton, 'take it or leave it'.[46]

When all else failed, 'ordinary' fucoid seaweed could be gathered in abundance on all shores and used as manure or sold on for that purpose. Different species had different qualities as fertiliser and were used selectively by farmers depending on the land crop being fertilised and the nature of the soil. The very many names that were used for different varieties of seaweed in various parts of Ireland indicate their widespread use in agriculture, and greatly enriched the vocabulary of the Irish language, as indicated in the invaluable work of Agnes Brennan:[47]

> [T]he present species [*Laminaria digitata*] is well known to the seaweed gatherers who never confuse the two [*L. cloustoni* and *L. digitata*]; in many cases they also distinguish between the two varieties of *L. digitata*. The commonest name for *L. digitata* in Connemara and Aran is *Coirrleach*; this name extends to north Clare. In west Clare the var. *stenophylla* is called *Leathach fada*. In the Carna district the var. *typica* is known as *Cupóga*. In Spanish Point Co. Clare, the name *Learach* is applied

> by some of the gatherers and *Leathrach* by others, to the variety *stenophylla*. In the Blasket Islands, Co. Kerry the species is called *Feamanach dubh*; at Ballyhoorisky, Co. Donegal, it is known as *Leath* ... In the Connemara and Aran districts the [*L. cloustoni*] fronds are known as *Ceann a' tslat* or *Feamuinn dearg*, and the stipe as *Slat mara*; the term *Sgothach* is also used for the frond in the Carna district. In north Clare the names *Múrach Bealtaine* and *Múrach Foghmhair* are applied to the Mayweed and the Autumn-weed respectively. In Spanish Point the name *Stúmpa* is applied to the stipes. In west Clare the stipe is known as *Feamain*, the frond of this as *Fadhairt*, and the entire plant as *Leathach dearg*. It is called *Feamanach buidhe* in the Blaskets. In Sligo the frond is called *Barrai* and the stipes *Slata mara*. At Ballyhooriskey, Co. Donegal, the frond is known as *Screadhbhuidhe*, and the stipes as *Slat mara*.

On some shores that lacked suitable rocky substrates – Aughinis in north Clare, for example – rocks were deliberately collected and distributed on the sandy shore so that weed would attach and grow on them. The distribution of the stones was not entirely random. They were laid in rows with passages left between them to facilitate the harvesting of the weed in due course.[48] Such areas are now called 'seaweed farms' and they are well documented in Co. Down and at Achill in Co. Mayo.[49] The seaweed farms of Aughinis have been recorded and described by Michael Gibbons (personal communication).

It might have been expected that the Congested Districts Board (CDB), when it commenced operations in 1891, would immediately step in to improve and develop the operations of the kelp industry, one of the most enduring occupations of the western seaboard, despite its harsh demands. In fact, the Board did nothing initially: not one single reference to seaweed or kelp was made in its Annual Reports before 1895. In its fifth Annual Report for 1895–6, it merely acknowledged that kelp making was 'undoubtedly important' and that 'very large sums of money, relatively speaking, are earned by people living on the seashore ... by burning kelp'.[50] The following year, the Board finally thought about building a special kiln for experiments in producing good unadulterated kelp but, on its own admission, 'the fall in price deterred us'.[51] The CDB would not mention kelp again until the 13th Annual Report in 1904. In that and in the Reports of 1905 and 1906, it confined its observations to recording simply the fluctuations in price and in amount produced. By then, kelp was achieving only £4–£5 per ton, and families in Connemara were selling from five to ten tons each.[52] Prices would go into further severe decline in the years immediately following.

In 1910, the Board, newly reorganised under the terms of the Birrell Act of 1909, activated once again the experimental burning of seaweed to produce good-quality, unadulterated kelp.[53] The new experiment was a success and was repeated the following year, when prices for top-quality product were up to £5 a ton. The Board wrote to the agents 'to ascertain whether the purchasers of kelp for chemical purposes wish the new method to be extended generally on the coast of the congested districts'.[54] The agents were unimpressed, claiming that they preferred to buy kelp in blocks rather than reduced to pebble-size packaged in sacks, as the Board's experimental product was. They alleged that they would need to empty every sack to perform tests for adulteration, and the emptying and refilling of the sacks would cause too much delay and expense.[55] Few things are more indicative than that of an unwillingness to see the Board become involved in the business, excellent kelp or not. In response, the Board dropped its kelp-making experiments. Robinson[56] records a kelp kiln 'of a type introduced by the Congested Districts Board early in this [twentieth] century', which he said was the only moderately well-preserved specimen of the special CDB kiln still standing in Connemara. It is located near *Céibh an tSáilín* on Gorumna Island.

Some buyers started to buy dried weed of the best kind for thirty to thirty-two shillings a ton, much more *pro rata* than it would cost to buy locally burnt kelp; again, one wonders if the buyers took this costly action in order to forestall any further plans by the Board to intervene more actively in the business? Whatever the reason, the CDB was intent, at last, on encouraging the kelp industry and started to buy, in its own name, the right to cut and gather seaweed on the foreshore between high and low water, and below low water, under the Crown Lands Act of 1865.[57]

Commencing with Inishnee and adjacent islands and shores in the Barony of Ballinahinch in 1910,[58] it bought the rights at Glengevlagh and Letterbricken in the Barony of Ross in 1914; at Inishbofin, Rossadillisk, Gorumna, Roeillaun and Dog Island in 1915; at various other islands, rocks and shores of the Berridge Estate of Ballinahinch in 1916; throughout the Barony of Dunkellin in inner Galway Bay, around the Aran Islands, and in south Connemara in 1915 and 1916.[59] It also acquired the rights for places in counties Kerry, Mayo and Sligo. Kelp making was one of the principal marine activities that the Board encouraged during the years of the Great War. Its action in this regard was timely: driven by circumstances connected with the war, the export of kelp from Ireland increased during those years and for a year after.

Despite competition from foreign sources, over 12,000 tons of kelp was exported from counties Galway and Clare alone, from 1914 to 1919; adding in smaller harvests from other congested counties brought the aggregate value of the national kelp exports to almost £135,000 for the years 1914 to 1919.[60] Smaller amounts of weed were sent to Dublin for the production of potash there.[61] By 1919, the Board held the seaweed rights for most of the shores in Co. Galway, and its intention to help the kelp industry seems to have been both firm and practical. Whether the rights purchased were exclusive, or shared in common with the local kelp gatherers, is not entirely clear, but there are no reports of the Board ever excluding anyone from gathering what seaweed they could.

The making of kelp for the production of iodine and potash dropped off after 1919; little weed was burned from then until 1924, when French chemical firms became interested in the trade and stimulated a revival in demand. The new Department of Fisheries of Saorstát Éireann directed the attention of Irish firms to the potential of kelp in the chemical industry, and the desirability of setting up an Irish processing plant 'for the extraction of fine chemicals on the spot'.[62] This was an innovative pharmaceutical proposal, long before its time. There had been few enough such maritime industrial beginnings under the old CDB regime and even fewer were to come under the newly independent national government.

Inshore Fishing

It is a truism, repeated in many references to fishing in Co. Galway in the late nineteenth century, that the men of south Connemara most definitely were not fishermen. Examination of the statistics given by the Inspectors of Fisheries[63] on the landings of lobsters from 1889 to 1900 (shown in Table 10.1) gives the lie to that truism. Landings were consistently large during that period, as they were in earlier years and would remain for many more. The fishing was carried out using pots manipulated from small boats and currachs (second- and third-class boats in the official records) working out of the many small landing places all along the coast from Killary Harbour to Black Head.

Landings of 'other shellfish', which included scallops, oysters and clams, were also significant, especially for the poor who dredged and sold them. The humblest maritime activity, gathering periwinkles in the intertidal zone for instance, was done mainly by women and children, and a sum of £355 was reportedly earned by them in the Lettermore

TABLE 10.1

Numbers and value of lobsters landed in all Ireland, 1889 to 1900, together with the numbers landed from the west coast alone.

	All Ireland		West coast only	
Year	**Number**	**Value**	**Number**	**Value**
	(x 1,000)	**£**	**(x 1,000)**	**£**
1889	415	£10,600	203	£4,200
1890	238	£7,400	81	£2,100
1891	212	£7,200	58	£1,800
1892	222	£7,600	54	£1,600
1893	248	£7,200	59	£1,600
1894	231	£7,100	43	£1,200
1895	276	£8,100	41	£1,100
1896	292	£8,500	69	£1,300
1897	259	£7,500	53	£1,400
1898	307	£8,100	72	£1,800
1899	311	£8,900	54	£1,300
1900	286	£8,300	45	£1,300

Source: data collated from Reports of the Inspectors of Fisheries for 1889–1900.

district of Connemara alone in 1912.[64] Such small sums must often have helped to pay for the necessities (if not some small luxuries) that relieved the grinding burden of poverty afflicting the most vulnerable families. Saved over years, such sums must often have helped to make up the fare for an emigrant passage to America for a promising youngster.

For a time there was only one company buying lobsters in Connemara and it paid prices so low that some fishermen refused outright to deal with it, selling instead to a buyer in north Clare who had extensive storage capacity near Finavarra. That was the depot of the Scovell Hamble Fishing Company, which had made a sea-water reservoir there at a cost of between £2,000 and £3,000 sometime before 1880.[65] The whole reservoir was about forty acres in extent, eight being used for lobsters and the remainder being designated an 'oyster ground'. The lobster enclosure was said to be suitable to hold and feed sixty to eighty thousand individuals. The place was in use as a storage pond until very recently and is still in existence. Other fishermen sold their lobsters directly to the owners of tank-boats arriving at Clifden quay. Tank-boats, sometimes called well-boats, had sea-water tanks or wells in which lobsters could be transported alive over long distances. Ironically, it was the Scovell Company that also ran these well-boats. Prior to the construction of the railway to Galway, such boats visited Clifden regularly and took away the lobsters to Southampton.[66]

Some fishermen consigned their catches to salesmen in Dublin and in England, to sell on commission in London, Manchester and Liverpool. Once the Galway to Clifden railway was operating (from 1895), lobsters packed in hampers with seaweed or heather to keep them moist and cool,[67] were brought by cart to the train at Clifden or Recess stations, consigned by rail from there to Galway and onwards to Dublin, from where they were shipped across the Irish Sea. By the time they arrived at market, many would allegedly have died – as many as half the number in some hampers – and on some occasions as few as only one or two individual lobsters allegedly survived. Needless to say, the fishermen had to rely on the honesty of the salesmen regarding the actual numbers allegedly surviving the long journey. It was not unknown for a fisherman to have to recompense a dealer for the cost of rail transportation due to the survival of insufficient numbers to cover that expense.[68]

Sales from Clifden were among the biggest in the whole country – big enough, indeed, to be quite incredible. The reported landings, especially in the last years of the nineteenth century are truly amazing, if they could be relied upon: in 1893, for example, there were 237,876 lobsters reportedly landed at Clifden and 99,000 at Galway according to one source.[69] Numbers were even greater in 1894.[70]

Fishing close to shore for white fish for home use and local distribution was widespread, even if it was carried out more sporadically and by fewer people. It was done using long lines, generally by the same men and boats that worked the lobster fishery. The Inspectors of the Congested Districts Board reported in 1892 that in the Carna district 'the number of *bona-fide* fishermen in the district are [*sic*] few, but ... a large number of families fish "off and on".[71] In the Clifden district 'the number of boats solely employed in fishing is practically *nil* except at Inishshark and Inishbofin Islands'; in Letterfrack 'a good many families fish pretty constantly when not engaged on their farms, but at other periods of the year not more than a dozen men ... fish regularly'; and in south Connemara 'there is a good deal of fishing'.

One South Connemara village, *Trábán* on Gorumna Island, with thirty currachs and ninety fishermen, depended heavily on local fishing. The boats, each crewed by three men, rowed out three to four miles in search of fish. The average catch, using long lines or spillets, was reported to be about three dozen cod or ling per boat per week, i.e. about one dozen fish per man per week.[72] No serious fish market or curing station could ever survive profitably on catches as small as that. The P&R Commission had built two huge entrance piers, almost five meters high, at the mouth of a small creek opening off the exposed beach at *Trábán*, making a sheltered haven 'for two hookers and some small boats'.[73] Boats could be moored or beached inside the harbour, which had a rude inner quay (RH 118). At low tide the entrance appears incongruously like a coastal fortification and it is not easy today to understand the justification of such a harbour for so small a fishing 'fleet' at that time. A road was also made from it to the village. Today the harbour does not appear to be in use.

The piers that were used most prominently for sea fishing (not necessarily all at the same time; centres would wax and wane as circumstances changed and the century advanced) included Inishbofin, Cleggan, Clifden, Bunowen, Roundstone, Mace, Crumpaun Carna, Ardmore, Kilkieran, Cashla (No. 137X1), Barna, Claddagh, Kinvara, Bush Harbour, Bournapeaka and Glenina. Generally speaking, these piers were sited on the open coast or at the entrances of the big bays. That meant they were closer to the known and reputed fishing grounds and the boats lost little time in sailing to their preferred fishing stations. Many of them are also among the more substantial piers, as might be expected from their more exposed locations.

In various periods, curing stations, fish stores and other land-based light industrial plants associated with a fishing industry were erected at a small number of the sites, but they rarely lasted more than one or two seasons. One such operation was that of the Connemara Industrial Company Ltd, which cured fifty barrels of mackerel at Kilkieran and Carna in 1889 for export to Manchester and Liverpool, but then abruptly ceased activity.[74] The supply of white fish was inadequate to sustain curing operations for long, and the much-vaunted herring and mackerel fisheries were too seasonal and too unreliable to support a permanent curing industry. Absence of large population centres and remoteness from established fresh markets also meant little opportunity existed for a fresh fish industry; unless the catch could be cured locally and quickly, there was little that could be done with it. As a result, when seasonal gluts occurred, as sometimes happened with herring and mackerel, the excess catch was spread on the land as fertilizer.[75] Some commentators saw this practice as wasteful, when in fact it was the only possible use of the catch in the absence of a market and curing facilities.

In 1887, at the height of the great decade of pier building, there were, in all of Ireland, curing stations of commercial size only in Dublin city, Dungarvan and Castletown Bere.[76] But on a domestic scale, local drying or curing was widely practised. The Galway and Clare fishermen salted their small catches at their own homes and sold them in Galway, Clifden and Westport.[77] Small villages had grown up around many of the piers – at Cleggan, Clifden, Roundstone, Carna, Kilkieran, Barna, Kinvara and Ballyvaughan, for instance – creating small local markets, and some were within reasonable distance of Galway city, the largest market (small though it was) for fresh fish in the west of Ireland. The north Clare piers had land communication with Ennis and Limerick, towns not entirely easy to access over the Burren, but providing possible outlets for fresh fish landed on the north Clare coast. Even in the Aran Islands' villages, which have reputations as fishing communities, fishing was meagre at very best, and was carried on under conditions of great difficulty as Brady reported in 1886:[78]

> There is a large number of open row boats and curraghs on the three islands of Aran, but that is their only mode of fishing; and they can only fish at short distances from the land, and cannot fish except in suitable weather. There is not a single first-class fishing vessel attached to the islands. The people are too poor to provide themselves with such, or obtain security for loans for such. There is one drawback to such vessels being kept, the want of proper harbour accommodation.

What fish the islanders did catch was salted and laid out to dry on the limestone pavement or on the roof thatch.[79] Some of the semi-dried fish was later sold in Galway city or bartered for turf with the men from Connemara. But prior to the work of the Congested Districts Board from 1892 onwards (chapter 11), the reported fishing endeavours of Aran provided little hard evidence to support the reputation the islands now have for extensive and sustained fishing from the earliest times.

In all parts of the islands and the mainland, families had their own potato gardens; small numbers of cattle, pigs and poultry were reared and sold for cash or credit; eggs were widely traded and shop goods were purchased or exchanged. A relatively small number of shopkeepers – families like the Conroys of Garafin and Kilkieran, the McDonaghs of Crappagh, the O'Donnells of Carraroe, the Connollys of Lettermullan and the Ridges of Mweenish[80]– serviced the rural communities, bringing goods like flour, sugar, tea, tobacco, salt and so on from Galway by sea, largely in their own hookers, taking back the eggs, fish, turf, kelp and other products produced by the local population. The old pier at *Sruthán* (Sruffaun) had steps midway along its wharf wall that were known locally as '*staighre na mine*' (the 'meal stairs'), where bags of flour and meal used to be unloaded. This information comes from an older local resident, encountered fortuitously at the site. It is probable that the steps were one place where relief provisions were landed from the relief vessel HMS *Valorous* when it visited the area in 1880. The relief squadron under the Duke of Edinburgh distributed 182 tons of meal, 52 tons of flour, 24 tons of oatmeal and 333 tons of potatoes, supplied by the Mansion House Relief Fund and the Relief Fund of the Duchess of Marlborough.[81] It may be from that episode of relief food distribution that the steps originally got their evocative name. Modernisation of *Sruthán* pier in recent times has meant *staighre na mine* are now long gone and no doubt even the memory of them, with its resonance of distress and the kindness of strangers, will fade away completely in time.

Other Uses of Small Piers and Quays

For the ordinary domestic routines of humble coastal people – going to Mass; attending school, markets and fairs; shipping cattle and other livestock; meeting and helping neighbours; and the general social intercourse of weddings, wakes, funerals, patterns and local pilgrimages common in the region – small, unsophisticated landing places sufficed to meet their needs. These activities necessitated communication between the many inshore islands and the mainland, and between mainland communities on opposing shores of the long inlets. Sea passage across the bigger bays like Greatman's Bay and Kilkieran Bay was shorter and could be less demanding than walking the long circuitous distances around the boggy heads of the inlets. Communication with Galway, north Clare and the Aran and Inishbofin Islands formed part of everyday life for some people. Movements of men and beasts were often difficult and hazardous, especially in winter, whether on foot at low tide or by boat when the tide was high. In following their mundane activities, island people either walked along the rough intertidal trackways that connected one island with another when the tide was out, or else waited until the tide filled and then rowed, or sailed, to their destinations. The nearer a landing place was to individual homes or settlements, the easier and more convenient such movements were.

Just as people today try to park their cars as close as possible to their homes, people in Connemara sought to moor or beach their boats as close as possible to where they lived; the currach or rowboat of nineteenth-century Connemara was the motor car, or more precisely the small white van, of its time and place. Even the smallest clear piece of shore sufficed as a landing place for the currachs or rowboats (*curachaí adhmaid*) of which most families possessed one at least. Countless numbers of such small natural landing places still exist and greatly outnumber the larger public piers and quays. Larger hookers (aquatic precursors of the ubiquitous road lorries of today) needed to be beached on a level keel for loading, and the boatmen would often make a semi-permanent berth called a *stad trí cloch* ('three-stone berth'). Two of the stones beside the gunnel amidship helped to hold a grounded hooker level, the third stone being a mooring stone to which the boat was tied.[82]

In some places where a landing site served a community of families, the inhabitants cleared away obstructing rocks and manoeuvred large stones to one side to form a rough breakwater. In that way, they cleared and improved the landing place while providing some rude shelter from the swell. In other places, small cuts were made into the boggy shore to provide sheltered lying-up areas, with or without building a rough, rocky 'quay' alongside. One of the smallest such quays recorded by Robinson[83] is that known as *Céibh an tSagairt* ('the priest's quay') in *Camas Íochtar*. This is a small cut into the boggy land with a rough, stone quay built along one side to facilitate embarkation and disembarkation. The priest from *Ros Muc* used to say Mass in a small church (now in ruins) located inland a little north of the quay. He was taken by boat from his home across *Cuan Chamais* to *Céibh an tSagairt* and from there he walked the rest of the way to the church. Such small quays and landing places were generally made by the people themselves but they sometimes received help and even small money contributions towards their construction.

Brady and Joynt, as we saw, dispensed small sums to help with slips and landing places of this kind, for which there appear to be no detailed records. James Hack Tuke was similarly generous with small sums for like purposes as he passed through distressed districts. When work was done by the OPW, under the *Relief of Distress 1880 Act*, a saving was made on the estimates by employing direct labour rather than contractors. The OPW put these savings to good use: according to the 52nd Report of the Board, the saving amounted to £932 which was spent 'on additional useful work at eleven piers' for which there is no further documentary information. In fact, most of the very small vernacular piers and quays have no public documentary records at all. Many of them appear to date from the 1879 to 1889 period when charitable bodies were operating throughout the West and people like Brady, Tuke, Joynt, Gaskell and others supervised small private relief works that were overshadowed by the official operations of the FPC, the FPH Commission and the P&R Commission.

Such, then, were the maritime occupations and livelihoods of the humble people whom these humble works served. It was not the rustic idyll of the Celtic revivalists, but neither was it a life prostrated by poverty or entirely mired in passivity, lassitude and indolence, as some accounts of the time seem to relish reporting. The absence of a big-cash economy does not necessarily equate to the direst poverty, and perhaps the plight of the western poor, severe as it undoubtedly was, was not much worse than that of their contemporary urban counterparts, who worked long hours in the mills and mines of industrial Britain, or crowded in desperation into tenements in the larger Irish cities.

The nature of humble people's lives on the Galway and north Clare coasts was shaped by their basic needs and the natural resources available to them, but their rhythms were determined by the tides and the seasons. People had evolved an 'amphibious culture'[84] based on local production and trade across and around Galway Bay, supporting a simple economic livelihood that utilised the natural features of the region, and depended on the sea for its main thoroughfare. W.S. Green described that life:[85]

> In some districts kelp is burned, and, at the beginning when we got boats for training crews, we found it sometimes difficult to get men to go away from their kelp burning. In South Connemara, too, the men have a constant source of income in carrying turf from Connemara to Galway, and whenever we have tried to get crews from South

FIGURE 10.4. Left: *Céibh an tSagairt.* Right: Canower Quay. Photos NPW.

> Connemara into the fishing boats, we have found they have always been hankering after the turf, and going back to carry turf whenever the fishing slowed down. They leave the fishing boats then and go back to the turf carrying.

The small piers and harbours were essential nodes in the network of trade, communication and social intercourse that held the fabric of the coastal communities intact. How, in such an environment, could they ever have been anything else?

The Allport Commission, and the Select Committee on Harbour Accommodation before it, had laudable ambitions to introduce deep-sea fishing to the west of Ireland, ambitions the Congested Districts Board would also share in the years to come. All were well aware of the existing human conditions and the maritime occupations of the coastal communities, but all failed to appreciate the real importance of these, and how they could be improved. The Allport Commission, for instance, stuck rigidly to its mandate (although, as a Royal Commission, it was free to direct its attention in whatever direction it wished), simply dismissing what it heard in evidence by remarking that 'we have received some evidence on the question of the capture of lobsters and of oysters; but as these matters do not form part of the reference to this Commission we will merely refer to the opinions on them expressed by Sir Thomas Brady and Mr Coleman, which no doubt are worthy of consideration.'[86] This arch comment, made almost as an aside, exemplifies the focussed perspective of the Commission's approach, which aimed to advance a ready-made solution – deep-sea fishing – to a situation that the Commissioners understood perhaps only poorly, or did not really wish to hear about.

Belief was also widespread at the time, with little or no secure empirical evidence, but based on anecdote and unsubstantiated opinion, that fish were in great abundance off the west coast and it only needed sufficiently large boats and large safe harbours in order to reap a bonanza.[87] Herrings in the earlier period, and mackerel in the later nineteenth century, certainly visited the coast annually and entered many of the big bays, where they were relatively easy to catch and were eagerly sought by local boats. However, they were only a seasonal resource and did not remain available all year round, or indeed continue for very many years. Consistency and sustainability are as important as abundance for a fishing industry to be viable, with associated markets that can develop and endure. The Allport Commission rightly dismissed the assumption that abundant, untapped fish stocks existed off the west coast, because it did not hear any hard, convincing evidence for it.

Throughout the 1880s, Inspectors Brady and Hayes had pleaded for the provision of a fishery research boat to survey the coasts, in order to establish what fishing grounds did actually exist, what fish stocks they held, how abundant and sustainable they were, and how best to encourage a fishery for them.[88] This was precisely the kind of information that the Allport Commission lacked. The Fishery Inspectors' repeated requests for a survey vessel to gather such information[89] had been consistently ignored or, worse still, they were accused of seeking a yacht for their own leisure and pleasure![90] Although the Allport Commission did not support the Inspectors' eminently practical and reasonable request for an investigation of the fishery potential of the west coast, the Royal Dublin Society would prove to be much more responsive: within two years, the Society would commence its own fishery surveys under William Spotswood Green. But most of all, it was the establishment of the Congested Districts Board in 1891 that promised greater development and a brighter future for the coastal communities of Galway. We will see in the next two chapters whether it delivered on that welcome promise.

Endnotes

1 A. Jarvis, 'Dock and Harbour Provision for the Fishing Industry since the Eighteenth Century', in D.J. Starkey, C. Reid, & N. Ashcroft (eds), *England's Sea Fisheries*, pp. 146–56.

2 R. Robinson, 'The Line and Trawl Fisheries in the Age of Sail', in Starkey, Reid. & Ashcroft (eds), *England's Sea Fisheries*, pp. 72–80.

3 Select Committee on Harbour Accommodation, BPP 1884 (290), queries 2103 and 2195.

4 Report of the Inspectors of Irish Fisheries for 1880, BPP 1881 [C. 2871], p. 5.

5 W. Le Fanu, in Report of the Committee appointed to enquire into the Board of Works, Ireland [The Crichton Committee], BPP 1878 [C. 2060], queries 1719–20.

6 Report of the Inspectors of Irish Fisheries for 1879, BPP 1880 [C. 2627], p. 9.

7 Data in Report of the Inspectors of Irish Fisheries for 1884, BPP 1884–85 [C. 4545], pp. 5–9.

8 *An Act to extend the Provisions of the Acts for the encouragement of the Sea Fisheries of Ireland, by promoting and aiding with Grants of Public Money, the construction of Piers, Harbours and other Public Works*, 29 & 30 Vict. C. 45.

9 Report of the Committee appointed to enquire into the Board of Works, Ireland [The Crichton Committee], BPP 1878 [C. 2060], pp. xlv–xlvi.

10 Royal Commission on Irish Public Works [The Allport Commission], Second Report, BPP 1888 [C. 5264-I], p. 9.

11 J. Donnell, in 1st Report of the Commission of Inquiry into the Irish Fisheries 1836, appendix 18, p. 121; T.F. Brady, in Select Committee on Harbour Accommodation, BPP 1883 (255), query 1793.
12 R. Manning, in Select Committee on Harbour Accommodation, BPP 1883 (255), queries 4027–28.
13 NAI OPW/8/109.
14 NAI OPW/8/106.
15 Report of the Committee appointed to enquire into the Board of Works, Ireland [The Crichton Committee], BPP 1878 [C. 2060], p. lxi.
16 W. Le Fanu, in Report of the Committee appointed to enquire into the Board of Works, Ireland [The Crichton Committee], BPP 1878 [C. 2060], query 1707.
17 W. Le Fanu, in Report of the Committee appointed to enquire into the Board of Works, Ireland [The Crichton Committee], BPP 1878 [C. 2060], query 1715.
18 Breathnach, *A Word in Your Ear*, p. 80.
19 T. Fraser, in Royal Commission on Irish Public Works [The Allport Commission], Second Report, BPP 1888 [C. 5264-I], written evidence, pp. 681–86.
20 Capt. A. Boxer, in Select Committee on Harbour Accommodation, BPP 1883 (255), query 2169.
21 W. Le Fanu, in Report of the Committee appointed to enquire into the Board of Works, Ireland [The Crichton Committee], BPP 1878 [C. 2060], query 1754.
22 J. Perry, in Royal Commission on Irish Public Works [The Allport Commission], Second Report, BPP 1888 [C. 5264-I], letter dated 6/11/1886 in appendix, p. 721.
23 Blake, *History and Position of the Sea Fisheries*, p. 24.
24 Blake, *History and Position of the Sea Fisheries*, p. 32. See quotation at head of chapter 6.
25 W.L. Joynt, in Minutes of the Select Committee on Harbour Accommodation, BPP 1883 (255), query 2841.
26 Rev. T. Flannery, in Report from the Select Committee on Industries (Ireland), BPP 1884–85 (288), queries 12662–5.
27 Ibid., queries 12684–88.
28 N. Kean, *South and West Coasts of Ireland Sailing Directions*, Twelfth Edition (Dublin: Irish Cruising Club, 2008).
29 C. Mac Cárthaigh (ed.), *Traditional Boats of Ireland. History, Folklore and Construction* (Cork: The Collins Press, 2008), pp. 161 ff.
30 Galway County Council Archive. Galway County Library, GC/CSO/2/56.
31 Galway County Council Archive. Galway County Library, GC/CSO/2/54.
32 Minute of the Engineer-in-Chief of the OPW, 1903. NAI OPW 5 3811 /15.
33 W.S. Green, in Minutes of the Royal Commission on Congestion in Ireland, BPP 1906 [Cd. 3267], query 4733.
34 Ibid., query 5164.
35 Ibid., query 4942.
36 Major W.P. Gaskell, report on the District of south Connemara, in Morrissey, *Verge of Want*, appendix G, p. 180.
37 Ibid., appendix J, p. 182.
38 M. Mongan, in Minutes of the Royal Commission on Congestion in Ireland, Tenth Report, BPP 1908 [Cd. 4007], queries 53993–96.
39 Mrs Blake in Minutes of the Royal Commission on Congestion in Ireland, Tenth Report, BPP 1908 [Cd. 4007], queries 52914–37.
40 M. Mongan, in Minutes of the Royal Commission on Congestion in Ireland, Tenth Report, BPP 1908 [Cd. 4007], queries 53990–92.
41 Rev. M. McHugh, in Minutes of the Royal Commission on Congestion in Ireland, Tenth Report, BPP 1908 [Cd. 4007], query 53270.

42 M. Mongan, in Minutes of the Royal Commission on Congestion in Ireland, Tenth Report, BPP 1908 [Cd. 4007], queries 53985–88.

43 C. O'Connor, in Minutes of the Royal Commission on Congestion in Ireland, Tenth Report, BPP 1908 [Cd. 4007], appendix, query 53875, p. 62.

44 T. Hazell, in Minutes of the Royal Commission on Congestion in Ireland, Tenth Report, BPP 1908 [Cd. 4007], queries 53664–67.

45 J.M. Synge, *In Wicklow, West Kerry and Connemara* (Dublin: Maunsel and Co., 1919), p. 190.

46 Rev. T. Flannery, in Report from the Select Committee on Industries (Ireland), BPP 1884–85 (288), queries 12972–74.

47 A.T. Brennan, *Notes on some common Irish Seaweeds* (Dublin: Institute of Industrial Research and Standards. Published by the Stationery Office, 1950), pp. 11–12.

48 M. Gibbons, personal communication, 2015.

49 E.E. Evans, *Mourne Country*, 2nd edition (Dundalk: Dundalgan Press, 1967); N. P. Wilkins, *Ponds, Passes and Parcs. Aquaculture in Victorian Ireland* (Dublin: The Glendale Press, 1989).

50 5th Annual Report of the Congested Districts Board for Ireland, BPP 1896 [C. 8191], p. 24.

51 7th Annual Report of the Congested Districts Board for Ireland, BPP 1898 [C. 9003], p. 34.

52 13th Annual Report of the Congested Districts Board for Ireland, BPP 1905 [Cd. 2275], p. 32.

53 20th Annual Report of the Congested Districts Board for Ireland, BPP 1912 Cd. 6553, p. 30.

54 19th Annual Report of the Congested Districts Board for Ireland, BPP 1911 [Cd. 5712], p. 38.

55 21st Annual Report of the Congested Districts Board for Ireland, BPP 1914 Cd. 7312, p. 31.

56 Robinson, *Connemara. Part 1.*

57 *Crown Lands Act 1866*, 29 & 30 Vict. C. 62.

58 Return to Parliament, BPP 1912–13 (338).

59 Return to Parliament, BPP 1919 (204).

60 Values aggregated from DATI Reports on Sea and Inland Fisheries for 1914 to 1919.

61 DATI Report on the Sea and Inland Fisheries for 1919, BPP 1921 [Cmd. 1146], p. xi.

62 Report of the Department of Fisheries, Saorstát Éireann, 1927, p. 23.

63 The information in Table 10.1 is collated from the annual Reports of the Inspectors of Irish Fisheries from 1891(BPP 1891 (6403)) to 1900 (BPP 1901 [Cmd. 718]).

64 21st Annual Report of the Congested Districts Board for Ireland, BPP 1914 Cd. 7312, p. 27.

65 Report of the Inspectors of Irish Fisheries for 1880, BPP 1881 [C. 2871], p. 15.

66 Report of the Inspectors of Irish Fisheries for 1892, BPP 1893–94 [C. 7048], p. 41.

67 DATI Report on the Sea and Inland Fisheries for 1900, BPP 1901 [Cmd. 718], p. xi.

68 C. O'Connor, in Minutes of the Royal Commission on Congestion in Ireland, Tenth Report, BPP 1908 [Cd 4007], queries 53832 and 53850.

69 Report of the Inspectors of Irish Fisheries for 1893, BPP 1894 [C. 7404], p. 23.

70 Report of the Inspectors of Irish Fisheries for 1894, BPP 1896 [C. 7793], p. 28.

71 Major R. Ruttledge-Fair, report on the District of Carna. Baseline Reports of the Inspectors of the CDB. In Morrissey, *Verge of Want*, p. 136.

72 Major W.P. Gaskell, report on the District of south Connemara. Baseline Reports of the Inspectors of the CDB, Appendix H. In Morrissey, *Verge of Want*, p. 181.

73 Report of the Piers and Roads Commission, BPP 1887 [C. 5214], p. 9.

74 Report of the Inspectors of Irish Fisheries for 1889, BPP 1890 C. 6058, p. 18. See also Morrissey, *Verge of Want*, pp. 141–47.

75 16th Annual Report of the OPW, BPP 1847–48 [983], p. 24.

76 Report of the Inspectors of Irish Fisheries for 1887, BPP 1888 C. 5388, p. 7.

77 Report of the Inspectors of Irish Fisheries for 1887, BPP 1888 C. 5388, p. 9.
78 Brady, T.F. to Burke, 5/12/1886, in O. J. Burke, *The South Isles of Aran* (London: Kegan Paul, Trench and Co., 1887).
79 See figure 10.50, p. 220 in J. Waddell, J. W. O'Connell and A. Korff (eds), *The Book of Aran* (Kinvara, Co. Galway: Tír Eolas Press, 2nd printing, 1999).
80 Mac Cárthaigh, *Traditional Boats of Ireland*, pp. 159–60.
81 D.G. Morant, Report to the House of Commons on Relief of Distress on the West Coast of Ireland, BPP 1880 [C. 2671].
82 Ó hIarnáin, Coilín. Personal communication, 2014.
83 Robinson, *Connemara Part 1*, p. 107.
84 Royal Commission on Congestion in Ireland, Final Report, BPP 1908 [Cd. 4097], p. 95.
85 W.S. Green, in Minutes of the Royal Commission on Congestion in Ireland, First Report, BPP 1906 [Cd. 3267], query 4942.
86 Royal Commission on Irish Public Works [The Allport Commission], Second Report, BPP 1888 [C. 5264-I], p. 4.
87 Select Committee on Harbour Accommodation, BPP 1883 (255), para 7 (e), p. v.
88 Report of the Inspectors of Irish Fisheries for 1882, BPP 1883 [C. 3605], p. 3.
89 Ibid. See also endnote 91.
90 Major J. Hayes, in Select Committee on Harbour Accommodation, BPP 1883 (255), queries 3812–21 & 3892–96.

CHAPTER 11

THE CONGESTED DISTRICTS BOARD

1891–1922

Some of those who have undertaken to reform the congested districts have shown an unfortunate tendency to give great attention to a few canonised industries, such as horse-breeding and fishing, or even bee-keeping, while they neglect local industries that have been practised with success for a great number of years.

J.M. Synge, 1919[1]

Maintaining the Piers and Managing the Sea Fisheries (Ireland) Fund, 1889–1902

On the termination of the FPH Commission in 1889, the OPW reverted to its normal role in the maintenance of the existing piers as prescribed in the 1853 Act.[2] With no new constructions in train or under consideration, its engineers surveyed all the recently built piers, which they found to be generally in good condition. However, the sea wall at Cleggan was badly undermined by scouring; the necessary repairs were initiated and completed by 1890.[3] Maintenance continued steadily at a small number of other named places (none in Co. Galway or north Clare) each year from then until 1897.

When surveys were carried out at Bournapeaka, Ard West and Cleggan in that year, the first two needed attention and the work was undertaken straight away. Ard West needed the relatively large outlay of £500 for unspecified repairs to the pier that had been built in 1886.[4] In 1900, Trawndaleen and *Spidéal* new pier were surveyed; the latter was then reported to be very dilapidated.[5]

With no new piers being built, there remained in the Sea Fisheries Fund a balance of about £8,000, which had increased to £10,960 by March 1897 and was growing every year as loan interest and repayments came in; one year later it stood at £13,614,[6] and at £16,411 in March 1899.[7] The OPW therefore called a meeting with the Fishery Inspectors to determine how best to utilise the rapidly growing sum. As a result of this, the engineers surveyed sixteen more sites, among them Portstewart in Co. Derry, Portavogie in Co.

Down, Ardmore in Co. Waterford and Kilronan on *Inis Mór*, Aran Islands. Reports and indicative estimates were drawn up for works proposed at all of them, using the balance in the SFI Fund.[8] Three were sanctioned by Treasury for funding in 1900: a new inner harbour at Portstewart (estimate £3,732); a breakwater and wharf with access road at Portavogie (estimate £7,590); and an extension of 100 feet to the main pier at Kilronan (estimate £3,494). Including administrative costs, these three would absorb over £16,600, virtually eliminating the entire balance.

In 1901, *Spidéal* pier, already known to be dilapidated, suffered again in storms and Treasury sanctioned up to £3,000 from the fund for immediate repairs. These repairs, the Kilronan pier extension (see later) and Portstewart harbour, were all completed by May 1902, leaving a new balance of £8,412, the fund having been boosted again by further repayments.[9] This source of money for repairs was proving to be very resilient indeed.

The Congested Districts Board (CDB)

The establishment of a Congested Districts Board was proposed by Arthur Balfour, Chief Secretary for Ireland in 1890, after his tour of the West, and two years after Col. Fraser's report on the successful 'experiment of an executive commission' that surely had influenced Balfour's thinking. Set up in 1891, under the *Purchase of Land (Ireland) Act*[10] of that year, it was intended to be a development board that would initiate and encourage broad economic expansion by 'aiding and developing [various named industries] ... and fishing (including the construction of piers and harbours, the supply of fishing boats and gear, and industries connected with and subservient to fishing)' in electoral districts where the land valuation was less than thirty shillings per head of population.[11]

'Congested' districts were not necessarily the most populous, but they were certainly, and by definition, the poorest districts in the country; almost all lay west of a line from Baltimore to Derry city. (To this day, such a line broadly divides Ireland into two different economic, social and resource-based regions – see, for example, *Irish Agriculture in Transition, A Census Atlas of Agriculture in the Republic of Ireland*.)[12] The CDB, which was never intended as a vehicle for administering short-term relief, was scheduled to remain in existence for twenty years initially, and to continue thereafter as Parliament would determine. Although the Chief Secretary for Ireland was a member, *ex officio*, the Board was completely independent of the Irish executive in Dublin and answered directly to the Government in Whitehall. Its funds were derived entirely from Irish sources – the Irish Reproductive Loan Fund, the Sea and Coast Fisheries Fund and the interest accruing to the Church Surplus Grant Fund – and amounted initially to more than £112,000 *per annum*. James Hack Tuke and Fr Charles Davis, PP of Baltimore, Co. Cork, were among its first permanent members, and William L. Micks was appointed Secretary. W.S. Green, by then an Inspector of Fisheries, was appointed an early, 'temporary' member. He, Davis and Tuke formed the fisheries committee of the Board; Davis died soon after appointment and was replaced by Rev. Patrick O'Donnell,

Catholic Bishop of Raphoe, later to be elevated to Archbishop of Armagh and Cardinal, who would remain a member until the Board was wound up in 1923. He, too, was a member of the Board's fisheries committee.

Within the congested districts, fishery piers fell exclusively under the CDB, and were funded initially from that Board's own resources. Its fisheries and fishery piers operations in Galway can best be considered in three broad chronological phases. The first was a start-up phase from 1892 to around 1896, during which it concentrated mainly on building small piers, landing places and other minor works (CDB Reports 1 to 9). The second, a transitional phase, was from around 1898 to about 1910, during which it paid more attention to larger piers and harbours in co-operation with other new funding agencies; in this phase also it concentrated on developing and encouraging the offshore, pelagic fisheries (CDB Reports 10 to 18). The third phase, from about 1911 to its winding-up in 1923 (Reports 19 to 29), saw it return to building small piers and landing places, but with rather less money and much less enthusiasm.

From the very beginning, WS Green's CDB membership and his continuance as Inspector of Fisheries gave him an influential and overlapping role respecting fishery piers and fisheries in all counties, both congested and non-congested. It seemed possible that the advice repeatedly given by the earlier Inspectors of Fisheries – that the west coast had particular needs and characteristics different to those of the east and south coasts; that development of deep-sea fisheries needed more than just bigger piers; that proper education and training of fishermen was essential; and that an investigation of the fish stocks along the west coast was urgently needed – might, at long last, be seriously addressed, and their recommendations be implemented.

State of the Piers before the CDB Commenced Operation

Just before the CDB commenced operation, the Inspectors of Fisheries published a large map of the whole country (scale 1:633,600), prepared by the Ordnance Survey of Ireland, showing the locations of all the piers and harbours built for the fisheries up to 1890.[13] Almost 200 piers and quays and minor works were shown in all the maritime counties, but this did not include the structures erected by the P&R Commission, or the very small structures funded mainly by the various charities. The map and its accompanying table of information acquainted the CDB with the existing structures and their state of usefulness, creating a baseline from which future pier development might realistically be measured.

The picture presented was not very optimistic: of the thirty-two piers and quays shown in Co. Galway, only one was suitable for deep-sea vessels (depth > 8 feet at LWST); fourteen were fishing piers 'dry at low water'; two more had less than eight feet of water at low tide; five were 'not much used for fisheries'; and ten were listed as mainly 'trading' piers. Of the four piers shown in north Clare, three were listed as 'for trading' and only one (Glenina) as a fishing pier, but dry at low water. In all, only twenty-four of the forty piers illustrated in

TABLE 11.1

Number and usefulness of the piers and quays in Co. Galway and north Clare as indicated by the Inspectors of Fisheries on their map entitled *Ordnance Survey Map of Ireland showing locality and expenditure on works etc connected with Fisheries 1891.*

Piers with more than 8ft of water at LWST:
N = 1. Bunowen.
Piers with less than 8ft of water at LWST:
N = 2. Cleggan; *Spidéal* new pier.
Piers dry at LWST:
N = 15. Rossroe; Leenaun; Glengevlagh; Inishbofin; Errislannan; Inishlacken; Ard West; Mason Island; Ardmore; Kilronan; Barna; Claddagh; Oranmore Castle; Bush Harbour; Glenina (north Clare).
Piers not much, if at all, used for fisheries:
N = 5. Smeerogue; Ballyconneely; *Caladh Thaidhg*; Carrowmore; Durrus.
Piers used mainly for trade:
N = 13. Ballinakill (Derryinver); Barnadearg (Ballinakill); Clifden; Roundstone; Carna; Kilkieran; Maumeen; Cashla; Trawndaleen; Tarrea; Burren New Quay (north Clare); Bournapeaka (north Clare); Ballyvaughan (north Clare).

Source: Report of the Inspectors of Irish Fisheries on the Sea and Inland Fisheries of Ireland for 1890. BPP 1890–91 [C.6403]).

Excavations (one at *Trácht Each*), slips (one at Inishshark) and rock clearances (two, at Gannoughs and *Baile Thíos*) are omitted here since depth of water alongside at LWST is less relevant in their case. Total number of piers and quays: 36. LWST = low water of ordinary Spring tide.

Co. Galway and north Clare were active as fishery harbours, even if they dried out at low tide. They are named in Table 11.1. The depth status of the piers in all counties of Ireland is shown in Appendix I.

Phase One, 1892–98

As a first activity, the Board appointed special Inspectors to assess the existing state of the congested districts and to report back.[14] Major Gaskell was one of these, charged with reporting from Connemara. Their reports are important, informed and informative accounts of the various districts examined. Among many matters, they described briefly the state of the piers and quays and made a number of suggestions as to how they could be restored or improved, or where new ones might be usefully placed. With this information, the Board could start to undertake small marine works, many of them actions that should have been done when the piers were being erected originally. Since many had been built as relief measures, niceties like dredging, rock removals, provision of lights and beacons, clearance of slips and making good access roads had mostly been neglected. As for new piers, the Board commissioned the engineer R.C. Parsons to design 'numerous *small* [emphasis added here] works on different parts of the coast' that would be constructed under his supervision. Parsons favoured the use of concrete in building these and the Board went along with that.[15] As a result, most of the structures erected in the 1890s were built using large concrete blocks, usually cast on site. By 1896, when all the designs had been supplied, and most of the works had been completed, or nearly so, the Board dispensed with Parsons' services, praising him for his diligence and good work. The engineers who worked under him, Messrs Frederick Gahan, Henry Keating, Mr Jellett and Charles Deane Oliver, were retained until everything was finally completed.

Frederick Gahan (1866–1955) continued on permanently with the CDB to become a Senior Inspector in Donegal and Mayo until its dissolution in 1923, when he transferred to the Land Commission. In his later years, he did valuation work for the Electricity Supply Board (ESB) in connection with the Poulaphuca hydro-electric scheme and the Portarlington peat-powered electricity-generating station.[16] Like many connected with the fishery piers, his proved to be a long and productive life, one that, in his case, was a bridge from the old, nineteenth-century technology of the CDB to the modern technology of the ESB.

The works done by the CDB resembled, to a great extent, those of the earlier P&R Commission: they were generally small, directed at facilitating the local inshore fisheries, but not exclusively so; they were widely scattered rather than concentrated in a single locality; they placed some emphasis on the needs of the islands and they improved existing facilities in small but important ways. A number were of uncertain value to fisheries but met the general needs of the places and communities they served. Derrynea quay (No. 136), for instance, situated at the very head of Cashla Bay, is a long distance from the open sea and the fishing grounds. There had been a landing place there from much earlier times and the small quay had been built in 1883. Because it was already a place well known for the export of turf, the new pier built by the CDB was most likely provided to facilitate that particular trade, rather than to assist the fisheries.

Altogether, one hundred and twenty-eight small marine works, principally landing places and slips, were completed in the six congested districts counties (Donegal, Sligo, Mayo, Galway, Kerry and Cork) by March 1900, at a total cost of £47,084. They are listed in appendix 33 of the ninth Report of the Board, published in 1900.[17] The average cost was £368 per site, a relatively paltry amount. Only sixteen had cost more than £1,000 each. Altogether, twenty-eight separate works were carried out in Galway (listed in Table 11.2), comprising four small piers, one slip and six landing places, along with dredging, lighting and other minor improvements at a number of sites.

The piers most relevant to the fisheries were those at Mace Harbour (No. 061) and Rossadillisk (No. 015). North Clare was not then a congested district, so nothing was done there. At other places, simple slips and landing places were provided – at Inishshark, Inishbofin and the Aran Islands (Kilmurvey, No. 174) on *Inis Mór* and on *Inis Meáin* – and at places on the mainland such as Doonloughan (No. 026) and Derrynaclogh (King's) (No. 003). The existing piers at Kilkieran and Kilronan were dredged, and lights

TABLE 11.2

Small marine works undertaken in Co. Galway by the Congested Districts Board from 1891 to March 1900. The number in brackets after each site is its number in the Ryan Hanley survey. The column headed 'date' indicates the first year in which the work is mentioned in the CDB reports. The costs indicated are approximate, but represent the final cost reasonably well. North Clare was not then in the congested districts. N = 21+.

New piers	Cost	Date
Built Derrynea Pier (No. 136)	£296	1892
Built Rossadillisk Pier (No. 015)	£625	1895
Built Mace Harbour (No. 061)	£2,216	1898
Feenish Isl. Improvements to pier (No. 168).	£105	1900
Slips		
Made *Inishmeáin* Slip (No. -) [AUTHOR]	£885	1895
Landing places		
Doonloughan (No. 026) LP	£374	1899
Inishshark (No. 164B) LP	£346	1895
Inishbofin (No. -) LP	£150	1895
Aughrusmore (No. 017) LP	£432	1895
Kilmurvey (No. 174 /5) LP	£292	1893
Derrynaclogh (No. 002B) LP	£96	1897

Other works	Cost	Date
Clifden (No. 021) beacons erected	£269	1895
Kilkieran (No. 702) dredged	£92	1893
Cashel (No. 046 or 047) pier improvements	£140	1893
Kilmurvey (No. 174/5) quay improvements	£21	1893
Kilronan (No. 181) dredged, repaired and lights installed	£152	1893
Cleggan (No. 014) improvements	£622	1896
Killary Harbour. Navigation marks set up	£430	1895
Roundstone (No. 040) rocks removed	£19	1898
Kiggaul (No. 105) improvements	£38	1899
Lights at various sites	£693	1896

were erected at a number of places to make the approach from the sea at night much safer. The aggregate spent on Galway piers and quays by the CDB from 1892 to 1900 was over £8,200, not including the contributions that Galway County Grand Jury approved for some of them.

By concentrating its attention on such small works, the Board appeared to be encouraging and facilitating inshore fishing, whose continuance it anticipated in a forecast it made in its second report in 1892:[18]

> Fishing in open boats or in canvas *curraghs* or canoes will probably continue for a very long time to be the kind of fishing most resorted to along the western coast of Ireland, and the inshore fisheries could not otherwise be fully availed of, as is known to those possessing practical experience of the coast.

At the Royal Commission on Congestion in 1906, in the words of W.S. Green, the Board would even take credit that:

> the landing piers we have made, the little piers, of which we have built a number, have been the immediate cause of new boats being got and new crews taking up fishing, men that had no chance of fishing before, and new centres of fishing being created. Each of those little slips and piers has been reproductive in that sense.[19]

Transformative Legislation and the CDB

A radical change of direction of the CDB became evident as early as 1896, a critical year for the Board. James Hack Tuke died in January and was replaced by Viscount de Vesci. The Board's fishery committee had now lost two (Fr Davis and Tuke) of its original three members. They were, arguably, the two Board members having the greatest personal familiarity with the activities that were actually carried on by local communities along the coasts of the congested districts. The year 1896 was also the year the CDB decided, for the very first time, to finance a large, deep-water pier: it agreed to grant £3,600 for one at Killybegs in Co. Donegal, provided the Government voted an additional £6,600 to its construction. When Treasury eventually agreed to provide the extra grant – the money was to come from the resilient SFI 1883 Fund – the Donegal Grand Jury stepped in with an additional contribution. This satisfactory financial arrangement caused the CDB to give thought to the likely future funding of all the marine works then under consideration. It recommended that legislation be introduced allowing the counties to take over the works once they were completed and to take on their maintenance thereafter. The legislation for this, the Board said, should be permissive in nature and not prescriptive, and their wish was soon fulfilled.[20]

In an important, wider political development, the old system of local government by landlord-dominated Grand Juries was swept away by the *Local Government (Ireland) Act* in August 1898, and replaced by a new framework of elected County, Borough and District Councils.[21] Section 18 of the Act provided that a County Council could agree to take over any marine works constructed or acquired by the CDB in its county. The provision was couched in the permissive manner that the CDB had hoped for: 'the council of any county may, if they think fit, agree with the Congested Districts Board to take over any marine work', which would then become public property and be maintained by the Council. The Board invited the Councils of Cork, Donegal, Galway, Kerry and Mayo to take over forty-eight of the marine works on which it had already spent about £30,000 (works comprising lights, beacons and rock removals did not count, since they did not constitute real property transferrable to the counties).

Galway and Mayo would not agree to take over any; Cork and Donegal were more forthcoming, but wanted renovations to be carried out on particular structures before they would finally commit themselves; Kerry took over only two small works. Owing to this generally poor response, the CDB resolved, for the future, to offer only a financial contribution for works they approved of, and to leave the business of planning, design and construction, and therefore the eventual maintenance, to the county councils.[22] This initiated a gradual but pronounced withdrawal of the CDB from the provision of small fishery piers.

The Local Government Act 1898 was the first of a number of Acts passed at that time that transformed the funding of fishery piers, changing it from a fishery and, in all but name, a 'relief' activity to a much more industrial exercise. Indeed, 1898 marks the end of the nineteenth century from a fisheries pier perspective: all future piers would be predominantly industrial undertakings driven by Acts like the *Department of Agriculture and Technical Instruction (DATI) Act, 1899;*[23] the *Marine Works (Ireland) Act, 1902;*[24] the *Ireland Development Grant Act 1903*[25] and the *Act to Promote the Economic Development of the United Kingdom and the Improvement of Roads Therein (1909).*[26]

The Department of Agriculture and Technical Instruction Act, 1899

This Act established a new Department of Agriculture and Technical Instruction (DATI) and transferred the Fishery Inspectorate, along with its staff, into the new DATI establishment. There the Inspectorate became the Fisheries Branch of the Department, with an annual budget of £10,000 to be applied exclusively to promoting the sea fisheries. Eleven years had passed since a recommendation to place the Inspectorate under another government department had been made by the Allport Royal Commission, which can now be seen to have cast a long shadow. Thirty years of independence of the Inspectorate, established in 1869 under the Fisheries (Ireland) Act of that year,[27] ended there and then and from 1899 onward (excepting the brief period 1923–25), fisheries would be a subsidiary branch of one Government department or another, right up to the present day. W.S. Green was appointed Chief Inspector of Fisheries and head of the Fisheries Branch, and he retained his 'temporary' membership of the Congested Districts Board. The Act also prescribed that the vice-president of DATI, Horace Plunkett, would be a permanent member, *ex officio*, of the CDB. Under section 18 of the Act, DATI, at the request of the CDB, could exercise and discharge any of the powers and duties of the latter Board in any congested districts county, provided the expense incurred was defrayed by the CDB, or from some other source.

Having already absorbed the Fishery Inspectorate, DATI seemed poised to take over all the CDB's fisheries responsibilities, something the latter might not have opposed very trenchantly, given the multiplicity of its other duties. Indeed, the Recess Committee,[28] which had been chaired by Plunkett, had been instrumental in the establishment of DATI, and had envisaged that DATI would take over the fisheries duties for the whole country in the course of time.[29]

The Marine Works (Ireland) (MWI) Act, 1902

The second significant and relevant enactment of the time, the *Marine Works (Ireland) Act* of 1902, gave a huge boost to fishery pier renovation during phase two of the CDB's activity. Under this Act, Treasury allocated £100,000 for grants to be given for marine works carried out by the OPW, where it was certified by the Lord Lieutenant that the works 'were necessary for the development of an industry or trade carried on by the inhabitants of a congested districts county and which, owing to exceptional circumstances, could not be executed without special State assistance'.[30] A condition of the awarding of grants was that the CDB, DATI and the relevant County Council were to agree in advance to contribute, either with finance or in kind, for example using their own staff, or by giving a free grant of land for the intended work. Local proprietors were also required to give all reasonable financial and other assistance to the objectives of the individual works proposed.[31]

The Councils of the maritime counties had, in addition, to covenant an annual sum, one half of 1 per cent of the cost of each relevant pier, to a general pier maintenance fund that would be set up and administered by the OPW. The Lord Lieutenant could fix appropriate tolls for the use of the piers erected, and the OPW could appoint harbour constables to control their use. For a period of ten years, county rates would be levied only on the value of the land actually taken up by any pier or quay. Most importantly, once they were certified for grants by the Lord Lieutenant, the piers were to vest in the OPW. Where piers were already public property under earlier Acts, the County Councils could agree to vest them in the OPW, who would then maintain or renovate them in the normal way.

The thrust of the Act was, therefore, to restore the role of the OPW in the fishery piers of the congested districts, a position it had lost under the 1891 Act.[32] It was an important restoration: under the influence of the above three named Acts, the provision of fishery piers was moving rapidly to become a twentieth-century industrial development activity and the OPW was by far the most appropriate body to implement this change and modernisation.

The Ireland Development Grant Act, 1903

This Act provided a special grant of £185,000 a year, to be used by the Lord Lieutenant for education and other developmental purposes in Ireland. It was in operation for three years before it was invoked in the construction of fishery piers. Because the *Marine Works (Ireland) Act* of 1902 applied only to the congested districts counties, the Development Grant Act balanced this by being applicable only to piers and quays in the non-congested counties. The CDB, however, did receive an annual subvention of £20,000 from the Development Grant Act fund which, at £1.2 million, was a far larger fund than that of the MWI Act.

In March 1907, the authorities in Portavogie, Co. Down, proposed major alterations and renovations to the pier and causeway built there only four years earlier. Fortunately for them, the Development Act fund was available to grant £6,216 for the extensive and expensive work contemplated.[33] Arklow harbour in Co. Wicklow then followed suit and was granted £14,000 under the Act, with £11,250 of this being directed through DATI;

Wicklow Harbour Commissioners joined in the benefit, receiving approval for a grant in excess of £20,000 for the harbour there.[34] Sligo Harbour Commissioners received sanction for a free grant of £20,000 for improvements at Sligo, for which plans were approved in 1908.[35] All these applications were for structures with commercial uses, rather than exclusively for fisheries, and the amounts granted were orders of magnitude greater than the sums administered by the CDB up to then. A major significance of these Development Act grants was their confirmation that an administrative change was underway in earnest: the County Councils were exercising their new powers – in some cases with the assistance of bodies like harbour boards and railway companies – by taking a much more forward role in the provision of large piers and quays.

By and large, it was the County Councils that initiated new proposals for piers and that contributed the matching local finance required by the Development Grant Act. This gave a decidedly more pronounced self-help, commercial aspect to all applications, which the legislation no doubt anticipated and certainly encouraged. The new funding was not aimed at rectifying previous deficiencies and omissions, but at facilitating and encouraging new, forward-looking developments and innovations. This would become even more evident some years later when new funds under a different Development Act would come on stream. But the *Ireland Development Grant Act* of 1903 was not insignificant in contributing about £1.25 million for education, transportation and piers in Ireland between 1903 and 1911. It is a much-overlooked piece of legislation that was repealed in the UK and in Saorstát Éireann in 1923.[36]

The Roles of the OPW, the CDB and DATI

In practice, therefore, from around 1901 onwards, fishery piers and quays in Ireland were constructed and improved using three different funding mechanisms:

> Small works in the congested districts, generally costing less than £1,000 each, were built and funded by the CDB from its own resources, with or without local County Council support (Table 11.3).
>
> Large works in the congested districts, costing in excess of about £1,000 each, were funded under the MWI 1902 Act, with minor contributions from the CDB (10 per cent of estimated cost), County Councils and some private contributions. The works were built, or their construction was supervised, by the OPW (Table 11.4).
>
> Large works costing in excess of £1,000 each in counties outside the congested districts counties, were funded initially from the remnants of the old SFI 1883 Fund, with minor contributions from DATI, the County Councils and private sources. The SFI Fund was replaced by the *Ireland Development Grant Act 1903* funds and, later, by the *Development and Road Improvement Act 1909* funds.

TABLE 11.3

Small marine works undertaken in Co. Galway by the Congested Districts Board from 1901 to 1909.

Small piers and breakwaters	Cost	Date
Trácht Each, Inish Oirr (No. 185) breakwater	£415	1905
Killeaney pier (No.178) repair works	£74	1909
Small slips		
Carraroe (--) slip made	£7	1908
Portnamuck (--) slip made	£313	1909
Landing places		
Inishshark (No. 164B)		1904

Other works	Cost	Date
Cleggan pier (No. 014) shed and repairs	£720	1908
Inishbofin (No. 164) sea wall made	£30	1908
Aughrusmore (No. 017) pier repairs	£37	1906
Kilronan (No. 181) shed and repairs	£1,291	1908
Killeaney pier (No. 178) rings	£5	1908
Inish Oirr (No. -) *[AUTHOR]* Cleared strand for slip at *Baile Thíos*	£327	1901

The number in brackets after each site name is its number in the Ryan Hanley survey. The column headed 'date' indicates the first year in which the work is mentioned in the CDB reports. The costs indicated are approximate, but represent the final costs reasonably well.

TABLE 11.4

Large marine works (cost > £1,000) undertaken in Co. Galway under the *Marine Works (Ireland) Act (1902),* 1903–11.

Site	Date	Final cost	MWI	CDB	Other
Kilronan (No. 179) Rebuild *an tSeancéibh.*	1903–6	£3,320	£3,080	£240	
Cleggan (No. 014) Make cattle slip and BW.	1903–7	£3,734	£3,474	£260	
Cleggan (No. 014) Extend the pier.	1907–11	£9,797	£8,797	£1,000	
Roundstone (No. 040X) Extend the north pier.	1903–6	£1,949	£1,700	£250	
Kinvara (No. 160) Restore the pier and wharf.	1905–6	£1,368	£1,200	£200	Co. Co., DATI

The dates indicate the period of construction. MWI = amount granted from the Marine Works (Ireland) Act 1902; CDB = amount contributed by the Congested Districts Board.

To facilitate comparison of the data in Table 11.4 (which applies to Co. Galway alone) with data for other counties in Ireland, appendix I lists all the large works proposed or completed in all of Ireland between 1902 and 1922. Part (a) of the appendix refers to the congested districts counties (excluding Galway); part (b) refers to the non-congested districts counties. Some of the works in part (b) were not completed due to the Great War of 1914–18.

Maintaining the Existing Piers

Keeping the earlier piers, previously vested in the counties, in good repair was quite a separate activity from building new works, and both DATI and the CDB avoided all involvement in such maintenance. From 1893 on, the Coast guard made annual returns to the OPW on the condition of the piers in the various locations. Where they reported damage or disrepair, the OPW brought the matter to the attention of the appropriate County Council, requesting that the County Surveyor attend to the matter. That was generally the first of many letters that would be exchanged before the work was actually undertaken. At Garafin pier, for example, the deck was in very poor condition in 1903. The OPW had to write five times to Galway County Council before £43 worth of work was done on it in 1905![37]

Generally speaking, the County Councils did the necessary repairs just before the Lord Lieutenant felt it necessary to instruct the OPW to undertake them under the provisions of the 1853 Act. Occasionally, the OPW engineer's exasperation with the Councils became apparent in his memoranda to his Board. When, for example, Galway County Council failed to repair Lettercallow harbour (No. 099) after many reminders, the engineer's memo described the situation. The original work had been done by the Piers and Roads Commission, he stated, at a total cost of £142. The repairs now needed would cost £100. As the harbour was not used for fishery purposes and apparently was little used for any other purposes either, he did not think it was worth raising it with the Lord Lieutenant and 'it appears futile to carry on the correspondence with the County Council any longer and I suggest that the matter be allowed to drop.'[38] The County Council did eventually proceed with the Lettercallow work, if only to save face. Monitoring and maintaining the piers and quays from 1892 to about 1915, therefore, was mainly a three-way affair between the Coast guard, the OPW and the County Councils. Many of the old piers were maintained in a minimal way although Galway County Council had its own piers and harbours committee to consider and approve the requisite expenditure.

From 1915 on, the Great War caused a disruption to normal activities; after the war, it seems that little maintenance work recommenced anywhere before the old OPW was replaced by a new Board of Works of Saorstát Éireann.

Winds of Change: The Royal Commission on Congestion in Ireland

By 1906 winds of change were blowing: the CDB was fifteen years established, and congestion did not appear to be easing significantly. Led by Sir Anthony McDonnell, Undersecretary to the Irish executive, opinion was gathering momentum that the CDB should not be extended beyond its scheduled termination in 1911. Parliament accordingly set up a Royal Commission in 1906, called the Royal Commission on Congestion in Ireland (RCCI), to make a thorough investigation of the CDB's activities and progress, and to advise on its future. It is sometimes referred to as the Dudley Commission, after its chairman, Lord Dudley.

Fisheries, along with the piers and harbours, were not the Commission's main focus, but it succeeded nevertheless in eliciting a considerable amount of information on these topics. McDonnell, who was a member of the Commission, was intent on having all the CDB's fisheries responsibilities transferred to DATI, so his examination of witnesses was understandably focussed, thorough and demanding.

In his evidence regarding the CDB's fisheries activities, Green explained what the Board had ostensibly set out to do and was currently engaged in:[39]

> We are trying to build up fishermen as fishermen, pure and simple, because we believe, when we look round to the Scotch boats, we see that the men, up, say, in Banffshire, can live in the greatest comfort out of the local fishing – but in order to do that, they have to wander over hundreds of miles of sea sometimes to get good fishing. So it comes to this: we have a certain amount of fishing on our west coast. We must try to train up men to take the full value out of the local fishing, and then when it is not to be had near their homes, to go to places where the fishing is really good, put in a season there and then come back to their homes; such men will make money out of fishing which will enable them to live in comfort.

Whether the CDB had really set out from the very beginning with this clearly enunciated plan to 'build up fishermen as fishermen' and 'follow the fishery' is moot, as evidenced by the Board's providing support exclusively for small piers and landing places during the early part of phase one (see quotations on p. 287 above). But it certainly was in pursuit of Green's stated goal that the Board decided for phase two to supply bigger boats and to provide training by Scottish fishermen in their use, and in the use of sea drift nets. Such bigger boats needed larger and deeper quays and piers and their introduction did bring some beneficial results, especially to the fishermen of Donegal. A relatively large number of substantial piers and harbours were constructed in that county using the SFI, MWI and Development Commission funds from 1902 onwards. There was considerable emphasis during the Royal Commission hearings, especially in Green's oral evidence, on deep-sea and pelagic fishing. He recounted in detail the Board's efforts to build bigger boats, to give loans for boats and gear, to obtain and train crews and in other ways to encourage fishing for herring and mackerel in the offshore area. These efforts proved a failure in Connemara, and had only a limited and transient success at Aran and Inishbofin Islands. Green attributed the lack of success in South Connemara to the habits and nature of the coastal population:[40]

> They are not as enterprising as the men in other places. There is a great difference in their character. They are not at all as pushful. They don't keep their houses or anything else like they do in other congested districts. They have not got the same ambition.

It was not that they were inadequate boatmen, unprepared to fish, but they had, he alleged, lost the skill of fishing. They needed to be taught how to do it, and be trained in the demands of a fishing industry, something that would take a very long time: it was not a matter of a year or ten years, he said, it was really a matter of at least a couple of generations to get 'a real maritime spirit' going with them.[41] On the other hand, he and others, local residents and Catholic priests, described to the Commission the daily occupations of the people, which gave no indication that they lacked energy or ambition, or were indolent or lacking in boat skills.

The Final Report of the Royal Commission on Congestion

The Commission on Congestion sat for two years, published a series of interim reports, and presented its Final Report in 1908.[42] It supported the continuation of the CDB into the future, despite the staunch opposition of McDonnell. As to the fisheries, and seemingly contrary to the evidence, it accepted that the creation of an offshore fishing industry was a task that had faced the Board *from its inception*, and was one that was still before them. But it recognised and cautioned that not every congested district was amenable to taking up and actively pursuing an offshore fishery:[43]

> In some places men make a fairly good living entirely from inshore fishing, especially from lobster fishing, if marketing facilities are convenient; but, for the most part, these men are partly dependent on their small holdings and must lead a somewhat amphibious existence. It would not seem desirable in general to attempt to effect any change in their condition ... In some places, especially in South Connemara, they do a very considerable export trade in turf, though supplies, conveniently situated, are diminishing, and to others, kelp-making affords a not inconsiderable source of income. To a large number, however, inshore fishing is the most important business. Except in a few places where there are good landing places for large boats and transit facilities, it does not seem desirable to attempt to introduce deep-sea fishing, unless surveys and investigations indicate that there are really good fishing banks in the vicinity. The most that can be done is to assist the men to better, though not necessarily larger, boats, and to improve, where absolutely necessary, boat slips and landing places, though it is obviously impossible to meet a large proportion of the applications made in this respect.

These observations reflected what the Commissioners had crystallised from the evidence of the various witnesses, and were broadly the same as those repeatedly made down the years by Brady, Blake, Joynt, Nolan and many others. The Commission's eventual recommendation based on these observations – that it was not desirable to impose deep-sea fishing on all coastal communities – was strikingly straightforward and sensible.

The actual experience of the CDB in South Connemara, where it had failed to attract crews to large boats;[44] where the men were unwilling to leave home to follow sea fisheries around the coast and to places abroad;[45] where loans for large boats went unpaid for lack of success in fishing;[46] where the coast was exposed and difficult, and land transport poor; and where the small holdings, the turf trade, the kelp trade, the lobster, oyster and inshore fishing were well established and actively pursued (even if some were small-scale), should have alerted the CDB much sooner that deep-sea fishing was never likely to be readily accepted and made viable in that quarter. In attempting to introduce deep-sea fishing, the CDB was bringing a pre-conceived remedy to an ailment it had failed to diagnose correctly, or understand properly. This was exactly what other Committees and Commissions – including the Allport Royal Commission – had already attempted without much notable success. In fact, of the many down the years who had addressed the difficulties of the Galway coast, Col. Fraser alone had accurately grasped the nature and circumstances of that coast and its communities and how they could begin to be addressed: [47]

> In certain respects, the natural features of the coast meet the wants of the people, and are otherwise favourable. The mass of the inhabitants are farmers, and are non-gregarious. Each man, even if a large harbour were available half-a-mile off, prefers to keep his boat beside or inside his cottage, and I doubt that the creation of large harbours would secure their use, except in a few instances ... The first and most pressing need, I think, is to enable the largest possible number of the coast population to get what fish they can catch to a profitable market, and so to increase the mean prosperity of this very poor chain of communities.

In simple terms, the CDB might more profitably have supported and developed the already existing maritime occupations in the region or district, rather than seeking to graft uncertain deep-sea fisheries on to unwilling communities. We will see in the next chapter how the CDB responded to the recommendations of the Royal Commission.

The Piers and Quays of the Congested Districts Board

Some Piers of Phase One, 1892–8

Mace Harbour

The new harbour at Mace, started in 1898 and estimated to cost £2,000, was the structure most obviously directed at facilitating the fisheries. A rough, manmade rocky platform at the mouth of a creek was already in use as a '*stad*' (exact meaning: a 'stop'), a rude landing place where goods were loaded and unloaded at low tide, but which was inundated at high tide and very exposed to the open sea. The old *stad* is still clearly visible on the west bank of the creek, just inside the substantial breakwater that the CDB first erected. The breakwater, which now has a steeply battered sea face, narrowed the mouth of the creek, providing good shelter for a harbour within. The creek itself was deepened by excavation for a few hundred yards inland and a new, straight wharf wall was built along the west bank, well inside the improved harbour space. It soon became a useful general landing quay and small fishery port, with access to the open sea and to the fishery grounds around the Skerd Rocks, as well as to the neighbouring islands, like Mason Island and Mweenish Island.

FIGURE 11.1. Mace Harbour. Top left: the old '*stad*' inside the breakwater. Top right: the sea slope of the breakwater in 2015. Bottom: the wharf wall (erected in 1898) in 2015. Photos NPW.

Rossadillisk Pier

Rossadillisk pier, situated in the townland of Silerna, north of Clifden, was built under the direction of Charles Deane Oliver.[48] It is located on a headland just outside Cleggan Bay, about two miles seawards from Cleggan village, at that time a well-established port with relatively good services and road connections to Clifden, Westport and Galway. The advantage to the fishermen of the new small pier at Rossadillisk was the easier access it

FIGURE 11.2. Rossadillisk pier today. Note the stone-built original pier with steeply stepped nose; the concrete jetty extending out from it is more recent, possibly early twentieth century. Picture taken from the slip, not visible. Photo NPW.

provided to the open sea compared with Cleggan pier. Hyacinth D'Arcy of Clifden had submitted a memorial for a small pier at Rossadillisk in September 1847,[49] but there is no documentary evidence that anything ever came of that early proposal. Apart from that, there is no mention of the place in any of the public documents examined (such as the comprehensive lists of the OPW),[50] and there is no indication of any structure existing there in the map of the Fishery Inspectors published in 1891.[51] Major Ruttledge-Fair, in his baseline survey for the CDB, advised that a pier at Rossadillisk would be a very useful work and was much required at the place.[52] We can infer from all these that there was no pier there when Ruttledge-Fair carried out his survey in 1892. The first modern mention of the place occurs in the records of foreshore permits for the period 1892 to 1896.[53] A permit, which acknowledged the ancient rights of the Crown in the foreshore, had to be secured when any interference with, or development of, the shore was contemplated, so that works must have been in mind for the place when the licence was obtained.[54]

As it stands today, there is evidence of three separate phases of construction. The main part of the pier is a well-built stone structure, about seven feet wide and about 150 feet long, steeply stepped at the nose. From the nose, a concrete arched jetty, narrow and low with a metal handrail, projects outwards towards the navigation channel. The jetty provides boat access beyond the pier nose at times of low water. There is a stone and concrete boat slip beside the pier on its east side.

The question arises whether the stone pier is the original CDB structure, with the concrete jetty constituting a later, twentieth-century addition. Alternatively, the stone pier could be an earlier structure to which the CDB added the concrete jetty. Taken together, the documents suggest that the stone section of the pier is the original CDB work, and the concrete section is a later addition. According to the CDB annual reports, work commenced in 1894[55] and was completed in 1896[56] at a total cost of £625. Up to the 1980s, the concrete jetty was decked with timber, laid across and bridging the concrete supporting pillars. The timber deck was later replaced with concrete slabs and the whole pier and jetty have been recently renovated. The old slip alongside, previously surfaced in stone flags, has now been covered entirely in concrete. Access to the site is along an unmade track. Strangely, Rossadillisk pier is not shown in the 1898 edition of the six-inch OS map.

Feenish Island (Caladh Feenish) pier

The work done at Feenish Island, close to Carna village in Connemara, involved modification of a small harbour located on the northwest coast of the island. There was an old, angled, rubble-stone pier with a low storm wall already in existence there, located on one side of a small inlet. An angled, longshore quay faced it on the opposite side, the two structures enclosing a tidal basin between them. The deck of the old quay was originally unpaved, but coped in limestone. The CDB lengthened this quay with concrete, adding a new jetty section, which projected out at right angles to form a return head. This made the original longshore quay L-shaped and narrowed the entrance, providing greater shelter to the harbour within.

All the older, pre-CDB, stone structures are of unknown date, their form and layout suggesting that they pre-date the Relief of Distress works of 1879–82. The island was part of the Leonard estate, and the original pier and quay may have been erected by the landowner and his tenants at an earlier period. The site is shown as a natural inlet in the first edition of the six-inch OS map (1839), with no evidence of any manmade structures existing there at that time. The original works almost certainly date from a time after 1839; most likely it was built with the aid of charitable funds during the Great Famine. The Leonard estate was purchased by the CDB 'at an exceptionally high price'[57] (£5,345) in 1896.[58] It was one of the poorest estates in the West and included the islands of Mason and Feenish. Feenish Island extends to 153 acres, on which twenty-five tenants and their families lived in the 1890s, twenty-one of them subsisting on only 78 acres.[59]

The island is now uninhabited and, to a large extent, sand-blown. The concrete section of the harbour was in poor condition until renovated in the last few years.

Some Small CDB Marine Works in Co. Galway, 1898–1909

Relieved of having to fund large expensive works after 1902, the CDB could return to supporting small fishery piers and slips in the congested districts. Those completed in Co. Galway cost £3,320 in total, and are listed in Table 11.3. They were built in co-operation with Galway County Council, whose contribution took various forms, such as cash, labour and engineering supervision. The Council also constructed small piers entirely at its own expense, for instance in 1912, at Blackweir (No. 152) (Corraduff) in inner Galway Bay,[60] so that £3,320 was only part of the total outlay at the time. One of the small CDB works of phase two was the construction, in 1905, of a concrete seawall on the western side of the P&R Commission's landing place at *Trácht Each*, near Ballyhees on *Inis Oirr*. The strand at the village nearby was also cleared of rocks and boulders in 1901 to form a village landing slip for currachs, at a cost of over £300. The two most costly works undertaken were at Cleggan in 1908, where repairs and construction of a new shed cost £720, and at Kilronan the same year, where repairs and a new shed came to £1466. These last two locations had received financial support from almost every funding scheme, public and private, since the beginning of the nineteenth century, a reflection of the extent and severity of distress that was virtually endemic on the islands. The full stories of some Aran piers and quays are deferred to later in this chapter.

Large CDB Marine Works in Co. Galway, 1903–13

The MWI 1902 Act was not intended to initiate any new industry or trade, but to assist industries or trade already underway by the inhabitants of any congested district. The majority of the works funded (Table 11.4) comprised deepening, extending and reconstructing existing piers and harbours in order to make them more suitable for large, modern fishing and trading vessels, especially for steamers.

Cleggan, Kilronan and Roundstone were three of only four places in Co. Galway that met the criteria to merit funding under the Act. In those places, the fishermen were taking up the larger boats introduced by the CDB for the pelagic herring and mackerel fisheries. The fourth location was Kinvara, on the south shore of Galway Bay, a place of significant local sea-borne trade. The initial decision as to which locations merited MWI funding was arrived at jointly by the CDB and DATI. With Green as the Chief Fishery Inspector in DATI and still an active member of the CDB main Board, both bodies were understandably *ad idem* on the places that were to be recommended for certification by the Lord Lieutenant.

The sites selected in Co. Galway were sanctioned in 1902 when the Act came into force, but construction was not started in any instance until 1903 or later. First, the exact sites had to be approved in discussion between the CDB, DATI, the OPW and the County Council; then the OPW had to make preliminary surveys and draw up indicative plans and estimates for each one; next, these reports were forwarded to the Lord Lieutenant, who certified the projects that could be undertaken as constituting 'works in aid of an existing trade or industry'. Only on issuance of this certificate could plans be finalised and construction commence, and the first certificates did not issue until 1903.[61]

It seemed as if history was repeating itself: the administrative burdens inherent in the MWI scheme almost rivalled those of the old Fishery Piers and Harbours Act of 1846. There was, however, nothing in the 1902 MWI Act that restricted it to funding only the fishing industry, and it may be seen as a failure of the CDB that it did not seek support under the Act to promote the other marine activities, such as, for example, the lobster and the kelp trades, which were actively being carried on in most of the congested districts. We will discuss these again in chapter 12.

The Gradual Evolution of Cleggan Harbour

Cleggan was a place in which the CDB showed a particular interest. Originally designed and constructed by Nimmo in the 1820s, it had dropped from the public records after the Great Famine, although it vested in the county in 1853 and its maintenance thereafter fell on the Galway County Grand Jury. In 1882, it appeared in the list of memorials as an application for 'improvement of a small quay'. Manning visited the place early that year and drew up a report, preliminary plan and specifications for its restoration.[62] Despite its dilapidated state, nothing was done immediately.[63] In December 1883, the FPH Commissioners requested a full survey and detailed plan, which was provided by the OPW in February 1884.[64]

The place must have been in a really decrepit state by then because it was one of only two Galway sites that were included in the 'top-priority' list No. 1 of works to be funded under the SFI 1883 Act; the estimate for its restoration, which involved repairs and extension outwards of the harbour, was £8,000. The free grant sanctioned was £6,000, the second largest single award ever made to a western locality (only Carrigaholt in Co. Clare had received a larger initial grant); Galway County Grand Jury took a loan of £2,000 to make up the residual contribution.

Work started in 1884 and continued until 1888. The following year, the new works were damaged by a storm and needed final repairs costing £100.[65] Cleggan harbour was becoming a busy place by then, and ongoing repairs and maintenance were attended to by the County without delay. Everything was finally in good shape by 1890.[66] Repairs were needed again in 1896, for which the CDB contributed almost £600 in that year and in 1897;[67] in 1899 the Board paid a further £132 for a light on Cleggan Head to make the sea approach safer.[68] In order to keep the harbour in good repair and to enforce published regulations for its proper use, the CDB leased it from Galway County Council for a period of five years, from 1901, on condition that the Board would not be responsible for any further major structural works.[69] With this arrangement in place, the CDB could ensure its regular upkeep, routine maintenance and regulatory control, using its own resources. Cleggan was an important harbour, not only for the fisheries (the Inishbofin and Inishshark boats and others landed their catches there), but for meeting the wider needs of the people of the islands and improving their access to Clifden, Galway and places beyond by road and rail.

With the harbour busy, well maintained and managed by an established authority, Cleggan was an ideal candidate for major development under the MWI Act of 1902. A plan for a new breakwater and a cattle slip, together with the removal of part of an existing groyne, estimated to cost £3,700 in total, was drawn up and received the certificate of the Lord Lieutenant in 1904.[70] Work began immediately and was finished essentially within budget by early 1907.[71] Provision of a cattle slip testifies to the port's increasing importance for purposes other than fishing, and greatly improved its general usefulness. Success breeds success, and within a year a new proposal, this time for an extension of the main north pier, estimated at £7,300, was certified for further MWI investment.[72] That year also, part of the harbour entrance had become dangerous and the OPW experienced difficulty in getting the County Council to undertake the repairs necessary. The CDB had surrendered its lease by then, and with the failure of the Council to act, the OPW indicated its intention to carry out the necessary work using its powers under the 1853 Act. At the very last moment Galway County Council relented and undertook the repairs under its own County Surveyor.[73] Meantime, the extension of the north pier, planned and funded under the MWI Act, started under contract in September 1908. Unexpected problems were encountered with the foundations, causing the estimated costs to increase to £9,450.[74] This new estimate was quickly approved and sanctioned, and everything was completed in 1911 at a final cost of £9,797;[75] the CDB had contributed only £230 of this.

Other small works at Cleggan, aided by the CDB in the years following, included the addition of a shed,[76] alterations to a cliff wall,[77] and the addition of navigation lights. Between 1913 and 1915, the CDB would also install a water supply to the pier at a cost over £627.[78] As a place deeply involved with fishing and local coastal trade, Cleggan had well merited investment under the MWI Act and the final aggregate amount it received from that source alone was almost £13,500.

The various works carried out from 1884 onwards brought the overall layout of the port to its present general size and shape, still largely conforming with Nimmo's original design, although considerably larger. The return head of the extended pier is a modern addition.

Roundstone Harbour Completed at Last

Roundstone, as we saw earlier, had first been proposed as a desirable site for a village by Nimmo as early as 1813. He designed and completed the wharf and south pier there in 1822–23, one of his earliest works in Co. Galway. A second pier situated to the north of

Nimmo's was reportedly started in 1830, but that amounted to little more than an attempt to build a short, rough mole. J.S. King, late fishery officer for the region, described it to the Fishery Inquiry of 1835: 'There are two piers at Roundstone. The south pier is in good repair. The north pier was built (to its present extent) in 1830, with a portion of the charity funds distributed for the relief of the poor. Its completion would render Roundstone Harbour the best and safest between Galway and Westport.'[79]

Deficient as it reportedly was, however, it was sufficient for Thomas Martin, who owned the place, to submit a memorial 'for the repair of the Harbour of Roundstone' in April 1846.[80] By then Roundstone had grown to a substantial village of around seventy houses, mainly slated rather than thatched.[81] We can surmise from the engineer's estimate (£1,200) for the alleged 'repair' work that the proposal was really for the transformation of the rudimentary mole into a proper new north pier: £1,200 was close to the full cost of building a pier completely from scratch during the Great Famine. At around the same time, Nimmo's south pier was in such good condition that it would require only £33 for repairs before being transferred to the county in 1854.[82] In November 1846, Treasury sanctioned a free grant of £600 for the north pier, with a further £300 to come from the district and £300 from the proprietor.[83] However, only £14 had been spent before work was reported as 'suspended' in 1848.[84] That terse observation masked a sad event that occurred just as things were getting underway: Thomas Martin of Ballinahinch, the memorialist, had succumbed to famine fever contracted, it is said,[85] as a result of his visiting some of his tenants in the poor house; he had a humane reputation for caring for his tenants in practical ways. (His late father, Richard Martin, was known as 'Humanity Dick' because of his concern for the welfare of animals and for bringing into Westminster in 1822 the

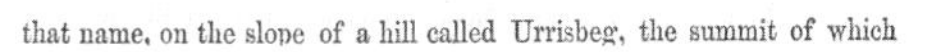
that name, on the slope of a hill called Urrisbeg, the summit of which

FIGURE 11.3. Two views of Roundstone, seen from the same approximate position. Left: engraving from a travel guide, first published in 1866 (Roney 1872), showing the edge of Nimmo's quay with two boats alongside it; there is no evidence of a north pier. Right: photograph taken from the same approximate vantage point in 2015, showing the north pier. Nimmo's 1822 quay is behind the houses to the right. Photo NPW.

FIGURE 11.4. Roundstone north pier, looking from the (cobbled) 1905 extension back towards the original 1881 section (flat stones with grass between). Photo NPW.

first Act outlawing cruelty to animals.) After Thomas's untimely death, all work ceased on the north pier, no one else came forward as a replacement local contributor, and the grant lapsed in consequence. Nimmo's south pier and quay vested in the OPW in 1848[86] and then in 1854, after small repairs, it was transferred to the county. The north pier was to remain uncompleted for almost thirty more years.

In 1879, a memorial was submitted for a 'pier and harbour extension and completion of breakwater' at Roundstone.[87] It was one of those selected by the Fishery Piers Committee (FPC) for funding of three-fourths of the cost out of the sum of £45,000 voted in the *Relief of Distress Act* of 1880. Staff Commander Charles Langdon, a member of the Committee, claimed that both the south and the north piers were in complete ruins when he visited there along with the young engineers engaged by the OPW. 'The north pier at Roundstone, the largest pier was down altogether', he told the Select Committee on Harbour Accommodation in 1883.[88] The truth was that the north pier had never been fully constructed. To build a new pier 313 feet long at the site, with a boat slip eight-five feet in length, and to clear the harbour bed of rock and sand, was estimated to cost £2,000.[89] Treasury sanctioned a free grant of £1,500, with added contributions of £500 to be made up locally.[90] The Canadian Fund Committee gave £150 and other unnamed sources were to provide £350. Work commenced in 1880, £1,603 being spent on it in that year.[91]

Manning visited in June 1881 and confirmed that the work had been completed to his satisfaction. It had been carried out by direct labour under the Board's supervision, with an average of twenty-five men employed daily. Labour alone was reported to have cost £1,178, suggesting that the daily wage was around two shillings per man per day, about the same as was paid at other sites at that time. The final cost of the completed work was £1,935.[92] After a further £132 was expended on it, the north pier transferred to the county in 1882.[93] When asked in 1883 whether it needed any further work, Langdon replied that Roundstone was then, after all the work, 'a very good harbour indeed'.[94] The north and south piers made a sheltered 'safety' harbour as it had a useful slip within and adequate wharf space.

Having been restored in 1881, Roundstone did not feature very highly in the distribution of the SFI Fund from 1883 onwards. It appeared in list No. 2, the piers of secondary importance, and then only as one of a group of 'alternatives':[95] it would be funded if the railway came there as a branch of the Galway to Clifden line; if not, the sum allocated, £1,800, would be used for piers elsewhere. The railway did go ahead eventually, but it did not extend quite as far as Roundstone, and it was other piers in the vicinity, like *Béal an Trá* (Ervallagh), that benefitted from the SFI 1883 money. In 1888, James Perry, the Galway County Surveyor, reported that Roundstone Harbour was 'in fair repair' and 'the old pier useful'.[96] It would remain like that for another decade.

In 1897, some rocks were cleared from inside the harbour by the CDB.[97] It was only a partial solution to the problem: the main problem was the excessive width of the harbour entrance due to the shortness of the north pier. A few years later, the OPW, at the request of the Irish executive, undertook surveys of various piers and harbours to determine their status and requirements for renovation and restoration.[98] The *Marine Works (Ireland) Act* had been enacted and it was necessary to establish which locations might qualify to benefit under it. Roundstone was one of the places selected. An extension of the north pier to cover a dangerous rock and to improve the approach and the accommodation within was approved and certified by the Lord Lieutenant for funding. Tenders were invited but none was received, so the OPW had to take on the task with direct labour.[99] Work was well underway in 1904 and was completed in 1905 at a final cost of £1,949.[100] It involved a partial rebuilding and extension of the north pier and some repairs to the quay and south pier. The CDB contributed £250[101] and Galway County Council most likely made up the balance needed: the Council certainly made a payment, in this case £29, to the maintenance fund as required by the MWI Act.[102]

In 1908, lighting, life-saving and protection appliances were installed.[103] When examined in October 1911 by the OPW engineer, all was found to be in good order and repair.[104] A final small expenditure of £27 occurred in 1912, leaving Roundstone essentially as it is today.[105] The extension made to the north pier by the CDB in 1905 is still distinctly visible and easily distinguished in the stonework of the pier face and the surface of the deck. Nimmo's south pier and quay were two of the earliest public marine structures built in the county in the nineteenth century and amongst the last to be mentioned in the public records before national independence one hundred years later. Roundstone deservedly remains one of the most popular villages of Connemara,

FIGURE 11.5. Roundstone as it is today: Nimmo's quay and pier are in the centre, and the north pier towards the right. Photo by M. Gibbons, reproduced by permission.

considerably enhanced by the harbour with its two piers, amply confirming Nimmo's good opinion of the place.

Jetty Piers Proposed at Clifden Harbour

Clifden never had a jetty pier; its port facilities have always consisted only of a longshore quay. However, no less than three different jetty piers were proposed there early in the twentieth century, none of them at the site, or in place of, the old quay. In 1902, the OPW put forward a preliminary plan for a jetty pier[106] at a site chosen by W.S. Green, about 2,000 feet west of Nimmo's original wharf wall at Doaghbeg. An indicative drawing, still extant, was prepared at the time by the OPW engineer Charles Deane Oliver.[107] It was to be a timber jetty pier projecting out from a stone-pitched approach road; the estimated cost was £3,000 for the jetty and £1,000 for the approach road. Although no site survey had been carried out, it was alleged that the place would be suitable to allow steam ships with a draft of nine feet to approach at all states of the tide. While this proposal was under consideration, W.E. Jackson, who owned a stone quarry in the locality, approached the Chief Secretary for Ireland with a plan for an alternative pier 'of special design'.

Negotiations were opened with Jackson, who forwarded to the Lord Lieutenant detailed drawings and specifications dated 9 January 1904, prepared for him by James C. Gilmour CE of London.[108] These show a trellised pier projecting outwards from an approach embankment pitched in stone. The approach road was 157 feet long and the jetty 200 feet. Jackson's preferred site was located east of Green's original site, somewhat closer to Doaghbeg. The special features were that all structural joists and piles were to be made of rolled steel, braced with steel joints. All piles were to consist of two lengths of steel connected together with 5/8-inch-thick plates above high water, and driven down to the bedrock. The pier head was T-shaped and a two-ton hand crane was to be mounted in one arm of the T. The entire deck was to be of pitch pine. Initially Jackson proposed to build the lot at his own expense; later, when the estimated cost reached £7,158, he changed his stance and said the County Council would not have to pay anything over £4,000.

At the outset, the chief engineer of the OPW was unable to evaluate the proposal properly because, he said, it was deficient in detail. Later, in March 1904, he issued a full assessment. In his view, the scheme was too light for the probable traffic, and the costs were under-estimated. Obviously he suspected, probably correctly, that the pier was mainly intended for the shipping of stone, which might account for its steel construction and the two-ton crane. James Gilmour, Jackson's engineer, responded two weeks later, claiming that Clifden could be developed as a seaside resort and that he had designed the pier to be as long as possible in order to make a promenade for visitors! There was, of course, no mention of the possibility of stone being shipped. He drew up an amended plan which Jackson forwarded to the Chief Secretary on 31 March.[109]

The chief engineer of the OPW considered the new plan and specifications to be 'just within the limit of safety' and the new estimate of £7,490 to be still too low. Nevertheless, on 25 July 1904, the Chief Secretary's office drew up heads of agreement to be finalised with Jackson. Considerable correspondence then ensued between all the parties. While Jackson's final reply does not appear to be extant in the records, we know the final outcome. Stated simply on the cover of the OPW file for Clifden are the final words: 'This file comes up to the stage where Mr Jackson's scheme was finally abandoned'.[110]

On 28 April the following year, 1905, the Under secretary wrote to the OPW inviting the Board to prepare new plans for Clifden pier, enclosing a copy of Green's and Oliver's original plan for a timber trellis structure. The OPW undertook a site survey in May and June and settled on a different site again, this one about 300 feet west of Green's original site, about half a mile west of Doaghbeg. As for Green's original plan for a timber structure, the OPW engineer regarded it as 'totally inadequate as a permanent structure' and went on to draw up a new plan of his own. This, too, comprised a straight jetty extending out perpendicular from the shore and approached by an embanked road.[111] He made some revisions to this a few months later.[112]

In his scheme, all construction was to be in stone, and would cost £7,000 for the pier and £2,000 for the road. He expressed concern, on balanced consideration, that Clifden was not really a suitable place for a fishery pier at all: it would need to be lighted and buoyed and would be liable to excessive siltation. A long series of letters and minutes had been exchanged since a jetty pier had been first mooted in 1902; the uncertainty and indecision this caused resulted in delays that condemned the project to repeated postponement. In the end the delays would prove fatal: rapidly moving events on the financial front outran the slow progress of Clifden pier. The Government had announced, in March 1903, that Galway County Council would receive a grant of £4,000 from the MWI 1902 fund to be used for Kinvara, or Salthill, or Clifden. Of this grant, £1,100 had already gone to Kinvara while Clifden dithered; when the final estimate for Clifden pier rose to £9,000, three times the already-reduced amount available, it could no longer be entertained, the plan was abandoned, and no jetty pier was ever built there. Clifden Harbour remains today largely as Nimmo envisaged it, although he never saw it completed. When the railway arrived in 1895, Clifden became important for the landing of lobsters and their onward conveyance by rail, but this did not last long.

Clifden is an attractive and popular town today but its harbour, located in a very scenic area, is now of little use for the fisheries.

Endnotes

1 J.M. Synge, *In Wicklow, West Kerry and Connemara*, p. 189.
2 *The Grand Jury (Ireland) Act 1853,* 16 & 17 Vict. C. 136.
3 58th Annual Report of the OPW, BPP 1890 [C. 6184]; 59th Annual Report of the OPW, 1890–91 [C. 6480].
4 67th Annual Report of the OPW, BPP 1899 [C. 9465], p. 5.
5 68th Annual Report of the OPW, BPP 1900 [Cd. 314], p. 23.
6 66th Annual Report of the OPW, BPP 1898 [C. 9029], p. 65.
7 67th Annual Report of the OPW, BPP 1899 [C. 9465], p. 75.
8 66th, 67th and 68th Annual Reports of the OPW, BPP 1898 [C. 9029]; BPP 1899 [C. 9465]; BPP 1900 [Cd. 314].
9 70th Annual Report of the OPW, B.P.P. 1902 [Cd. 1261], p. 73.
10 *Purchase of Land (Ireland) Act*, 54 & 55 Vict. C. 48.
11 1st Annual Report of the Congested Districts Board for Ireland, BPP 1893–94 [C. 6908].
12 S. Lafferty, P. Commins and J.A. Walsh, *Irish Agriculture in Transition. A Census Atlas of Agriculture in the Republic of Ireland* (Dublin: Teagasc, 1999).
13 'Ordnance Survey Map of Ireland showing locality and expenditure on works etc connected with Fisheries 1891', in Report of the Inspectors of Irish Fisheries on the Sea and Inland Fisheries of Ireland for 1890, BPP 1891 [C. 6403].
14 The reports of the CDB Inspectors are archived in Trinity College Dublin. They have been reprinted in Morrissey, *Verge of Want*.
15 2nd Annual Report of the Congested Districts Board for Ireland, BPP 1893–94 [C. 7266].
16 O'Donoghue, *County Surveyors*.
17 9th Annual Report of the Congested Districts Board for Ireland, BPP 1900 [Cd. 239], appendix XXXIII, p. 95.
18 2nd Annual Report of the Congested Districts Board for Ireland, BPP 1893–94 [C. 7266], p. 24.
19 W.S. Green, in Minutes of the Royal Commission on Congestion in Ireland, First Report, BPP 1906 [Cd.

3267], query 4738.

20 5th Annual Report of the Congested Districts Board for Ireland, BPP 1896 [C. 8191], p. 26.

21 *Local Government (Ireland) Act 1898,* 61 & 62 Vict. C. 37.

22 10th Annual Report of the Congested Districts Board for Ireland, BPP 1901 [Cd. 681], p. 46.

23 *Agriculture and Technical Instruction (Ireland) Act 1899,* 62&63 Vict. C. 50.

24 *Marine Works (Ireland) Act 1902,* 2 Edw.VII C. 24.

25 *Ireland Development Grant Act 1903,* 3 Edw. VII C. 23.

26 *Act to Promote the Economic Development of the United Kingdom and the Improvement of Roads Therein,* 9 Edw. VII C. 47.

27 *An Act to amend the Laws relating to the Fisheries of Ireland 1869,* 32 & 33 Vict. C. 92.

28 T. West, *Horace Plunkett: Co-Operation and Politics, an Irish biography* (Gerard's Cross: Colin Smythe, 1986), pp. 42–54.

29 T. P. Gill, *Report of the Recess Committee* (Dublin: Browne and Nolan, new edition, 1906), p. 109.

30 2 Edw.VII C. 24, section 1.1.

31 2 Edw.VII C. 24, section 1.1.

32 *Purchase of Land (Ireland) Act 1891,* 54 & 55 Vict. C. 48.

33 76th Annual Report of the OPW, BPP 1908 [Cd. 4228], pp. 7–8 & 53–54. The sum paid under the Act is extracted from the Lord Lieutenant's reports dated 1908, 1909 and 1910.

34 Report of the Lord Lieutenant under the Ireland Development Grant Act 1903 for the year 1908, BPP 1908 [Cd. 4086]; for the year 1909, BPP 1909 [Cd. 4659]; for the year 1910, BPP 1910 [Cd. 5149].

35 76th Annual Report of the OPW, BPP 1908 [Cd. 4228], p. 54.

36 *The Land Act, 1923, Saorstát Éireann,* No. 42 of 1923.

37 NAI OPW5/3806/15.

38 Memo of J.C.S. to the OPW, 12 October 1905. NAI OPW5/3812/15.

39 W.S. Green, in Minutes of the Royal Commission on Congestion in Ireland, First Report, BPP 1906 [Cd. 3267], query 4944.

40 W.S. Green, in Minutes of the Royal Commission on Congestion in Ireland, First Report, BPP 1906 [Cd. 3267], query 5167.

41 W.S. Green, in Minutes of the Royal Commission on Congestion in Ireland, First Report, BPP 1906 [Cd. 3267], queries 5223–28.

42 Royal Commission on Congestion in Ireland, Final Report, BPP 1908 [Cd. 4097].

43 Royal Commission on Congestion in Ireland, Final Report, BPP 1908 [Cd. 4097], p. 95.

44 16th Annual Report of the Congested Districts Board for Ireland, BPP 1908 [Cd. 3767], p. 25.

45 W.S. Green, in Minutes of the Royal Commission on Congestion in Ireland, First Report, BPP 1906 [Cd. 3267], query 4710.

46 17th Annual Report of the Congested Districts Board for Ireland, BPP 1908 [Cd. 4340], pp. 28–29.

47 Written Evidence of Col. Fraser, in Royal Commission on Irish Public Works [The Allport Commission], Second Report, BPP 1888 [C. 5264-I], appendix, p. 683.

48 6th Annual Report of the Congested Districts Board for Ireland, BPP 1897 [C. 8622], p. 26.

49 NAI OPW/8/221.

50 54th Annual Report of the OPW, BPP 1886 [C. 4774], Appendix E, pp. 68–72; Royal Commission on Irish Public Works [The Allport Commission], Second Report, BPP 1888 [C. 5264-I], lists handed in by General Sankey.

51 Report of the Inspectors of Irish Fisheries for 1890, BPP 1891 [C. 6403].

52 R. Ruttledge-Fair, 'Report of Major Robert Ruttledge-Fair to the Congested Districts Board on the District of Clifden, 26 August 1892.' Reprinted in Morrissey, *Verge of Want.*

53 Return to Parliament, BPP 1897 (285).

54 Today, the Irish foreshore is vested in the State and the placement of structures there is governed by the Foreshore Acts.
55 4th Annual Report of the Congested Districts Board for Ireland, BPP 1895 [C. 7791], appendix XXVII, p. 53.
56 6th Annual Report of the Congested Districts Board for Ireland, BPP 1897 [C. 8622], Appendix XXXII, p. 64.
57 6th Annual Report of the Congested Districts Board for Ireland, BPP 1897 [C. 8622], p. 13.
58 5th Annual Report of the Congested Districts Board for Ireland, BPP 1896 [C. 8191], p. 13.
59 7th Annual Report of the Congested Districts Board for Ireland, BPP 1898 [C. 9003], p. 21.
60 Return to Parliament, BPP 1919 (204).
61 72nd Annual Report of the OPW, BPP 1904 [Cd. 2229].
62 50th Annual Report of the OPW, BPP 1882 [C. 3261], p. 68.
63 Return to Parliament, BPP 1884–1885 (266).
64 52ndAnnual Report of the OPW, BPP 1884 [C. 4068], p. 28.
65 58th Annual Report of the OPW, BPP 1890 [C. 6184].
66 60th Annual Report of the OPW, BPP 1892 [C. 6759].
67 7th Annual Report of the Congested Districts Board for Ireland, BPP 1898 [C. 9003], appendix XXXIII, p. 80; Minutes of the Royal Commission on Congestion in Ireland, fourth report, 1907 [Cd. 3509], appendix 1, p. 159.
68 9th Annual Report of the Congested Districts Board for Ireland, BPP 1900 [Cd 239], appendix XXXIII, p. 94.
69 11th Annual Report of the Congested Districts Board for Ireland, BPP 1902 [Cd 1192], p. 40.
70 72nd Annual Report of the OPW, BPP 1904 [Cd. 2229], p. 10. Estimate of the costs is given in 74th Annual Report of the OPW, BPP 1906 [Cd. 3109], p. 9.
71 75th Annual Report of the OPW, BPP 1907 [Cd. 3693], p. 8.
72 76th Annual Report of the OPW, BPP 1908 [Cd. 4228], p. 9.
73 76th Annual Report of the OPW, BPP 1908 [Cd. 4228], p. 54.
74 78th Annual Report of the OPW, BPP 1910 [Cd. 5301], p. 8 & p. 32.
75 80th Annual Report of the OPW, BPP 1912 [Cd. 6356], p. 7 & p. 26.
76 16th Annual Report of the Congested Districts Board for Ireland, BPP 1908 [Cd. 3767], appendix XXII, p. 78.
77 21st Annual Report of the Congested Districts Board for Ireland, BPP 1914 Cd. 7312, appendix XVIII, p. 71.
78 23rd Annual Report of the Congested Districts Board for Ireland, BPP 1914–16 Cd. 8076.
79 J.S. King, in 1st Report of the Commission of Inquiry into the Irish Fisheries 1836, p. 223.
80 NAI OPW8/326.
81 Mr Wainwright, 'The first section of the Galway Estates of the late Thomas Barnewell Martin Esq. [The Martin Estates Connemara] to be sold in August 1849 by Auction.'
82 23rd Annual Report of the OPW, BPP 1854–55 [1929], appendix 8.
83 15th Annual Report of the OPW, BPP 1847 [847], appendix, p. 8.
84 17th Annual Report of the OPW, BPP 1849 [1098], p. 280.
85 S. Lynam, *Humanity Dick. A Biography of Richard Martin, M.P. 1754–1834* (London: Hamish Hamilton, 1975), p. 282.
86 17th Annual Report of the OPW, BPP 1849 [1098], appendix E2, p. 280.
87 48th Annual Report of the OPW, BPP 1880 [C. 2646], p. 26.
88 Select Committee on Harbour Accommodation, BPP 1883 (255), queries 3082–84.
89 Royal Commission on Irish Public Works [The Allport Commission], Second Report, BPP 1888 [C. 5264-I], list No. 4 handed in by General Sankey, p. 708.

90 49th Annual Report of the OPW, BPP 1881 [C. 2958], p. 29.
91 50th Annual Report of the OPW, BPP 1882 [C. 3261], p. 51.
92 Return to Parliament on expenditure on piers under the Relief of Distress Acts 1880, BPP 1882 (203).
93 51st Annual Report of the OPW, BPP 1883 [C. 3649], p. 26 & p. 47.
94 Select Committee on Harbour Accommodation, BPP 1883 (255), query 3179.
95 53rd Annual Report of the OPW, BPP 1884–85 [C. 4475], p. 26.
96 Royal Commission on Irish Public Works [The Allport Commission], Second Report, BPP 1888 [C. 5264-I], appendix, p. 721.
97 7th Annual Report of the Congested Districts Board for Ireland, BPP 1898 [C. 9003], appendix XXXIII, p. 80.
98 71st Annual Report of the OPW, BPP 1903 [Cd. 1748], p. 6.
99 72nd Annual Report of the OPW, BPP 1904 [Cd. 2229], p. 10.
100 74th Annual Report of the OPW, BPP 1906 [Cd. 3109], p. 9.
101 14th Annual Report of the Congested Districts Board for Ireland, BPP 1906 [Cd. 2757], p. 36.
102 76th Annual Report of the OPW, BPP 1908 [Cd. 4228], p. 90.
103 77th Annual Report of the OPW, BPP 1909 [Cd. 4941], p. 53.
104 80th Annual Report of the OPW, BPP 1912 [Cd. 6356], p. 27.
105 21st Annual Report of the Congested Districts Board for Ireland, BPP 1914 [Cd. 7312], appendix XVIII, p. 71.
106 72nd Annual Report of the OPW, BPP 1904 [Cd. 2229], p. 10.
107 NAI OPW/5/10733/07.
108 Plan of pier proposed at Clifden by James Gilmour, dated 9 January 1904. NAI OPW/5/10733/07.
109 Plan of pier proposed at Clifden by James Gilmour, dated 28 March 1904. NAI OPW/5/10733/07.
110 NAI OPW/5/4076/5.
111 Plan dated 3 May 1905 for pier at Clifden. NAI OPW/5/10733/07.
112 Plan dated 29 July 1905 for pier at Clifden. NAI OPW/5/10733/07.

CHAPTER 12

Yet Another New Beginning: Not So Little ... But a Little Too Late

The aspirations aroused exceeded the ability to satisfy them.

J. Lee, 1973[1]

Important Steps in Modernising the CDB and DATI

It was the engineers of the OPW who designed and supervised the various SFI and MWI piers and quays, so that the need for the CDB to have its own engineers was considerably reduced after 1903. By then they were engaged more often in assisting DATI with the design of fish passes and fish hatcheries than in working on piers and other marine works for their own Board. In 1904, therefore, the CDB engineers were formally transferred to the staff complement of the Fisheries Branch of DATI. It is ironic that this resulted in the Fisheries Branch being given the engineering expertise that the Inspectors of Fisheries had sought so assiduously but failed to get more than twenty years earlier. That failure had contributed to the discord that marred the great decade of pier building in the 1880s.

The post of engineer created in DATI in 1904 would remain established in the Fisheries Branch of successive departments until the late twentieth century. Ironically, in 1989, the post was amalgamated with the marine engineering section of the OPW to form a new Marine Engineering Division within the Department of Agriculture and Fisheries: in a sense, the post of Fishery Branch engineer was being reunited with its OPW ancestor, while remaining within the fisheries portfolio. However, any balanced and fair assessment of the OPW's own contribution to the earlier fishery piers will acknowledge that, given the burden of diverse duties it had to perform, the amount and the purpose (relief of distress) of much of the funding, and the geographical remoteness and topographical difficulty of many of the pier sites, it had done a truly remarkable job for almost 100 years. Many of its structures have stood the test of time, tide and severe weather, a credit to the men who designed, built and looked after them.

On another front, both the CDB and DATI had acquired their own supply and survey vessels in 1901. Such vessels had been sought repeatedly by the earlier Inspectors but their requests had gone permanently unanswered. SS *Granuaile* became the CDB supply vessel, used to convey goods and equipment to various pier sites and fishing stations around the coast, and SS *Helga* was acquired by the Fisheries Branch of DATI. The Branch purchased this steam cruiser as a survey and research vessel to facilitate studies on the biology and distribution of marine fishes, activities that now, at last, were coming to be regarded as essential. Times, and resources, were indeed changing.

However, the future of the CDB and of DATI would be shaped, not by these new, much needed resources alone, but by events drawn on a much broader and bloodier canvas, beginning in 1914. But before then, two further enactments – the *Development and Road Improvement Act* and the *Irish Land Act*, both passed in 1909, were to transform the whole pier-funding scene and, along with the report of the Royal Commission on Congestion, constitute another new beginning in the history of the Irish fishery piers, however short-lived that would prove to be.

The Development and Road Improvements Act, 1909

The construction of large piers in phase two had been driven by the three important Acts, the SFI 1883 Act, the MWI 1902 Act and the Development Grant (Ireland) 1903 Act. In phase three, from about 1911 to 1923, an entirely new Act – the *Act to Promote the Economic Development of the United Kingdom and the Improvement of Roads Therein*[2] – would become the principal driving force. It passed in 1909 and was amended the following year by an Act of 1910.[3] It is appropriate therefore, before considering the implementation of the recommendations of the Royal Commission on Congestion, to give some detail of the effect of this Act in a wider context, even though Galway and north Clare did not benefit from it at all. Those interested only in the Galway and north Clare piers may wish to bypass this section and move forward directly to the wider significance of the Act given on page 317.

The principal Act, known in short as the Development Act 1909, set up a new fund of £2.5 million, to be dispensed over five years, promoting economic development in the whole of the United Kingdom. Works already in receipt of public funds were specifically excluded from support under the Act, except in exceptional cases. Advances would be given by way of free grants or loans to selected enterprises that gave an increase in labour, and from which a direct remunerative advantage could be expected, sooner or later.[4] Applications were to be made directly to Treasury. Should an application come from a Government department or public authority (like the OPW), Treasury sent it directly to a Development Commission which it had set up. If an application originated elsewhere – from, say, a business, a county council or a harbour board – Treasury sent it to a relevant Government department which, having considered it, would forward it along with its own observations, to the Development Commissioners.

The Commissioners, appointed on 12 May 1910, examined all the applications and made their own recommendation regarding each one to Treasury, which made the final determination in every case. For their part, the Development Commissioners set out principles that would guide their own deliberations: applications would be evaluated in a wide context, rather than based on merely local considerations; ideally, applications should be part of coherent and comprehensive schemes.[5] For example, any harbour proposal would be considered in the context of a regional or national strategy for harbours and ports. Since the Act was aimed primarily at supporting agricultural and fishery developments, applications for purely commercial ports would not be considered; applications were also expected to show significant financial support from local interests, so that the Development Fund would need to meet only part of the estimated costs.[6]

Up to 31 March 1911, Treasury received no less than 170 applications for grants under the Act. It forwarded to the Commissioners twenty-four of these, involving annual expenditure estimated at more than £210,000 and over £200,000 in capital costs.[7] The Department of Agriculture and Technical Instruction, Ireland (DATI) had been particularly quick off the mark, applying on 18 July 1910 for £50,000 to develop and improve the Irish fisheries; that was the earliest application received by the Commissioners. Following discussions with Fishery Inspector W.S. Green and T.P. Gill, the Secretary of DATI, the Commissioners asked for a general account and overview of the Irish fisheries so that they might be better able to judge the total amount likely to arise in claims for support. Of the £50,000 requested in July, £32,500 was for six named harbours on the east coast, the remainder being for other small projects of diverse kinds. The Commissioners wished to know the principles on which the six specified harbours had been selected when, it appeared to them, there were others no less worthy of support. DATI responded with a revised proposal, this one costing £59,250, with changes to the harbours selected.[8] The Commission again refused to deal with this application in isolation from requests expected to come from other Irish harbours. Showing themselves to be well aware that the east coast fisheries came under the jurisdiction of DATI, whereas the west coast came under the CDB, the Commissioners wanted to know the principles under which the Irish fisheries *as a whole* were to be developed, rather than proceed with the harbour requirements of the east coast in isolation. Remarkably, and well worthy of note, this was the first time that a coherent, national plan for fisheries and fishery piers in Ireland was sought or even raised formally in any official body or document. The Development Commissioners wrote to the Irish administration with their request in January 1911 and awaited a reply, postponing in the meantime an application they had received from the CDB for a pier at Dingle.

The following year, 1912, nine more Irish harbour applications arrived, this time from 'unofficial sources' (that is, not from Government departments), seeking a total of over £83,000. They included the following: Down County Council seeking £26,000 for seven sites; Donegal County Council, £35,000 for five sites (including Buncrana and Rathmullan); Sligo Harbour Commissioners, £50,000 for harbour improvements there; and Newry Port and Harbour Trust, £12,500.[9] The Dublin Port and Docks Board also made an application for £1,200 for Skerries harbour, but this was rejected.[10] These applications were additional to the ones submitted by DATI and the CDB that amounted to almost £130,000 – over £42,000 for DATI and over £87,000 for 'landing places' by the CDB.[11] None of the applications were from Co. Galway or north Clare.

Almost overwhelmed by applications from Ireland, the Commissioners engaged in extensive correspondence and consultation with DATI and the CDB. As a result, they decided to recommend a total grant of £50,000 to DATI, to be expended as follows: £19,000 for a harbour at Helvick Head in Co. Waterford; £20,000 for improvements at Kilkeel, Co. Down; £2,500 for general harbour dredging; and £1,000 to restock oyster beds and build oyster holding ponds.[12] A decision on the use of the remaining £7,500 was postponed to the following year when they recommended £4,500 for clearing the channel from Helvick Head to Dungarvan, and £3,000 for the maintenance and repair of steam and motor engines for fishing boats.[13] Turning to the CDB application, the Commission sent a delegation to visit some of the sites nominated for grants. There followed a meeting with the Board at which 'some considerable differences of opinion manifested themselves particularly ... on how far the Board's own income of some £230,000 a year was available and ... might defray the cost of the schemes proposed'.[14] Eventually, the Development Commissioners recommended a grant of £40,000 to the CDB from the Development Fund, or £50,000 partly from that fund and £10,000 from 'other sources'.

In total, therefore, the Commissioners had recommended that DATI and the CDB should each receive £50,000 for the entire five-year life of the Development Fund. The failure of DATI and the CDB to act in unison probably damaged the Irish case for greater support. Had they acted in concert to produce a rational, integrated scheme for the development and management of the Irish fisheries on a national scale they might have met with even more success; it was another missed opportunity for Irish fisheries development. Still, the outcome was not altogether unsatisfactory and it would permit a number of piers to go ahead. There was, however, a sting in the tail of the Commissioners' decision, stated in their own words:[15]

> The schemes of the Department and the Congested Districts Board will absorb the funds which the Commissioners can devote to Irish fisheries and fishery harbours for some time to come ...

The Commissioners had, in effect, given both bodies £50,000 each and told them to 'now go away'. But Treasury retained the requirement that the Commissioners' approval was needed for every individual project to be supported with the funds before it, Treasury, would sanction the project and advance the money.

Some proposals from 'unofficial sources' had come from County Councils for works already featured in the DATI and CDB schemes: Kilkeel, for example, was proposed by Down County Council but was also included in the DATI scheme. Similarly, proposals for Buncrana, Rathmullan and Burtonport made by Donegal County Council were included in the CDB scheme, and Baltimore in Co. Cork was within the same scheme.[16] Again, this appears to confirm a lack of co-ordination in approach between the various Irish bodies.

In each case, costs over what the Development Commissioners would grant were to come from other sources. Buncrana and Rathmulllan, for instance, were allocated £25,000 from the Development Fund, provided that Donegal County Council gave £10,000; Burtonport was offered £8,000 with the County Council to provide £1,500[17] and agreeing to pay the extra costs should the work go over budget. At these three places (Buncrana, Rathmullan and Burtonport), Donegal County Council requested, in addition, a tiny contribution of just £20 to be advanced by the OPW out of the old SFI 1883 Fund[18] (and in the case of Portavogie in 1907, the request had been for a mere £10). These miniscule SFI contributions were far from unimportant: they were essential in order to establish that, technically, the proposed works were sanctioned and carried out under the *Sea Fisheries (Ireland) Act* of 1883. That way, the Councils were able to raise loans under the 1883 Act to make up their contributions, and the works when completed would vest in, and be maintained by, each Council as that Act specified. By 1914 all the *Development and Road Improvement Act* allocation to DATI had been committed except for £4,500[19] and most of the CDB allocation had been committed as well.

When tenders for the agreed Donegal piers were received by the OPW in 1914, work could not commence, because advances from the Development Fund were suspended due to the outbreak of the First World War.[20] For the same reason, all CDB small grants for piers, landing places and roads ended in 1915.[21] No new allocations had been made under the MWI Act since 1911, leaving almost £46,000 unallocated and unused in that fund. From 1914 onwards therefore, no new pier works would be started anywhere in Ireland, and the Annual Reports of the OPW for the years 1915 to its termination in 1922 made no mention at all of fishery piers and harbours.

In 1918, the Development Commissioners expressed the hope that advances from the £1 million still remaining in the Development Fund would soon recommence and their ninth annual report for the year to March 1919 was indeed much more positive: the works at Burtonport, postponed since 1914, were now to be taken up.[22] The OPW estimated that costs had risen from £9,500 to £16,500 in the interval, in light of which Donegal County Council agreed to raise its contribution from £1,500 to £3,500 and the Commissioners responded by raising their grant from £7,980 to £12,980.

A proposal for Dingle harbour, first made in 1915 and sanctioned in 1918, was also taken in hand. The OPW estimated that, by then, costs had increased by £5,300 above the pre-war estimate. The Commissioners agreed to raise the original grant to £12,000 on condition that, as part of the works, ancillary water and sewage facilities would be installed at the expense of the Rural District Council and the CDB.[23] Harbour works at Helvick Head and Ballybrack, Co. Waterford and Kilkeel, Co. Down, approved before the war, were also taken to conclusion.[24] The CDB made a new, post-war application for an extension of the railway and alterations to the pier at Renard Point, Valentia Island.[25] Treasury agreed that the £7,000 remaining from the CDB's £50,000 allocation could be applied to that purpose, on condition that Kerry County Council, the CDB and the Great Western and Southern Railway Company made up the balance of the estimated cost. But in the end, Valentia,

Buncrana, and Rathmullan, all of which had been offered very considerable grants from the Development Fund, failed to come up with the local contributions after the war and the works never went ahead.[26] What had seemed so positive and encouraging had proved in the end to be a great disappointment.

The final involvement of the Development Commission in Irish fishery harbours was in March 1920 at three locations in Co. Antrim: Port Ballantrae, Dunseverick and Portmuck. The aggregate amount Antrim County Council requested for these places was £12,800; the Commissioners recommended a grant of two thirds of this, on condition that the final one-third came from the Council.

The Wider Significance of the Development and Road Improvement Fund Act 1909

The above extended account of the *Development and Road Improvement Fund* Act is presented for the light it throws on the dramatic change in funding that was taking place from 1909. Under the Act, large piers were funded as services to existing industries already showing potential for commercial success, rather than as amenities aimed at improving the general conditions of local communities: the funding rationale had shifted from rectifying past deficiencies to facilitating promising new advances in ongoing favourable commercial activities.

The Act applied to ports both within and outside the congested districts and irrespective of whether they were designed exclusively for fisheries or had other relevant agricultural purposes. It had also encouraged applications from County Councils and from commercial entities, not just from Government departments. Given these changes, the role of the CDB in funding large piers declined very rapidly to a subsidiary one. The Development Commissioners had, in reality, become the ultimate advisors to Treasury on the matter of Irish piers and harbours, and final decisions regarding them were centralised in London. Adapting the earlier metaphor of the late Fishery Inspector J. Redmond Barry,[27] the CDB was never the driver, and now was no longer even the engine, but merely one cog in a very complex process of harbour development. The whole business of piers and harbours, and the way they were to be funded, had entered the post-war world of the twentieth century, a vastly different landscape to that of the nineteenth. For the future the old, nineteenth-century driver – relief of distress – was no longer a consideration; 'industrialisation' was the new driver and 'development' the new engine. After the Great War, the case for building fishery ports with public money had taken on a completely different complexion: it lost its predominantly 'community' or supportive focus and, for the future, applications were expected to show evidence of commercial relevance and viability.

The digression also shows that Co. Galway and north Clare missed out on the greatest source of potential funding since the SFI 1883 Fund of almost thirty years earlier. They might have hoped to benefit from Development Commission funds eventually, but before the war there was no convincing evidence of any large, sustainable stocks of fish off the

Galway coast which commercial fishermen could access profitably with their existing boats and fishing technology. Because of that, no Galway or north Clare fishery piers warranted investment as, for example, the Cork and Donegal fishery piers did. Galway and north Clare failed the test of a real involvement in, or commitment to, sea fisheries that could provide 'an increase in labour, and from which a remunerative advantage could be expected within the foreseeable future'. It would be the second half of the twentieth century, with the introduction of enormous seine nets and deep-water trawling technology using large factory vessels, before deep-sea fishing off the Irish west coast, out to and well beyond the 100-fathom line, would become a realistic, commercially viable proposition.

The Congested Districts Board (CDB) and DATI expressed themselves reasonably satisfied with the way that funding had changed: 'Experience of the satisfactory completion of this important engineering work [Baltimore Harbour] in pursuance of the enactments enabling County Councils to carry out such important engineering works with a limited contribution has caused us to approve of carrying out such works through County Councils'.[28]

The Irish Land Act of 1909: A Dramatic Change of Direction

Returning to 1909, all the pieces were falling into place for a radical reorganisation of the CDB, precipitated by the findings and recommendations of the Royal Commission on Congestion. Resulting partly from the Commission's recommendations, Parliament passed the Irish Land Act of 1909, sometimes called the Birrell Act, brought in by Augustine Birrell, then Chief Secretary for Ireland.[29] It brought radical reorganisation to the CDB: it increased the membership of the Board to fourteen; removed the three 'temporary' members, of whom W.S. Green was one; reorganised its operation to make it a corporate body, transferring some of its agricultural responsibilities to DATI; increased its annual income to £231,000; brought Ballyvaughan in north Clare, and Claddagh in Galway city, into the congested districts; and it widened the definition of congested districts to include all parts of every county that had at least one congested district within it.

Under the Act, the primary function of the CDB for the future was to create as many economic land holdings as possible in the congested districts. Regarding fisheries and fishery piers, a consultative committee was set up, composed of three members drawn from the CDB and three from DATI, to ensure continuance of the co-operation already existing between them; Green was made one of the DATI members on the consultative committee.

A New Beginning ...

Implementation of the changes meant that the reorganised CDB took a new direction in its marine activities. It would, for the future, pay greater attention to hitherto largely neglected marine occupations such as lobster fishing and kelp making and to the small piers and landing places that serviced them. This signalled a far more profound change of purpose and direction than might seem, at a casual glance. In his evidence to the Commission on Congestion, Green (quoted earlier in chapter 11) had described the fishery objective of the CDB in the words 'we are trying to build up fishermen as fishermen, pure and simple.'[30] However, in 1913, one year after Green's retirement as Chief Inspector of Fisheries, the 21st Annual Report of the reorganised CDB expressed its objective in directly opposite terms:[31]

> 'Our main object in promoting fishery development is not to turn farmers into fishermen but to promote fishing of a kind that will enable the fisherman-farmer all along the western coast to supplement his insufficient earnings from the land by receipts from sea-fishing ...'

The Report went on in explanation:

> The boats that should be used in such cases ought to be inexpensive and light ... the use of boats of the local type for such occasional fishing is very often the wisest policy: they are sure to be cheap and safe; and it is unwise to lead fishermen-farmers to incur heavy liabilities for boats and gear if they can rarely be used and if they cannot earn enough to provide for interest on capital, depreciation, and replacement in addition to profits.

It was a truly remarkable about-turn from the previous emphasis on large boats, deep-sea fishing and 'fishermen as fishermen pure and simple.' It shows that the recommendations of the Commission on Congestion had struck home deeply with the reorganised Board. Its new direction involved, for instance, addressing fishing activity at the Claddagh in Galway, encouraging more part-time inshore fishing along the Connemara coast, and venturing once more into the kelp business.

The Claddagh in Galway, most of whose inhabitants depended traditionally on part-time fishing for their livelihood, was not a congested district before 1909, but that was changed by the Birrell Act. The Board could now take steps to improve the condition of what it called 'this primitive and interesting community' by providing them with better boats and equipment.[32] The Claddagh had potential as a good fishery station and there was already an active local market for fresh fish in Galway city. More important was the railway, with daily services to Dublin, which ensured that fish landed in the evening could be delivered fresh to the Dublin market by the following morning, and even sent onwards within the same day to the English markets. There was a balance exceeding £46,000 still unallocated in the MWI 1902 Fund, and Claddagh was well positioned to benefit from the application of some of this, even if it was too late to be funded by the Development Commission.

Along the Connemara coast too, the Board decided to start a 'fishery experiment in the small local boats that are now used for traffic in turf and seaweed.' It proposed to restore some of the best maintained of these boats to perfect order and provide herring nets for their use, the cost of which would be repaid by arrangement with the fishermen. Phase three of the Board's activity, therefore, started out as a genuine new beginning. From 1910 to 1913, the CDB erected seven more small piers, two slips, and a number of other minor works in Co. Galway. An entirely new pier was built at Tawin in inner Galway Bay, and modifications were carried out at other existing piers in Inishbofin, Doonloughan, Glinsk, Canower and Cleggan. They are listed in Table 12.1.

The choice of Tawin for a fishery pier seems almost inexplicable: returning from the fishing grounds, boats headed for Tawin would need to pass Galway port, the largest fresh fish market in the region and the principal railhead. Who would benefit, and in what way, by proceeding to land fish catches at Tawin? It seems reasonable to conclude that

TABLE 12.1

Final works of the CDB in County Galway, 1911–20.

RH No.	Small Piers	Cost	Date
066	Carna Pier (improvements)	£267	1911–13
164	Inishbofin Pier (improvements)	£203	1913–14
184	*Inis Meáin* Pier (new)	£1,348	1912–14
189	*Inis Oirr* Pier	£118	1912–13
150	Tawin Pier (new)	£609	1913–15
050	Glinsk Pier	£198	1912–14
175	Portmurvey Pier (new)	£522	1911–13
092	Muckinagh Pier (new)	£621	1914–20
077/078	Derryrush Pier	£173	1915–16
064	Cuileen Pier	£56	1918–19
-	Lettermullan Pier	£47	1919–20
042 B	Canower Pier (repairs)	£40	1914
-	Loughidur Pier	£175	1914
	Small Slips		
184	*Inish Meáin* slip (reconstruction)	£885	1912–13
126A	*Béal an Daingin* slip (*Tí Darby*)	£91	1919–20
	Other Works		
014	Cleggan (cliff wall and lights)	£114	1912–13
138	Cashla (install light)	£148	1912–13
026	Doonloughan (pier repairs)	£96	1911–13
061	Mace Harbour (improvements)	£128	1911–12

FIGURE 12.1. Tawin pier in 2012. Photo by Paul Gosling, reproduced with permission.

the pier was really meant to service the agricultural and domestic needs of families in the immediate locality, rather than be a convenient amenity for the fishery and there is evidence that such activities were indeed the mainstay of Tawin pier. The pier is a straight jetty with a short return head. Built as a concrete case filled with stone, it was dangerously dilapidated by 2001 and does not appear to be in use any more. A slip was made on *Inis Meáin* and piers were erected on each of the three Aran islands. These, and the other new measures implemented, were aimed at revitalising local inshore fisheries.

The Board also started experiments in the capture of cod, haddock and other demersal species with nets (presumably using some form of bottom trawl, or more likely, a trammel net) with a view to providing bait for the long-line fishery. Initial results were sufficiently encouraging to justify continuing the experiments, although it is difficult to understand the rationale of using net-caught cod and haddock as bait to catch the same or similar species on long lines; few fishermen would be impressed by the extra work and waste such a plan entailed. The reorganised Board also came around to an earlier proposal of Colonel Fraser's when it showed an interest in encouraging the sale of fresh fish for cash, which was already ongoing in a small way throughout the county. If surprise was expressed at the existence of a ready-money market for fish in a district supposedly reduced to semi-starvation,[33] the Board's response was to attribute this market activity to the capture and sale of lobsters and other fish by the men, and the collection and sale of periwinkles by women and children.

That was the first time the CDB had taken any notice at all, official or otherwise, of the shellfish and lobster fisheries. The original Board had never, at any time, regarded lobsters, or any other crustaceans or shellfish, as 'real' fish or 'real fisheries'. Their colleagues, the Inspectors of Fisheries, had stated in 1894, that the lobster fisheries constituted 'a flourishing industry',[34] yet the CDB in its Annual Reports, from the first in 1892 to the sixteenth for 1906, did not make one single mention of the lobster fishery and export trade, in which between 40,000 and 80,000 lobster were reportedly captured each year on the west coast, yielding £1,000 to £2,000 annually to the fishermen (see Table 10.1).

According to the official fishery reports issued by DATI – signed by W.S. Green – there were no fewer than 327 boats, engaging 1,055 men, in 1902 and 260 boats, engaging 761 men, in 1907 fishing for lobsters in the Galway and Clifden fishery districts alone.[35] Large numbers were engaged on the north Clare coast of Galway Bay as well. Were these not fishing activities supportive of the local communities and meriting encouragement? For the CDB to have overlooked them as not constituting 'real' fisheries may seem most peculiar today but in fairness to the Board, their neglect at the time was understandable and in accord with the general tenor of the age: over the second half of the nineteenth century the term 'fisheries' had come to signify predominantly the offshore pelagic and demersal fisheries that constituted the fishing industry as pursued by the Scottish and English fleets in mid- and distant waters; prevailing expert opinion, including that of Green himself,[36] held the deep-sea fisheries to be inexhaustible. In light of this, the inshore lobster fisheries simply did not count. Calculation from the data in Table 10.1 confirms that the average price obtained for lobsters was indeed very small – about £25 per thousand, i.e. just over six pence each (between 4.5 and 7.4 pence in different years, average 6.2 pence each over the twelve years listed). Small as that appears today, it compares reasonably well with a wage of less than two shillings (twenty-four pence) for a full day's labour on relief works; in such circumstances, the capture of two or three lobsters could make a real difference to a family's income. All things considered, the early neglect of the lobster industry by the CDB is inexplicable except in the context of deep-sea fishing alone being the canonised aspect of fisheries, as Synge so pointedly noted.

... A Little Too Late

While phase three, from 1910 to 1923, had started out well and promised to be a real new beginning, the promise had come a little too late and proved to be short-lived. In its 23rd Report for April 1914 to March 1915, the CDB recorded:[37]

> At Galway nearly all the crews of fishing boats were broken up, as the great majority of the young men of 'the Claddagh', the fishing quarter of the town, joined either the Navy or the Army, and comparatively few remain at home except those who are either too old or too young for military service. The few Galway boats that have been fishing with rearranged crews have had a profitable fishing ...

Fish catches (excluding shellfish) in the western area, from Loop Head to Erris Head, had averaged about 160,000 cwt yearly in the three years, 1912 to 1914; in the three years 1915 to 1917 it averaged only about 50,000 cwt yearly, so a falling-off was indeed noticeable (it was also evident in the fish catches landed elsewhere in Ireland, which were in decline from about 1910 onwards).[38] The decline was compensated for by a large increase in the monetary value of fresh fish sold. The CDB therefore resolved to make ice available in the congested districts to assist the transport of fresh fish to market in good condition.

However, fish prices fell again in 1919, and the cost of transport from the west of Ireland to Britain increased by 120 per cent, negating the Board's efforts.[39] After the war, the increase in Britain of home-based steam trawlers released from war duties resulted in fish gluts that greatly depressed the monetary value of the catches landed. This caused a sense of gloom to descend on the Irish post-war fishing industry. Things were even worse in Britain: in the wake of the gluts and depressed prices, fish markets virtually collapsed, causing the laying-up of over one half of all steam trawlers in the English fleet.[40]

The drop-off in the fishing effort in Ireland in 1914 was accompanied by the suspension of the Development Fund[41] and, in consequence, the postponement of the construction of large piers under its aegis. The CDB, too, suspended funding for new small marine works in 1915 but fortunately most of those already approved since 1910 were already well underway and some were even approaching completion. Work on these was continued, but only three completely new works – small piers at Muckinagh, Loughidur and *Cuileann* – which had been under consideration in 1914, were commenced in Co. Galway. From 1914 on, the reports of the CDB on piers took on a more perfunctory tone and after the war there was little enough enthusiasm to restart small pier works; the 27th Report of the Board suggests a noticeable disengagement had set in regarding them:[42]

> During the year [April 1918 to March 1919] a sum of only £1,699 was spent upon the maintenance of landing-places and other marine works; but considerable projects are before the Development Commission that were postponed temporarily at the outbreak of war.

After 1919, DATI continued to fund piers and harbours in the non-congested counties in co-operation with the Development Commission and this helped to complete some important structures at Kilkeel (Co. Down), Helvick (Co. Waterford), Arklow (Co. Wicklow) and Annagassan (Co. Louth). But the ports of Valentia, Buncrana and Rathmullan – the only sites in the congested districts that had grants from the Development Commission already guaranteed – all proved unable to come up with the local contributions and the works did not go ahead.[43] In a way, that was the final straw for the CDB. The Board did, however, go on to complete the piers at the three latest Galway sites mentioned above.

On 1 January 1923, the functions of the CDB and DATI were amalgamated and transferred to a new Department of Fisheries of Saorstát Éireann. An assessment made by that Department of the status of the fishery harbours in the whole jurisdiction at the time of transfer was brief:[44]

> The number of first class harbours having railway service, deep water landing facilities and safe shelter is small, being 12 only. There are 35 harbours which may be rated as second class. These are harbours having railway connections with shallow berthage at low water, or deep water harbours having no railway connection. In addition to these harbours there are upwards of 300 small harbours with piers or landing places scattered along the seaboard at which the fish caught in small boats is landed.

The figures quoted refer to the whole coast of Saorstát Éireann and the total number of fishery harbours they indicate is approximately 347, far short of the real number. Galway city was the only fishing port in Co. Galway and north Clare that had first-class status, mainly because of the existence of its commercial dock and railway terminal. Apart from that, not one place between Carrigaholt in Co. Clare and Ballyshannon in Co. Donegal had been named in the 'List of more important fishery piers and harbours affording facilities for fishing and general trading vessels' handed in to the Select Committee on Transportation in 1918 by Ernest Holt, then the Chief Inspector of Fisheries.[45] A century of fishery pier building in the West had left little or no impression on Irish officialdom, although 149 piers, quays and harbours were recorded in Co. Galway alone in the six-inch OS maps of 1899.

By erecting more than thirty-two small piers and quays in Co. Galway and north Clare in its thirty-two years of operation, the CDB had contributed appreciably to the vibrancy of the 'amphibious existence' of people in the region. Notwithstanding the failure to induce men to embrace full-time deep-sea fishing, the piers that were built or renovated were useful additions to the maritime infrastructure, and essential in the lives of the people who availed of them. They added to the store of marine transport facilities in remote places where few other means of communication existed hitherto. However, even their usefulness in this regard was set to decline with the completion of the inter-island raised causeways, first built by the P&R Commission and perfected by the CDB and, in the early years of Saorstát Éireann, the opening of new roads and the arrival of motorised road transport.

Indeed, it was the causeways more than the piers or the roads that marked the final opening-up of Connemara to the outside world. Today the small piers and quays constitute valuable assets in the maritime heritage treasury of counties Galway and Clare, even if they have much reduced suitability for modern fishery purposes.

Some Piers and Quays of *Inis Mór*, Aran Islands

There are four separate pier and quay structures located in Killeany Bay, *Inis Mór*: three are relics of three separate famine episodes, 1822, 1830–31 and 1846–49; the other is a twentieth-century boat slip.

Killeany Pier and Quay

The origin of this pier and quay during the famine of 1822 has been described already in chapter 2. Despite some known disadvantages, it served the Aran fishermen well from its completion in 1826, remaining a fishery station and valuable shelter harbour until it was eclipsed by the developments that took place at Kilronan during the Great Famine. Vested in the OPW in 1847,[46] Killeany transferred to the county in 1853 under the *Grand Jury (Ireland) Act* of that year without requiring any pre-transfer renovation. In 1881, Robert Manning, on instructions from the OPW, undertook unspecified 'repairs' at a total cost of £345.[47] At the rates of pay prevailing at the time, thirty shillings would have paid the wages of about twenty men and two supervisors for a single day's work. Therefore £345, less £80 for materials and other costs, would have provided a workforce of twenty-two men for a period of twenty weeks, sufficient to complete significant restoration. The work was carried out under the 1853 Act, the cost to be recouped by later presentment of Galway Grand Jury.

Nimmo's original structure already incorporated a roadside quay extending along the northern shore of the site. At some stage in its later history, a second roadside quay was added, this one extending southwards from the pier root, but the date of this has not been fully ascertained. Walled in cut stone, like the pier and the original north quay, it is most likely that this quay was part of the reconstruction designed by and carried out under Manning in 1881.

While these works improved the landing facilities, Killeany's main disadvantage was the approach to it by sea. Access from the open sea was through a channel surrounded by rocky ledges, which narrowed to about fifty feet at low water. This was a major hazard for visiting boats and an inconvenience even to boatmen with good local knowledge. Sailing boats approaching and leaving the pier could not pass one another safely at the channel's narrowest part. Seeing this, the Piers and Roads Commissioners spent £465 in 1886 on widening the channel to 162 feet by excavating 2,320 cubic feet of the surrounding rock.[48]

The following year, Col. Fraser continued the widening process, blasting away another 1,400 cubic yards of rock, deepening the channel and erecting five marker beacons at a cost of £450.[49] These were major, very practical works that greatly improved access to the pier and quay. The full extent of the rocky ledge and the size of the cut made through it can be clearly identified in Google Earth images of the place today. Killeany pier underwent further repairs in 1908–09 under the CDB, although by then it was not being used very much for fisheries. In spite of the widening of the approach channel, the anchorage, which dries out at low tide, did not suit many boats, especially the bigger ones that came into use as the century ended. The pier and its two quays are constructed of fine cut limestone, but more recent renovations at the site are made of concrete.

An tSeanchéibh, Kilronan

Famine, never far away in the early nineteenth century, struck Aran in 1830–31. Those were years of severe weather and consequent poor harvests, which resulted in localised famines in many places along the western seaboard. As a relief measure on Aran, a small jetty pier was erected at Kilronan village, where none existed previously: according to the local Coastguard Officer, E.N. D'Alton, it was built in 1831 'by the inhabitants who were remunerated for their labour with oatmeal sent for the relief of the poor'.[50] That the jetty would be useful for the fisheries was perhaps only a secondary consideration, although it was the first manmade quay convenient for the fishermen of the village and it pre-dates the main Kilronan pier by almost twenty years. Having received only charity support and nothing from the fisheries administration, it was a poor enough structure. Nonetheless, it lasted a long number of years without any further documented expenditure on it. Its value and importance were largely eclipsed by the new main pier that was erected near it during the Great Famine. After that, the small jetty came to be known as *an tSeanchéibh* (the old 'quay').

With the expansion of the fishing fleet attending Kilronan during the 1890s, space alongside the main pier became inadequate and *an tSeanchéibh* became important once more. The Lord Lieutenant therefore approved and certified it for renovation under the MWI 1902 Act. The plan was partly to rebuild, and partly to extend it so as to increase the berthage, at an estimated cost of £3,000.[51] Work commenced in 1905 and was completed in 1906.[52] The final cost was £3,320, of which the CDB contributed £240; the MWI fund contributed the bulk of the finance needed.[53]

Kilronan Main Pier

On 3 April 1846, Rev. J. Digby, whose estate included the whole island of *Inis Mór*, submitted a memorial for a pier at an unidentified location stated simply to be 'in Killeany Bay'.[54] The engineer Barry Gibbons, examined the locality and, in November 1846, an indenture was drawn up, signed by J. Radcliff and William Mulvany, two of the Commissioners for Public Works, granting approval and funds under the Piers and Harbours Act for a new pier. The Government would give a grant of £1,600 and Digby would make a contribution of £800 to meet the estimated cost of £2,400 for one to be constructed at Kilronan village. Digby's contribution was one-third rather than one-fourth, so the authorities must have deemed the pier to be as important for general communication, which of course it was, as for fisheries. A map accompanying the indenture confirms the exact location as that of the present main pier.[55] In time this structure would become the principal pier of the island. Its location near the mouth of Killeany Bay obviated the need to navigate the narrow passage leading to Nimmo's pier at Killeany. Kilronan new pier was designed as a substantial, straight, cut-stone structure, 440 feet in length, built along a rocky outcrop extending out into Killeany Bay, to a planned depth of five feet at low water of Spring tide. Rock had to be extensively cut away along the shore to accommodate the work, and the seabed on its western side was further excavated to provide a clean, reasonably deep berthage in the relatively sheltered space alongside. Thomas Ganley was in charge of operations, reporting directly to Barry Gibbons of the OPW.

Sketch plans of the stoneworks are still extant in the records.[56] While under construction in 1848, the works were damaged by a storm, which slowed progress, but everything was eventually completed satisfactorily by 1849. Kilronan pier soon surpassed Killeany in importance as a fishing station, but it would not appear in the public records again until the Congested Districts Board came on the scene.

W.S. Green favoured *Inis Mór* as a deep-sea fishing station ever since his own coastal survey in 1890. In consequence, the major development and expansion of Kilronan pier occurred during the time of the CDB, when it became a busy fishing station for local and visiting boats, especially those engaged in the seasonal herring and mackerel fisheries. The improvements commenced as early as 1893 when the CDB dredged the berthage and installed approach lights.[57] Kilronan was selected for further improvement at the meeting in 1899 of the CDB, the Fishery Inspectors and the OPW. On foot of its selection, Treasury sanctioned a grant of £4,394 from the SFI 1883 Fund for the required renovations and an extension, 100 feet long, to the existing pier;[58] the CDB agreed to give a maximum of £1,000 towards the cost, and the work started in 1901. It was completed in April 1902 at a final cost of £3,853,[59]and the whole harbour was transferred back from the OPW to the County Council in 1903.[60]

The usefulness of Kilronan harbour was further enhanced by the works carried out under the MWI 1902 Act on the nearby *seanchéibh* about that time (see above). In 1908, the main pier was improved again by the provision of a large shed at the relatively enormous cost of £1,291 and in 1916, a curing platform was constructed to serve the fishing industry that was then decidedly on the increase.[61] Kilronan was a popular port for boats of all types, and later a hulk was moored close to the pier for salting, curing and barrelling mackerel and herring. The mooring space alongside the pier soon became insufficient for the number of boats using it.

Kilronan did not remain an important fishery station after the Great War when the Aran fishermen took to landing their catches at Galway (and later at Rossaveel), where there was ready transport available to Irish and foreign markets. Kilronan was extensively improved and further extended later in the twentieth century. In more recent years, it was renovated to cater for the tourist business and it is today a busy port for local trade and tourism.

The fourth marine structure at Kilronan is a small slip, located in the harbour between *an tSeancheibh* and the main pier. Originally built in 1906, and repaired in 1914 by the CDB, it is now entirely a concrete structure having been extended in 1974; it is used principally by the local lifeboat.

Kinvara Pier and Quay

For most of its life, Kinvara harbour has been in a legal limbo. According to Lewis, James Ffrench constructed a quay, about fifty yards long, and a pier at Kinvara in 1773. These simple structures were raised, lengthened and the whole site modified by the addition of an open dock in 1807–08 when Richard Gregory of Coole bought the Ffrench estate.[62] On the other hand, Nimmo said that the original quay had been built by county presentment and was 'in rather bad order' by 1826.[63] He did nothing to improve the place, possibly because of its uncertain ownership.

By 1841, there were said to be 160 houses in Kinvara town, with almost 1,000 inhabitants, and a further 5,500 persons living in the neighbourhood, so that the harbour was serving a significant population even then. No maintenance or renovation was ever

FIGURE 12.2. Kinvara harbour at low tide. The dock is on the extreme left with the beached yacht. The quay is along the grassy section and the pier is where the other boats are beached. Photo NPW.

done to it with public money because of the uncertainty as to its true owner. A claim that it was inadvertently conveyed to a private owner under the Encumbered Estates Act[64] after the Great Famine further compounded the legal uncertainty. By the late nineteenth century it was largely derelict and it had never been transferred to the county, confirmation that it was considered by officialdom to be in private hands in the early 1850s. Even in the 1880s, Manning dissociated himself thoroughly from it: 'The harbour works there', he said, 'are not in a good state, but they were no child of mine'.[65]

With the harbour seriously dilapidated, in spite of (or maybe because of?) its huge turf trade with Connemara, the CDB realised that something had to be done to restore it to a safe condition. The Board indicated its willingness to contribute £200 towards remediation, even though Kinvara was not at the time in a congested district. The Government agreed to allocate £1,100 to it from the MWI 1902 fund via the OPW; Galway County Council offered £800 and DATI gave £1,100.[66] It was most unusual for all the parties to combine in this way to fund a single location; indeed, Kinvara is the only place in the West to have enjoyed this plurality of benefice. They all obviously agreed on the urgency of appropriate action and they left the vexed question of ownership and responsibility to one side temporarily to enable the commencement of the reconstruction.

The work was carried out in two stages. In the first, the OPW, using the MWI and the County Council contributions, restored, lengthened and raised the pier. The second stage, involving renovation of the quay and dock, was deferred until 'certain legal difficulties' – the ownership issue, no doubt – could be overcome. That stage was funded by DATI and the CDB, with an extra £300 subscribed by Galway County Council. (As we saw earlier, the county's allocation of £1,100 to Kinvara depleted the amount available for Clifden, to the latter's great disadvantage.). The actual construction work of the second stage was put in the hands of Galway County Council and its County Surveyor. The CDB agreed to take over the whole harbour from the Trustees of the Sharpe Estate (successors in title to the Ffrench/St George/Gregory estates)[67] when the work would be completed, and to hold it until it could be vested in the county. There is no evidence that it was ever transferred subsequently to the county and, even in 2001, its status was still uncertain.[68]

When all work was completed in 1909, notice was posted of the rules and bye-laws governing use of the pier and quay. These prohibited the disturbance, removal or injury to any stones, coping, bollards or rings.[69] A schedule of tolls was introduced in 1912. Trading boats, loading or discharging, paid three pence per ton of registered tonnage, and one halfpenny per ton per week if staying over. Ferry boats arriving from ports in Co. Galway paid six pence per visit in lieu of the tonnage charge. Fishing boats paid three pence per registered ton, valid for a year; currachs with turf or seaweed paid a 'licence fee' of one shilling *per annum* or part thereof. On top of that, a wharfage fee was charged, comprising one penny per ton loaded or discharged; each horse or head of cattle cost one penny; sheep, lambs, donkeys, and pigs cost one halfpenny each.[70]

Kinvara, although situated in Co. Galway, is in an electoral district (Ballyvaughan district of Co. Clare) that was not declared congested until 1909. Overlooked by Dun Guaire castle and backed by the village main street, the harbour is one of the most picturesque in Co.

Galway and its present-day leisurely ambience belies its busy past. In the early twentieth century, it even had a steamer connection to Galway city that was popular with tourists. That and the turf and seaweed trades, the staples of the port, are now long gone, but the turf trade is recalled and celebrated every year by a festival called *Cruinniú na mBád* ('the gathering of the boats') at the harbour.

Other Large Works (Outside the Congested Districts Counties) During Phase Two

No works were done in Co. Galway or north Clare with SFI 1883 funds after 1902. The extension at Kilronan and the repair works at *Spidéal* sanctioned in 1900 had been finished by then, and thereafter the SFI 1883 Fund was restricted to works in non-congested counties. There was a sum of almost £10,000 (including arrears, and anticipated future repayments) still available in the fund in 1903, when Treasury sanctioned £3,000 for a slip, pathway and breakwater in Ardmore, Co. Waterford.[71] Later it sanctioned £6,000 for a breakwater and deepening of the harbour at Passage East (Cheekpoint) in the same county.[72] Waterford County Council was to give £1,000 and DATI the same amount for Ardmore; both were also to give £2,000 each to the works at Passage East. When these works were fully paid for, the balance in the SFI Fund was finally exhausted.

Endnotes

1 J. Lee, *The Modernisation of Irish Society 1848 to 1918* (Dublin: Gill and Macmillan, 1973).
2 *Act to Promote the Economic Development of the United Kingdom and the Improvement of Roads Therein*, 9 Edw. VII C. 47.
3 10 Edw. VII C. 7.
4 9 Edw. VII C. 47, section 1.
5 1st Report of the Development Commissioners, BPP 1911 (199), p. 7.
6 2nd Report of the Development Commissioners, BPP 1912–13 (305), p. 34.
7 1st Report of the Development Commissioners, BPP 1911 (199), pp. 13–14.
8 1st Report of the Development Commissioners, BPP 1911 (199), p. 29.
9 80th Annual Report of the OPW, BPP 1912 [Cd. 6356], pp. 27–29.
10 2nd Report of the Development Commissioners, BPP 1912–13 (305), p. 41.
11 2nd Report of the Development Commissioners, BPP 1912–13 (305), p. 34.
12 2nd Report of the Development Commissioners, BPP 1912 –13 (305) , p. 46.
13 3rd Report of the Development Commissioners, BPP 1913 (273), pp. 33–34.
14 2nd Report of the Development Commissioners, BPP 1912–13 (305), p. 46.
15 3rd Report of the Development Commissioners, BPP 1913 (273), pp. 33–34.
16 82nd Annual Report of the OPW, BPP 1914 [Cd. 7563], pp. 23–24.
17 82nd Annual Report of the OPW, BPP 1914 [Cd. 7563], pp. 23–24.
18 82nd Annual Report of the OPW, BPP 1914 [Cd. 7563], pp. 23–24.
19 5th Report of the Development Commissioners, BPP 1914-16 (408), p. 8.
20 83rd Annual Report of the OPW, BPP 1915 [Cd. 8119], pp. 17–18; 24th Annual Report of the Congested Districts Board, BPP 1916 [Cd. 8356], p. 24.

21 24th Annual Report of the Congested Districts Board, BPP 1916 [Cd. 8356], p. 24.
22 9th Report of the Development Commissioners, BPP 1919 (214), p. 14.
23 9th Report of the Development Commissioners, BPP 1919 (214), pp. 14–15.
24 9th Report of the Development Commissioners, BPP 1919 (214), pp. 15–16.
25 8th Report of the Development Commissioners, BPP 1918 (118), pp. 10–11.
26 10th Report of the Development Commissioners, BPP 1920 (230), p. 234.
27 J.R. Barry, in Select Committee on Sea Coast Fisheries (Ireland) Bill, BPP 1867 (443), queries 3977–83.
28 26th Annual Report of the Congested Districts Board for Ireland, BPP 1918 Cd. 9139], p. 14.
29 *Irish Land Act 1909*, Edw 7 C. 42.
30 W.S. Green, in Minutes of the Royal Commission on Congestion in Ireland, First Report, BPP 1906 [Cd. 3267], query 4944.
31 21st Annual Report of the Congested Districts Board for Ireland, BPP 1914 [Cd. 7312], p. 18.
32 19th Annual Report of the Congested Districts Board for Ireland, BPP 1911 [Cd. 5712], p. 32.
33 21st Annual Report of the Congested Districts Board for Ireland, BPP 1914 [Cd. 7312], p. 10.
34 Report of the Inspectors of Irish Fisheries for 1893, BPP 1894 [C. 7404], p. 23.
35 DATI Report on Sea and Inland Fisheries for 1902 and 1903, BPP 1905 [Cd. 2535]; DATI Report for 1907, BPP 1908 Cd. 4298, Appendix10.
36 W.S. Green, 'Survey of Fishing Grounds on West Coast', in Report of the Inspectors of Irish Fisheries for 1891, BPP 1892 [C. 6682], pp. 43–59.
37 23rd Annual Report of the Congested Districts Board for Ireland, BPP 1914–16 [Cd. 8076], p. 14.
38 Tables of fish landings in DATI Reports on Sea and Inland Fisheries for 1915 to 1919: BPP 1916 [Cd. 8392]; BPP 1918 [Cd. 9018]; BPP 1919 [Cmd. 44]; BPP 1920 [Cmd. 601]; BPP 1921 [Cmd. 1146].
39 D. Hoctor, *The Department's Story – A History of the Department of Agriculture* (Dublin: Institute of Public Administration, 1971), p. 112.
40 Report of the Department of Fisheries, Saorstát Éireann, 1927, p. 25.
41 24th Annual Report of the Congested Districts Board for Ireland, BPP 1916 [Cd. 8356], p. 24.
42 27th Annual Report of the Congested Districts Board for Ireland, BPP 1920 [Cmd. 759], p. 40.
43 10th Report of the Development Commissioners, BPP 1920 (230), p. 234.
44 Report of the Department of Fisheries, Saorstát Éireann, 1927, p. 8.
45 First and Second Reports of the Select Committee on Transport, BPP 1918 (130), pp. 334–38.
46 17th Annual Report of the OPW, BPP 1849 [1098], Appendix E2, p. 281.
47 50th Annual Rcport of thc OPW, BPP 1882 [C. 3261], p. 27.
48 Report of the Piers and Roads Commission, BPP 1887 [C. 5214], p. 8.
49 Col. T. Fraser, Report to the Lord Lieutenant, BPP 1889 [C. 5729], p. 9.
50 E.N. D'Alton, in 1st Report of the Commission of Inquiry into the Irish Fisheries 1836, evidence, p. 223.
51 72nd Annual Report of the OPW, BPP 1904 [Cd. 2229], p. 10.
52 74th Annual Report of the OPW, BPP 1906 [Cd. 3109], p. 9.
53 80th Annual Report of the OPW, BPP 1912 [Cd. 6356], p. 7.
54 NAI OPW/8/11.
55 Indenture for a pier at Aran Islands, 16/11/1846. NAI OPW8/11.
56 NAI OPW8/11.
57 4th Annual Report of the Congested Districts Board for Ireland, BPP 1895 [C. 7791], appendix XXVII, p. 53.
58 9th Annual Report of the Congested Districts Board for Ireland, BPP 1900 [Cd. 239], p. 42.
59 70th Annual Report of the OPW, BPP 1902 [Cd. 1261], p. 6.
60 71st Annual Report of the OPW, BPP 1903 [Cd. 1748], p. 6.
61 24th Annual Report of the Congested Districts Board for Ireland, BPP 1916 Cd. 8356.

62 Lewis, *Topographical Dictionary*.
63 Nimmo, A., Report in 8th Report of the Commissioners of Fisheries, BPP 1827 (487), p. 45.
64 *The Encumbered Estates Act, 1849*, 12&13 Vict. C. 77.
65 R. Manning, in Report of the Select Committee on Harbour Accommodation, BPP 1883 (255), queries 4157–58.
66 15th Annual Report of the Congested Districts Board for Ireland, BPP 1906 [Cd. 3161], pp. 28–29.
67 Melvin, *Estates*; Landed Estates database, www.landedestates.ie/.
68 Colleran and Hanley, *Assessment of Piers, Harbours and Landing Places.* Available on the website: http://data-galwaycoco.opendata.arcgis.com/datasets/.
69 Poster of Bye-laws, Regulations and Rules of Kinvara Harbour, 1909. Galway County Library, GC/CSO/2/35.
70 Schedule of Tolls and Rates at Kinvara, 13/6/1912. Galway County Library, GC/CSO/2/35.
71 72nd Annual Report of the OPW, BPP 1904 [Cd. 2229], p. 35.
72 73rd Annual Report of the OPW, BPP 1905 [Cd. 2657], p. 43.

Epilogue

Let us return now to the unnamed Connemara man mentioned in the very first page of this book. He was right: there is a landing place at the end of every *boithrín* to the shore in Connemara and west Galway – and also along the Galway Bay shore of north Clare.

Of the manmade structures whose date of origin we know precisely, most were erected during periods of famine and distress: twenty-three in the famine of 1822, nine in the Great Famine, ten in the famine of 1879–80, and twenty-five in the distress of 1885–8. Built ostensibly for the fisheries, they were funded, whether by Government or by charities, with money provided primarily as a way of relieving the distress of famine by providing paid work; for all practical purposes, encouraging and endowing the fisheries was rarely more than a secondary intent.

We have encountered some of the dedicated engineers who designed or helped to build the various structures, men like Nimmo, Killaly, Donnell, Gibbons, Forsythe, Roberts, Manning and Parsons; we have met Commissioners for Public Works like Burgoyne, Mulvany, McKerlie, Le Fanu and Sankey; and the chairmen of various important Commissions and Committees.

Fishery Inspectors like Barry, Brady, Green and Holt have played a large part in the story as it unfolded. Almost forgotten philanthropists and humanitarians, societies and individuals (Irish and foreign) such as the London Tavern Committee, the Duchess of Marlborough Committee, the Mansion House Committee, the Canadian Fund, W.L. Joynt, J.H. Tuke, C.T. Redington, Major Gaskell, Col. Fraser and Lady Burdett-Coutts, have graced the pages of the story of the fishery piers and quays. Priests, too, like Fr T. Flannery, Fr R. Quinn, Fr T. Kelly and Fr Linskey made their appearances in interesting ways.

Least prominent of all in this historical account have been the men and their families who suffered the distress and destitution, built the structures and carried on their lives as best they could in the prevailing circumstances. As always, they constitute the silent majority. Because of the absence of their voices, accounts like this can at best only scratch the surface of the topic. The full *scéalta* (stories) of the people, places and events need to be further researched and voiced if the full picture is to be revealed.

However, by telling, as we do here, some of the stories that made it into the dry documentary record, hopefully someone else with different skills will carry the venture on and be stimulated to seek out what is still hidden elsewhere, fleshing out the fuller picture. From what has been learned so far, we can safely say that the abundance of piers and quays in Co. Galway and Galway Bay stands as a record of recurrent famine and abiding poverty rather than as an indicator of vibrant fishing activity. That is what the stones themselves cry out.

The multiplicity of piers and quays is central to the maritime heritage treasure of Galway and north Clare. We are indeed fortunate that it was so comprehensively recorded in 2000–01, commissioned by Galway County Council and, in north Clare, by Clare County Council in 2006–07. Since then many of the structures, having been severely damaged by the storms of recent years, have been repaired and made safe with the liberal application of concrete, the addition of new safety features and in some instances, protected by robust rock armouring.

Renovation we must accept as natural and requisite; the history of the piers is replete with recurring episodes of such repair and renewal. Indeed some of the instances where we hear clearest the genuine voice of the ordinary people are the memorials, some written in support of such maintenance and restoration works. Many of these are signed with the spidery, fragile signatures of men familiar with hardship but unfamiliar with writing; some signed with only their simple, witnessed mark.

Now that we know something about the origin and the history of the piers and quays, the generosity of those who contributed money to them, the skill of the engineers who designed them and the quality of the handwork of the starving men who built them, perhaps we can look forward for the future to a more sympathetic appreciation of these humble works for and by humble people.

Appendix A

A guide to the maps, showing the locations and alternative names of the piers. (For other information the reader should consult the following website: http://data-galwaycoco.opendata.arcgis.com/datasets/)

RH NO.	NAME	OTHER NAMES	DATE OF ORIGIN
1	Ashleigh	Glengevlagh; Glengimlagh; Leenane	1881
2	Leenaun north pier	Leenane	1822
2X	Leenaun south pier	Leenane	1867
2A	Derrynacleigh 3	-	-
2B	Derrynacleigh 2	-	1897
3	Derrynacleigh 1	King's	Pre-1899
4	Rosroe	-	1852
4B	Little Killary 2	-	-
5	Glasillaun	Glasillaun	1886–7
7	Renvyle	Gurteen; Smeerogue; Tully; Poulahly	1881
10	Letter (B'kill)	*Leitir*	1886–7
11	Derryinver	-	1823
12	Dawros 1	-	-
12A	Dawros 2	-	-
13	Dooneen	Keelkyle; Letterfrack; Ballinakill	1848
13X	Small Quay, W of 013	-	-
14	Cleggan	-	1823
15	Rossadillisk	-	1895
16	Aughrus Beg	-	1886–7
17	Aughrusmor	-	-
17X1	Skeaghduff	Aughrusmor	Pre-1839
17X2	Aughrus Mor	(Gannoughs)	1883–90
18	Claddaghduff	-	Pre-1899
18X	Kingston Glebe	*Cúl an Chlia*	Pre-1839
19	Kingstown	-	-
19A	Kingstown slip	-	-
20	Fahy pier	-	-
20X1	Fahy slip	-	1886–7
20X2	Doaghbeg wall	Doaghbeg	1822
20A	Clifden Beach Slip	-	1886–7

20B	Beach Rd	-	-
21	Clifden Quay	-	1822
21X	Salt Lake Pier	-	Pre-1839
21A	Faul Pier	-	Pre-1899
22	Drimmeen	Errislannan; Boat Harbour	1869
23	Curhownagh	Errislannan; *Laughán Leá*	1848
24	Derrygimla	-	Pre-1899
25	*Cuan na Luinge*	-	-
26	Dunloughan	-	*1899*
27	Pollrevagh	-	Pre-1899
28	Slackport	-	1869
29	Aillebrack	Dolan's; *Trá Mhíl*	-
30	Bunowen	-	1846–50
32	Ballyconneely quay	Doleen; *Dóilín*	1881
33	Dohulla BW	-	1887
34	Emlaghmore	-	Pre-1899
34A	Paul's Quay	-	Pre-1899
35	Callagh	*Caladh*	-
35A	Murvey	Mannion's; *Caladh Bháid*	-
36	Dolan	Whale Harbour; Foyle Harbour	-
37	Ervallagh	Bealantra	1886–7
37X	Ervallagh small pier	-	Pre-1899
38	Monastery 2	-	Pre-1899
39	Monastery 1	-	Pre-1899
40	Roundstone S pier	-	1823
40X	Roundstone N pier	-	1880–1
40X2	Inishnee	*Caladh na loinge*	-
40A	Aillencally	-	1886–7
41	Cloonisle	Derryadd West	1823
42	Scrahalia	-	-
42A	Rosroe	-	1886–7
42B	Canower	-	-
43	Illauncreevagh	The Dock	-
44	Lettercamus 1	Wallace's quay	Pre-1839
44A	Lettercamus 2	-	Pre-1899
44B	Dunreaghan 2	-	Pre-1899
45	Dunreaghan 1	-	-
46	Cashel 2	-	Pre-1899

47	Cashel 1	-	1893
48	Bunnahowen	-	-
48A	Bunnahowen old	-	Pre-1839
48X	*Oileán Gorm*	-	*1886–7*
48B	Glinsk school	-	-
49	Glinsk 2	*Céibh Caladh Mháire*	-
50	Glinsk 1	*Caladh Mór*	*1912–14*
51	*Leitreach Árd Thoir*	*Litir Árd Thoir* 1	-
51A	*Leitreach Árd Thoir*	*Litir Árd Thoir* 2	-
52	*Gleann an tSruthán*	Bertraghtboy; *Céibh Aill an dhá Bhinn*	-
53	*Cuan Caortháin*	McDonagh's Pier	Pre-1899
54	*Béalcarra*	*Cé Bairéid*	Pre-1899
55	Dawros	-	-
56	Moyrus 1	*Céibh an Phortaigh*	Pre-1899
57	Moyrus 2	*Céibh an Droighne*	Pre-1899
58	Coradan	*Dubhithir; Céibh an Choradáin*	Pre-1899
59	*Caladh an bháid*	-	-
60	Half Mace	-	-
60A	*Oil. an Bhromaigh*	-	-
61	Mace	*Céibh an Mhasa*	Pre-1838
62	Ard West	*Árd Thoir*	-
63	Ard East	*Árd Thiar*	-
64	Cuileen	-	1918–19
65	Carna	-	-
66	*Crompán Carna*	-	1885
66A	Carna school	-	-
66X	Quay on Mainis road	-	Pre-1839
67	*Aill an Eachrais*	-	-
68	Rusheenamanach	*Bun an tSrutháin*	-
69	*Caladh Feenish*	*Caladh Mháinse; Cé an tSean Tí*	Pre-1899
70	*Árd Mór*	Pollnadhu; Ardmore; *Cé na hÁirde*	1880
71	*Aill an Bhroin*	*Caladh na gCúg*	Pre-1899
72	Kilkieran	*Cill Ciarán*	1848
72A	Rosduggan	*Tí Sweeney*	-
72X	*Caladh na gClimini*	-	*1886–7*
73	*Sruthán na mbracai*	-	*1886–7*
74	*Aill Uaithe*	-	*1886–7*

75	Flannery Bridge quay	-	*1886–7*
76	*Loch Conn Aortha*	*Aill na dTurnóg*	-
77	Derryrush 1	-	-
78	Derryrush 2	*Cé an Liocháin*	-
79	Inver	-	-
80	*Aill an Mhianigh*	-	-
81	Turlough	*Céibh an Cora; Ros Locha*	-
82	*Rosdubh*	*Cé an Oileáin*	Pre-1899
83	*Sruthán Bhuí* 1	Rosmuck	1886–7
84	*Sruthán Bhuí* 2	*Caladh Cill Briocháin*	Pre-1899
85	*An tOileán Mór*	*Céibh Ceann an Bhothair*	1886–7
86	*An Aill Bhui*	Kilbricken; *Céibh Nua*	-
86A	*Crumpán*	Rosmuck; Crumpaun	-
87	*Aill Ghamhain*	*Ros Céide;* Roskeeda	-
88	Garafin	-	Pre-1886
89	Silera	Rosmuck	Pre-1886
89A	*Dún Manusa* 1	-	Pre-1899
89B	Screeb	-	-
89X1	E side, Rosmuck pen.	-	Pre-1899
89X2	Screebe Lodge	-	-
90	*Dún Manusa* 2	-	-
90X1	Quay S of above	-	-
90X2	*Cé an tSagairt*	-	-
90X3	*Cora Chláir*	-	-
90A	Camus Br. slip	-	-
90B	*Camus Iocthtar*	*Caladh Domhain*	-
91	*Cinn Mhara*	-	-
91X	*Cinn mhara* 2	-	-
92	Muckina	-	*1914–20*
92A	*Aibhnín*	*Cé an Chrumpáin*	-
92B	*Baile Lár*	*Cé an tSeanbhaile*	-
92C	*Cé na Relige*	-	-
92X1	Bunakill	-	Pre-1839
92X2	*Céibh Nua*	-	-
93	Annaghvaan	*Eanach mheáin*	-
93A	*Bun an tSrutháin*	-	-
94	Brandy Harbour	-	1886–7
95	*Sruthán Bhuí*	Lettermore	Pre-1899

96	*Aillin an Arainn*	-	-
96A	*Inish an Ghainimh*	-	-
96X	Q. W of Cashen Pt.	-	-
97	Garanta	*Caladh an Chuaile*	-
98	Lettercallow School	*Bun an tSrutháin.*	-
99	Lettercallow	*Céibh na gCaisle*	1886–8
100	Ross	-	Pre-1899
100A	Murragh Quay	*Cé an Roisín*	-
100B	Murragh slip		-
101	Rusheen Maumeen	*An Caladh beag*	-
102	Glantrasna	-	1886–7
103	Knock	-	-
104	*Baile na cille* 1	-	-
105	Kiggaul	-	1883–90
105A	Cé Barrett	*Cé Bharéid*	-
106	*Furnais* N	Paddy Jack's	-
106X	Quay on Dinish Isl.	-	Pre-1899
107	*Furnais* S	-	-
107A	*Furnais* slip	-	-
107B	Crappagh	Cnappagh	Pre-1899
108	*Crumpán*	Lettermullan	-
109	*Caladh Gholaim*	-	-
110	*Cora Bhuí*	*Cé na hÁirde*	-
111	*Stad Truscain*	-	-
111X	Q S of Kiggaul Br	-	-
112	*Baile na cille* slip	*Clann na Fionnoige*	Pre-1899
113	*Baile na cille*	-	Pre-1899
114	*Poll Ui Mhuirinn*	-	-
115	*Seanachomheas*	-	-
116	Drom	-	-
117	Aircin	*Cé an tSáilín*	-
118	Trawbaun	*Trá Bán*	1886–7
119	Derreendarragh	*Cé an Dóirín*	Pre-1899
120	*Clais na nUan* Quay	Teernea; *Cé Clais na nUan*	Pre-1839
120X	Teernea BW	Teernea; *Clais na nUan*	1887
121	Maumeen pier	-	1823–6
122	Sconnsa	-	1886–7
123	Lettermore	-	Pre-1839

124	*Cé na Leice*	Slate Harbour	-
125	*Cora Chiola*	-	1886–7
125A	*Oileán na dTrachta* Quay	-	1886–7
126	*Béal an Daingin* quay	Bealadangin; Bealadaingean; Tithe na Cora	1836
126A	Tí Darby slip	Bealadaingean slip	1919
126B	*Crumpán an Chaonaigh*	-	Pre-1899
127	Tooreen	*Tuanín.*	1886–7
127A	Rossroe Island	-	1886–7
127B	*Doire Fhartha* pier	Cornamona	-
127C	*Doire Fharta* slip	-	-
128	*Lathaí Dubha*	-	Pre-1899
129	*Caladh Thaidhg*	Collaheigue	1842
130	*Doilín*	Doleen; Lettermore	1883–90
130A	Keeraunbeg	-	Pre-1839
131	*Pointe*	*Cuan caol*	-
132	Glenmore	*Caorán Mhór*	-
132X	*Cé na bhFaocanna*	-	-
133	*Sruthán*	Shruffaun	Pre-1839
134	*Aill an Chúl chaladh*	Garryroe; *Céibh an Ghreasaí*	Pre-1899
135	Clynagh	*Cé an Croisín*	Pre-1899
136	Derrynea	-	1892
137	Rossaveel	-	Post- 1900
137X1	Rossaveel old	*An tSeancéibh*; Cashla; Costelo	1823
137X2	Quay N of 137X1	*Caladh na muice*	Pre-1899
137X3	*Cé an Tí Mhóir*	-	Pre-1839
138	*Casla (Nua)*	*Céibh an Chaoráin; Cé Ceann an Bhothair, Cé an tSalain;* Cashla	1886–7
138X	Quay S of 138	-	Pre-1839
140	Ballynahown	*Baile na hAbhann*	1886–7
140X	Travore	*Trá Mhór*	Pre-1839
144	Spiddal new	*Spidéal*	1871
144A	Spiddal old	*An tSeancéibh*	1823
144X1	Furbo	Furbogh; *Na Forbacha*	Pre-1839
144X2	Forramoyle slip	*Caladh na Famona*; *Céibhín na gCurach*	1886–7
145	Barna	*Bearna*	Pre-1800
145X1	Recorders Quay	-	*Pre-1820*

145A	Claddagh & Slate piers	-	1822–8
145B	Eyre's Dock	Mud dock	Pre-1800
145C	Moneen, L. Atalia	Lynch's; L. Atalia quay	*Circa 1824*
146	Roscam	*Pollachuslán*	Pre-1839
146X	St Marys Port	Pollvarla; St. Mary's Quay	Pre-1839
147	Oranmore Castle pier	-	Pre-1839
148	Rinville	New Harbour; Ardfry	1824
149	Prospect	-	Pre-1839
150	Tawin	-	1913
151	Carrowmore	Lynch's; Ballynacourty	1884
151X	Ballynacourty old	Nimmo's	1824–8
152	Black weir	Corraduff	Pre-1899
153	Keave	-	-
153X	Tyrone House	-	Pre-1839
154	Clarinbridge new pier	-	1842
155	Clarinbridge old quay	*Tobarnagloragh*	Pre-1840
156	Kilcolgan Quay	Moran's Quay; The Weir	*Circa 1800*
156X1	Stradbally	-	-
157	St. Kitts	Killeenaran	1824
158	Pollagh	-	-
158X	Mulroog	-	-
159	Tarrea	-	1849
160	Kinvara	-	1773
160X1	Carr's Island quay	-	Pre-1839
160X2	Duras Demesne	-	Pre-1839
161	Bush Harbour-	Parkmore; *Caladh na Sceiche*	1881
162	Duras pier and quay	Lynch's; Newtown Lynch; Doorus; Durrus	1824
163	East end I'bofin	-	Pre-1899
164	Old pier Bofin	Main Harbour	1881
164A	New pier I'bofin	-	Post-1900
164B	Inishark slip	-	-
165	Inishturk (near Omey)	-	Pre-1899
166	Inishturbot	-	1875
166A	Inishlacken	-	Pre-1899
166B	Inish Macdara	-	-
166C	Mason Isl.	*An tAircín*	*Exists 1886*
167	*Portach*, Mweenish	*Cé an Aifrinn; Portach Mháinse*	1886–7

167A	*Meall Rua*	Mweenish Island	1882
168	Finish Island	*Caladh Feenish*	Post-1839
169	*Caladh O Dheas* on Inis Travin	-	-
170	*Caladh an tSliogáin* on Inis Travin	-	-
171	*Caladh O Thuaidh* on Inis Travin	-	-
172	Inishbarra Island	Drinagh	1886–7
172X	Inishbarra Island	*Caladh na Scanaimhe*	-
173	Inisherk Is.	-	-
174	Kilmurvey small pier	-	1886–7
175	Kilmurvey main pier	-	*1911–13*
176	Bungowla, Aran	-	-
177	*Gort na gCapall, Inis Mór*	*Port na dtonn*	-
178	Killeany, *Inis Mór*	*Cill Éinne*	1823–6
179	Kilronan jetty	*An tSeancéibh*	1830
180	Kilronan slip	-	*Circa 1899*
181	Kilronan main pier	-	1848
182	Trawndaleen, *Inis Meáin*	*Caladh Mór*	Pre-1839
183	*Port na gCurrach, Inis Meáin*	-	-
184	Cora point, *Inis Meáin*	-	-
185	*Trácht Each, Inis Oirr*	*Trá Teach*	1886–8
185A	Lghthouse jetty, *Inis Oirr*	-	*1912–14*
186	*Corrráit, Inis Oirr*	-	-
187	*Poll na gCaorach, Inis Oirr*	-	-
189	Main pier, *Inis Oirr*	-	1886–7
189X	Ballyhees slip	*Baile Thíos*	*circa 1900*
CS01	Corranroo Pier (ruin)	-	Pre-1839
CS02	Corranroo pier & quay	-	*Pre-1900*
CS03	Munnia pier (ruin)	-	*Pre-1900*
CS05	Aughinis pier	-	*Pre-1900*
CS07	Burrin New Quay	New Quay	1823
CS07X	Burrin Old Quay	-	Pre-1822
CS08	Skerretts Quay	Rine Quay	*Circa 1770*
CS09	Parkmore Quay	-	*Pre-1820*
CS10	Muckinish W pier	-	*Pre-1900*

CS11	Muckinish E pier	-	*Pre-1900*
CS13	Muckinish S Pier	-	*Pre-1900*
CS14	Bell Harbour	Bealaclugga	*Circa 1780*
CS18	Muckinish W Quay	-	*Pre-1900*
CS21	Bishops Quarter Quay	-	*Pre-1839*
CS23	Loughrask Pier & quay	-	Circa 1840
CS24	Loughrask Pier		Pre-1900
CS27	Bournapeaka Pier	-	1879
CS30	Ballyvaughan Quay	Old quay; Quay at CGS	*Pre-1800*
CS 31	Ballyvaughan Pier	-	1846–7
CS31X	Culliagh Quay	-	Pre-1839
CS32	Harbour Hill House Quay	-	*Circa 1800*
CS33	Glenina	Coolsiva	1881

Appendix B

One hundred and eleven applications for piers and harbours in all coastal counties of Ireland received by the Commissioners for Irish Fisheries in 1820–9

Co. Antrim
Carrickfergus: not proceeded with due to excess cost
Carrickfergus (Scotch Corner): not commenced
Cushendun: not commenced
Mill Bay: rejected
Mill Island: rejected
Port Ballintrae: not completed
Portmuck: completed
Stephens Port: rejected on grounds of cost

Co. Down
Ardglass: improvement to existing pier rejected on grounds of cost.
Ballywalter: nothing done, no local contribution.
Bridget's Cove: nothing done.
Cultra, Belfast Lough: rejected.
Kimmersport, Ardglass: works completed.
Killileagh, Strangford Lough: rejected.
Kircubbin, Strangford Lough: rejected.
Newcastle, Dundrum Bay: works in progress.
Portaferry: rejected.
Rossglass, Dundrum Bay: not carried further.

Co. Louth
Clogherhead: works in progress.

Co. Dublin
Balbriggan: works completed.
Lambay: works in progress.

Co. Wicklow
Greystones: work suspended.

Co. Wexford
Arthurstown: works in progress.
Carna Bay: grant refused.
Courtown: further grant declined.
Crossfarnogue: no local contribution, nothing done.
Killake: rejected.

Co. Waterford
Annstown (Drumbrattan): rejected.
Ballinacourty Wyse's Point: works not commenced.
Ballingaul: nothing done, local contribution too small.
Fishcove, Tramore Bay: rejected.
Grandison Harbour, Dungarvan Bay: no action taken.

Co. Cork
Baltimore: works completed.
Bantry: nothing done.
Castletown Bere Haven: works completed.
Cape Clear: works in progress.
Clonakilty (Ring): works completed.
Courtmacsherry: works completed.
Coolagh (Ballycroane): works completed.
Glandore: works completed.
Knockadoon: grant not approved, nothing done.
Laurence's Cove, Bere Island: works in progress.
Monkstown: rejected.
Tragarriff, Bantry Bay: rejected.

Co. Kerry
Ballinskelligs: works completed.
Barra: rejected.
Brandon: works completed.
Cahersivine: works completed.
Dingle: works completed.
Kenmare: works in progress.
Kilfenora: no action taken.
Valentia: works in progress.

Co. Clare
Ballyvaughan: nothing done.
Burrin New Quay: works completed.

Carrigaholt: works completed.
Doonbeg Bay: no report.
Dunbeg: works completed.
Kilbaha (new site): works completed.
Kilkee: no action taken.
Kilrush: works in progress.
Liscannor: works in progress.
Querin: rejected.
Seafield: works completed.

Co. Galway

Ardcastle: rejected.
Ballinacourty: works completed.
Barna: works not completed, inadequate local contribution.
Berturboy: rejected.
Blackrock: rejected.
Bunowen: no action taken.
Claddagh: works in progress.
Cleggan: works completed.
Clifden: works unfinished, still in progress.
Costello: works completed.
Duras: works completed.
Greatman's Bay (*Clais na nUan*): rejected.
Inish Bofin: no report.
Inishshark: nothing done.
Killeany, Aran Island: works completed.
Kilkieran: rejected.
New Harbour: works completed.
Roundstone: works completed.
St Kitts: works completed.
Spiddle: works completed.

Co. Mayo

Achill: not considered.
Belmullet: works completed.
Broad Haven: grant refused, nothing done.
Bunatraher: nothing done.
Bundurragh: works completed.
Carrowmore: grant revoked, nothing done.
Ennisturk: nothing done.
Killala: works completed.
Old Castle: works completed.
Old Head: works completed.
Portacuille: grant revoked.
Saleen: works completed.
Tarmon (Blacksod): works completed.

Co. Sligo

Mullaghmore: works completed.
Pollagheeny: not completed due to cost.
Raughley: works completed.

Co. Donegal

Ballyotherland, Donegal Bay: nothing done, no local contribution.
Craigbuoy, Lough Foyle: no grant made.
Creven, Inver Bay: rejected.
Doaghbeg: dangerous rock removed.
Green Castle, Lough Foyle: works in progress.
Killybegs: rejected.
Moville: rejected.
Portnoe: works in progress.
Portmore: works in progress.
Portnacross, near Killybegs: nothing done.
Teelin: nothing done.

Co. Derry

Port Stewart: nothing done, due to excess cost.

Appendix C

Piers and harbours in all coastal counties of Ireland sanctioned under the Piers and Harbours Acts of 1846 and 1847

Site	Works	Applicant	Estimate	Actual Grant spent
Co. Antrim				
Cushendall (Red Bay)	Pier	J. McNeill	£5,000	£4,100
Co. Down				
Annalong	Wharf and BW	Inhabitants	£2,400	£1,654
Ballywalter	Pier and road	A. Mulholland	£4,500	£3,071
Newcastle	Pier and BW	Lord Annesley	£9,400	£6,319
Co. Louth				
Carlingford	Pier and road	R. McNeill	£3,350	£2,583
Clogher Head	Harbour improvements	Landowners	£750	£368
Dunany	Pier extension	A. Bellingham	£500	£416
Co. Dublin				
Rush	Rebuild pier	Sir R. Palmer	£1,400	£894
Co. Wicklow				
Greystones	Wharf	W. La Touche	£1,200	£896
Co. Wexford				
Cahore	Sundry			£76
Courtown	Screw-pile pier	Earl of Courtown	£4,570	£2,283
Duncannon	Pier and approach	A. Stephens and Others	£6,000	£3,344
Fethard	Sundry			£29
Kilmore	Pier	Inhabitants	£6,000	£4,640
Slade	Pier	Inhabitants	£1,400	£803
Co. Waterford				
Ballinacourty	Extend pier	T. Wyse	£500	£20
Ballinagaul	Pier and slip	J. Strangman	£1,800	£1,346
Helvick Head	Slip and improvements to basin	Lord Stewart de Decies	£450	£225
Co. Cork				

Ballycottin	Pier	M. Longfield	£2,840	£1,119
Burren	Pier and slip	Lord Bernard	£1,100	£481
Cape Clear	Pier and landing place	Sir W. Beecher	£460	£267
Glandore	Harbour improvements	J. Barry	£220	£252
Skull	Pier and approach	Inhabitants	£2,400	£1,409
Courtmacsherry	Improvements			£80
Castletown (Bere Haven)	Improvements			£90
Baltimore	Improvements			£18
Co. Kerry				
Brandon	Rebuild pier	Inhabitants	£600	£450
Blackwater	Pier and road	Rev. D. Mahony	£900	£526
Caherciveen	Sundry			£45
Dingle	Extend pier	Inhabitants	£3,000	£1,495
Greenane	Pier	J. Mahony	£300	£235
Kenmare	Sundry			£1
Knightstown	Extend pier and breakwater	Knight of Kerry	£2,700	£1,713
Kilmackillogue	Pier	Marquis of Lansdowne	£2,200	£930
Tarbert*	Pier			£5,000
Ventry	Pier and slip	J. Hickson and others	£2,200	£1,135
Co. Clare				
Ballyvaughan	Improvements			£119
Liscannor	Improvements			£82
Kilbaha	Improvements			£19
Co. Galway				
Kilronan	Pier and slip	J. Digby	£2,400	£1,607
Bunowen	Pier	J.A. O'Neill	£4,800	£4,256
Ballinakill (Keelkyle)	Pier and slip	F.J. Graham	£900	£410
Claddagh	Quay and piers	Rev. J.R. Rush and others	£4,000	£2,658
Curhowna (Errislannan)	Dock	J.S. Lambert	£800	£572
Kilkieran	Pier	Waste Land Society	£3,000	£1,004
Rosroe	Quay, slip and road	Gen. Thompson	£800	£565
Roundstone (north pier)	Pier repairs	T. Martin MP	£1,200	£35 work abandoned

Tarrea	Pier and harbour	M. Blake	£2,100	£1,481
Ardfry (new Harbour)	Sundry			£15
Barna	Sundry			£30
Clifden	Sundry			£18
Co. Mayo				
Belmullet	Canal	W. Carter	£9,000	£5,032
Belmullet	Pier			£10
Broadhaven	Pier			£21
Newport	Quay wall and deepen channel	Sir R. O'Donnell	£1,200	£780
Roigh	Sundry			£73
Saleen	Sundry			£14
Tarmon	Sundry			£15
Co. Donegal				
Ballyness	Pier	W. Olpherts	£640	£487
Bunatrahane	Slip	Rev D. Tredennick	£500	£435
Moville	Pier	W. Hazlett	£5,900	£4,490
Portnablahy	Pier	A. Stewart	£1,900	£993
Mount Charles (Sea View)	Pier	L. Cornwall	£2,675	£1,035
Greencastle	Sundry			£26
Bruckless	Sundry			£15
Killybegs	Sundry			£5
Teelin	Sundry			£20

Source: data from OPW19 and a Return to Parliament dated 1859. [BPP 1859 (Sess 1) (119)]. The total number of works undertaken was 67 including Belmullet canal.
* Tarbert is listed in the Return to Parliament but not in the OPW Reports.

Appendix D

Ninety piers and harbours in all coastal counties of Ireland named in the Act 16 & 17 Victoria, C. 136 vesting them in the relevant counties

Co. Antrim	Courtown	**Co. Kerry**	Liscannor	Tarrea	**Co. Sligo**
Carrickfergus	Duncannon	Brandon	Kilbaha	Clifden	Raughley
Cushendall (Red Bay)	Fethard	Blackwater	**Co. Galway**	Cleggan	**Co. Donegal**
Portmuck	Kilmore	Caherciveen	Ardfry	Leenaun (east jetty)	Ballyness
Co. Down	Slade	Castlemaine	Doorus		Bunatrahan
Annalong	**Co. Waterford**	Dingle	Errislannan (Laughawn Lea)	**Co. Mayo**	Moville
Ballywalter	Balllinacourty	Greenane	Keelkyle ('Ballinderrig')	Achill Sound	Portnablahy
Newcastle	Ballynagane (Ballynagaul?)	Kenmare	Kilcolgan	Belmullet	Mount Charles (Sea View)
Co. Louth	Helvick Head	Knightstown	Kilkieran	Broadhaven	Greencastle
Carlingford	**Co. Cork**	Kilmackillogue	Claddagh	Clare Island ('2 small piers')	Bruckless
Clogher Head	Ballycotten	Ventry	Barna	Dooniver	Killybegs
Dunany	Cape Clear		*Spidéal (Seancéibh)*	Inishturk Island	Rathmullen
Co. Dublin	Coulagh Bay	Greenane	Maumeen	Newport	Teelin
Lambay Island	Glandore		Killeany	Dooniver	Newport [*sic*]
Rush	Skull		Bunowen		
Co. Wicklow	Courtmacsherry		Kilronan pier and slip		
Greystones	Castletownbere		Bealandangan Pass	Old Head	
	Laurence Cove (Bere Haven)				
Co. Wexford	Baltimore	**Co. Clare**	Rosroe	Roigh	
Arthurstown	Kinsale (Cove)	Ballyvaughan	Roundstone (south pier)	Saleen	
Cahore	Kinsale (Worlds end)				

APPENDIX E

Piers and marine works in all coastal counties of Ireland sanctioned and carried out, 1879–84, under the Fisheries Piers Committee (FPC), with their sources of funding

Site	Special Allocations 1879[1] + 1883[4]	Relief of Distress Funds 1880[2]	Contribution From the Canadian Fund[3]	Total all sources, incl. contributions up to 1882[5]
Donegal				
Tawny Pier		£675	£135	£900
Teelin Pier		£6,000	£1,000	£8,000
Downies Bay Pier	+ £600	£4,350	£900	£5,800
Rannagh Pier, Arranmore	+ £150	£900	£200	£1,200
Mallinbeg Slip		£825	£238[2]	£1,100
Bunatruhan Harbour	£1500	£735	£118[2]	£980
Burtonport Pier or Quay	£375			
Ballysaggart Slip and Harbour	£900			
Poulhurrin Slip	£900			
Sligo				
Inishcrone Pier (BW and Slip)		£1,583	£215	£2,110
Pullendiva Pier	+ £150	£3,900	£1,050	£5,200
Mayo				
Tontanvalley Pier	+ £900	£1,515	£255	£2,020
Mallaranny		£2,700	£55	£3,000
Roonagh Point Landing		£1,275	£75	£1,700
Glenlara Slip		£600	£250[2]	£800
Lacken Pier		£1,830	£310	£2,440
Lecanvey Pier		£3,000	£600	£4,000
Galway				
Bush Harbour Pier		£2,348	£581	£3,130
Renvyle (Smeerogue) Pier	£3,000 (£2821)		£566*	
Leenane Pier		£600	£200	£800
Glenagimlagh Pier		£488	£63	£650

Ardmore Pier		£1,050	£350	£1,400
Roundstone Pier		£1,500	£150	£2,000
Inishlacken Pier		£450	£150	£600
Doleen Pier		£1,500	£350	£2,000
Oranmore Castle Pier		£435	£100[2]	£580
Ballyhees Slip	£188[4]			
Carrowmore Pier	£210[4]			
Clare				
Goleen Ross	£225	£450		£600
Glenina Pier		£900	£300	£1,200
Goleen Tullig Pier		£188	£63	£250
Kerry				
Gleesk Pier		£1,013	£150	£1,350
Portmagee Slip	£150[4]			
Cork				
Cape Clear	£3,500[5]			
Gerahies, Bantry Bay		£165		£200
Dunmanus Bay Pier		£458	£163[2]	£610
Milcove Pier		£570	£100[2]	£700
Baltimore Pier		£3,000	£200	£4,000
Wexford				
Carnsore pier	£1,650[4]			
TOTAL (N = 39)	**£10,400 + £ 3998**	**£39,807**	**£9,384**	**£55,720**

* This allocation includes the sum of £189 allocated from the *New York Herald* fund, but administered by the Canadian Fund Committee.

[1] Data from 49th Report of the OPW, p. 25. BPP 1881 [C. - 2958].

[2] Data from 49th Report of the OPW, p. 29. BPP 1881 [C. - 2958]. (Also in 50th Report, p. 28, BPP 1882 [C. - 3261] and 51st Report, p. 26, BPP 1883 [C. - 3649], both of which include the dates the works were transferred to the relevant counties.)

[3] Data from Addendum, pp. 33–34, to the Report of the Canada Fund Committee, BPP 1881 (326). The Committee funded a total of 29.

[4] Data from 52nd Report of the OPW, p. 26. BPP 1884 [C. - 4068].

[5] Data in Return to Parliament, 1882. BPP 1882 (203).

Local contributions other than those of the Canadian Fund are not included.

Appendix F

Sums spent under the Sea Fisheries (Ireland) Act, 1883, on marine works in all coastal counties of Ireland up to 1902

Location	Structure	Total spent £	Free Grant	Loans, contributions, etc.
Co. Antrim				
Ballywilliam	Slip	1,342	1,000	342
Co. Down				
Annalong	Harbour	3,794	2,833	961
Ballyhalbert	Pier	5,746	4,260	1,486
Kilkeel	Harbour	7,726	5,810	1,916
Portavogie	Pier	5,371	5,371	-
Co. Louth				
Carlingford	Harbour	14,604	10,950	3,654
Clogher Head	Harbour	19,322	14,535	4,787
Co. Dublin				
Loughshinny	Pier	1,977	1,701	276
Co. Wicklow				
Greystones	Pier	20,689	19,189	1,500
Co. Wexford				
Kilmore	Pier	7,785	5,839	1,946
Co. Waterford				
Ballynagaul	Pier	1,490	1,304	186
Boat Strand	Harbour	6,456	6,456	-
Cheek Point	Pier	3,335	2,585	750
Passage East	Harbour	4,120	3,055	1,065
Co. Cork				
Ballycotton	Harbour	20,491	16,491	4,000
Baltimore	Lighthouse	1,399	1,399	-
Baltimore	Slip	157	107	50
Castletown Bere	Pier	2,640	2,640	-

Knockadoon	Slip	1,456	1,456	-
Ross Bar	Pier	4,332	3,832	500
Union Hall	Pier	4,911	3,761	1,150
Co. Kerry				
Anascaul	Slip	759	759	-
Ballydavid	Slip	1,627	1,220	407
Brandon	Pier	4,960	4,960	-
Dingle	Lighthouse	589	589	-
Co. Limerick				
Glin	Small improvements	23	23	
Co. Clare				
Carrigaholt	Harbour	13,761	11,213	2,548
Kilkee	Slip	1,526	1,155	371
Liscannor	Harbour	5,184	3,888	1,296
Seafield	Pier	3,378	2,515	863
Co. Galway				
Ardwest	Pier	759	759	-
Bealandangan	Pier and Harbour	435	435	-
Bunowen	Pier	2,465	2,465	-
Cashla	Pier	2,036	2,036	-
Cleggan	Harbour	8,022	6,016	2,006
Collaheigue	Harbour	1,704	1,704	-
Crumpaun Carna	Pier	2,455	2,455	-
Gannoughs	Landing Place	255	255	-
Inishbofin	Small Improvements	162	162	-
Kilkieran	Small Improvements	773	773	-
Kilronan	Pier	4,640	4,640	-
Mason Island	Pier	590	590	-
Rossaveel	Approach Road	332	332	-
Trawndaleen	Pier	1,715	1,715	-
Co. Mayo				
Achill Viaduct	Small improvements	1,629	1,629	-
Belmullet East	Harbour	3,984	3,984	-
Belmullet West	Small Improvements	421	421	-

Carrowkeeran	Pier	2,894	2,644	250
Killerduff	Slip	1,164	1,164	-
Lackan	Pier	5,681	5,681	-
Lecanvy	Small Improvements	1,007	1,007	-
Pollnamuck	Landing Place	276	276	-
Co. Sligo				
Aughris	Pier	2,345	2,345	-
Easky	Pier	4,255	4,255	-
Inishcrone	Pier	6,318	6,318	-
Co. Donegal				
Bundoran	Harbour	3,249	3,249	-
Culdaff	Pier	3,842	3,352	490
Malin Head	Pier	10,038	7,730	2,308
Malinmore	Slip	1,255	1,255	-
Port Ochre	Slip	1,844	1,674	170
Port Salon	Pier	6,111	4,858	1,253
Co. Derry				
Portstewart	Harbour	3,887	2,947	940
Portstewart	Harbour	4,222	4,222	-
TOTALS		**261,715**	**224,244**	**37,471**

Source: data from return to Parliament, BPP 1902 (355).

Appendix G

All works done by the Piers and Roads Commission (P&RC) in counties Galway and Mayo, May 1886–May 1887

Co. and Structure	Net Cost	Works Done
County Mayo		
Inishturk Isl.	£213	Concrete breakwater, 120 ft. protecting boat slip
Portacoola, Clare Isl.	£502	Two cement breakwater arms, 185 ft., forming small harbour. Beacons put at entrance
Lighthouse Cove, Clare Isl.	£651	Two cement breakwater arms formed with slip roughed out of rock. Approach road 500 yards long
Carrowmore Harbour	£479	Breakwater of stone and cement formed at old pier. Site cleared and rocks blasted
Kilsallagh Landing	£39	Rocks in entrance channel blasted
Murrisk Slip	£63	Made concrete landing slip, 90 ft. long
Killeenakoff, Knappagh	£147	Built two-span concrete bridge, each span 15 ft. Approach road 1,300 yards laid down
Derryribeen, Kilmaclasser	£201	Improvement of existing Derryribeen Road
Carnaclea, Clogher	£151	Improvement of existing Carnaclea Road
Kilmeena	£24	Fahy Road built, 500 yards
Lettermogheera, Srahmore	£71	Built pier 120 ft. for landing seaweed brought by river into lake
Coraun, Achill	£428	Doughbeg and Coraun Road, five miles long, completed to Achill Sound
Dooega, Achill Isl.	£267	Rock clearance for 166 ft. to improve landing place. Small bridge, two spans of 12 ft. each made between the villages
Slievemore, Achill Isl.	£1,229	Keel Harbour excavated to LWST with 200 ft. of wharf. Entered by passage 170 x 24 ft. cut through rock. Two concrete breakwaters planned, one completed
Slievemore, Achill Isl.	£97	Keel Road, one half-mile from Keel harbour to main road, constructed
Slievemore, Achill Isl.	£126	Dooagh landing. Rocks blasted and faced and place improved. Not suitable for a harbour
Bunacurry, Achill Isl.	£212	Replace old pier with larger, higher substantial pier, 140 x 24 ft., strengthened with concrete
Ballycroy	£215	Tallaght Pier. Rocks blasted at entrance to small creek. Small stone pier built at site chosen by inhabitants
Bundoola, Belmullet	£509	Breach in main road of the peninsula made good by making causeway, 320 ft. long, faced with strong masonry in cement

Inishkea Isl.	£631	Made slip, 250 ft. in length, sheltered by outer breakwater in concrete
Aughadoon, Belmullet	£59	Rocks cleared from dangerous landing
Tip Harbour, Belmullet	£105	Blasted rock at centre of entrance channel. Further rock clearing at Muingereena Bay
Knocknalower, Belmullet	£311	Grauphil Landing. Outer rock cleared, landing place floored with concrete and small slip formed
Knockaduff, Belmullet	£180	Coraunboy slip made in concrete. Landing place for ferry
Muingnaboo, Belmullet	£202	Porturlin Slip, 120 feet long made and entrance cleared by blasting rocks
Co. Galway		
Furbo	£83	Made Derryloney turf road, 700 yards
Spidéal	£59	Made Ballintleva Road, 550 yards
Killaunin	£51	Made road to shore for seaweed, 530 yards
Selerna	£98	Made Boherbhui turf road, 2,000 yards
Kilcummin	£119	Made Minna turf road
Kilcummin	£107	Made Cloghmore road to existing landing place, 950 yards
Kilcummin	£73	Existing track from Rossaveel Harbour to main road made passable, about three-quarters of a mile
Kilcummin	£60	Completed Keeraun Road (laid out in 1848) for seaweed traffic
Oughterard, Screeb	£1,517	*Spidéal* to Screeb road completed for 61/2 miles. A girder and concrete bridge of 22 ft. span built near Screeb
Oughterard, Gorumna Isl.	£290	Intertidal causeway at Corrigaluggaun Pass replaced as far as possible by a 14 ft. road of heavy granite slabs with many gullets
Oughterard, Gorumna Isl.	£33	Road to Trabane Harbour from main road improved
Oughterard, Gorumna Isl.	£30	Made Pollawirra Road for transport of seaweed
Oughterard, Gorumna Isl.	£503	Pass at Rankin's House. Made a granite causeway 240 ft. long into Lettermullan Isl. with an iron revolving bridge on a central pivot, with two spans of 12 ft. each to allow passage of boats
Oughterard, Letter-mullan Isl.	£12	Furnace Bridge. A girder and concrete bridge was arranged for, but owing to interference with and threats to destroy the work, loose wooden planks had to be substituted
Oughterard, Turlough	£416	Made granite and concrete bridge arch of 12 ft. span at Inverbeg. Made causeway through the lake with girder and concrete bridge of 22 ft. span, with piers of granite in cement
Clifden, Owengowla	£133	Kylesalia Road made from Inverbeg to Invermore, and rocks blasted from Invermore to Kylesalia (to be completed by county)

Clifden, Owengowla	£1,184	Made Kylesalia Bridge across estuary. Six 30 ft. spans of 18-inch girders and concrete roadway on piers, 20½ ft. high, made of granite and cement with concrete abutments. Made Acoonera Bridge, 9 ft. long with girders and concrete on granite and cement piers
Clifden, Knockboy, Mweenish Isl.	£375	Made a dry-stone granite causeway, 1,400 ft. long above HWST, with a raised girder and concrete bridge, 18 ft. span, raised on piers of granite in cement over the main channel separating Mweenish Island from the mainland
Clifden, Roundstone	£921	Road, ¼ mile long made to Inishnee Island. Channel bridged by granite causeway 352 ft. long with girder and concrete bridge of four 30 ft. spans supported on concrete piers and abutments

Source: Report of the Piers and Roads Commission. BPP 1887 [C. 5214].

Appendix H

Depth of water at the pier head of the fishery piers of all coastal counties of Ireland as recorded by the Inspectors of Irish Fisheries in 1891

The number and depth of water at the pier head of all the fishery piers in Ireland as recorded by the Inspectors of Fisheries on a map entitled *Ordnance Survey Map of Ireland showing locality and expenditure on works etc connected with Fisheries 1891*, published in the Report of the Inspectors of Irish Fisheries on the Sea and Inland Fisheries of Ireland for 1890. BPP 1891 [C-6403]. Excavations and slips are omitted here since depth of water alongside is less relevant in their case. The total number of piers and quays in all Ireland recorded on the map is 160, 109 west of the Baltimore/Derry line and 51 east of it. A further thirty-five other works had also been recorded that did not affect the depths, e.g. rock removals, lights, buoying, etc. (Note that the figures for Co. Clare are for the whole county, not just for north Clare.)

County	No. of piers and quays	> 8 ft LWST	<8 ft LWST	Dry LWST	Not used for fisheries	Mainly for trade use	In ruins
Donegal	21	3	1	10		6	1
Sligo	5		3	1			1
Mayo	19		3	8	5	2	1
Galway	32	1	2	14	5	10	
Clare	11		1	4	1	4	1
Limerick	1					1	
Kerry	13		4	5	1	3	
W. Cork	7	1	2	3		1	
Sub-total	**109**	**5**	**16**	**45**	**12**	**27**	**4**
E. Cork	13	2	3	4	1	3	
Waterford	7		3	3			1
Wexford	8		3	4		1	
Wicklow	2	1	1				
Dublin	6	2		4			
Louth	4	1		2		1	
Down	7	2	2	1		1	1
Antrim	3		1			1	1
Derry	1		1				
Sub-total	**51**	**8**	**14**	**18**	**1**	**7**	**3**
TOTAL	**160**	**13**	**30**	**63**	**13**	**34**	**7**

LWST = low water of Spring tide; < 8ft = less than eight feet of water; > 8ft = more than eight feet of water.

APPENDIX I

Large fishery piers and marine works (> £1,000) carried out in all coastal counties of Ireland,1902–22, with the proposed source of funding

The *Marine Works (Ireland) Act* (MWI) was restricted to the congested districts counties only and was to be supplemented by contributions from the CDB and local County Councils. The *Sea Fisheries (Ireland) 1883 Act* was used for non-congested districts counties and was later largely bolstered by the *Ireland Development Grant Act, 1903* fund (IDG) and the *Development and Road Improvements Act, 1909* fund (DRI). The outbreak of the Great War caused all funding to cease, affecting a number of the proposed works. These did not restart after the War; generally, the County Councils were unable to raise the required local contribution.

Note how the estimated cost of the structures in the congested districts (part (a) of the table) are generally much lower than those in the non-congested counties. The funding of Kilronan and Spidéal under the SFI fund in part (b) was a special allocation approved by Treasury.

Part (a)

Works carried out in the congested districts counties, funded under the *Marine Works (Ireland) Act* (MWI), with support from the CDB and County Councils (Coco).

Site	Dates	Work	Est. cost	Funding
Liscannor, Clare	1906 –12	Deepen for steamers; extend groyne	£8600	MWI, DATI
Cape Clear, West Cork	1903-07	New pier head & entrance	£5,000	MWI, CDB.
Gortnasate, Donegal	1904–16	Deepen for steamers	£5,700	MWI, CDB, Coco
Cladnageeragh, Donegal	1904–06	Timber extension	£1,250	MWI, CDB, Coco
Portnoo, Donegal	1904–07	Rebuild old pier	£3,100	MWI, CDB, Coco
Downies Bay, Donegal	1904–07	Extend existing pier	£4,370	CDB, Coco
Falcarragh, Donegal	1907–08	Make boat slip	£1,000	Coco
Renard Pt., Valentia, Kerry	1907–10	Improve existing slip; erect storm wall	£4,800	MWI, CDB, local contrib.

Part (b)

Works carried out in counties outside the congested districts, funded under the *Sea Fisheries (Ireland) Act 1883*, the *Ireland Development Grant Act*, DATI and County Councils.

Site	Dates	Work	Est. cost	Funding
Ardmore, Waterford	1902–16	Slip, path, short breakwater	£4,181	SFI, DATI, Coco
Passage E., Cheekpoint Waterford	1906–7	Deepen and make breakwater	£6,874	SFI, DATI, Coco
Baltimore, West Cork	1914		£7,500	DRI
Kilronan, Galway	1899–1902	Extend main pier	£3,853	SFI, CDB
Kilmurvey, Galway	1909	Plan to repair and extend pier		Not done
Spidéal, Galway	1899–1902	Emergency repairs	£3,111	SFI, Coco
Rathmullan Donegal	1914	Major development	£17,000	SFI, DRI, Coco Not done. Stopped by war
Burtonport, Donegal	1914	Major development	£16,500	SFI, DRI, Coco. Not done. Stopped by war.
Buncrana, Donegal	1914	Major development	£17,000	SFI, DRI Not done. Stopped by war
Portstewart, no. 1, Derry	1900–02	New inner basin	£3,072	SFI, DATI, Coco
Portstewart, no. 2, Derry	1907–14	Deepen, enlarge, raise n. pier etc.	£1,982	SFI, DATI, Coco
Portavogie, no. 1, Down	1900–02	New pier and viaduct	£8,301	SFI, Coco
Portavogie, no. 2, Down	1907–11	New quays, pier and approach; new layout	£6,686	SFI, IDG, Coco

Bibliography

Parliamentary Papers

The full reference data to parliamentary papers are given in the endnotes to each chapter – see the note in the List of Abbreviations.

Acts of Parliament of the United Kingdom;
Bills introduced into Parliament;
Returns and Further Returns to Parliament.
Reports of Select Committees of the House of Commons;
Reports of Royal Commissions;
Reports of Commissions of Enquiry;
Reports of the Piers and Roads Committee;
Reports of the Lord Lieutenant under the Ireland Development Grant Act;
Reports of the Commissioners for Public Works Ireland 1831 to 1922;
Reports of the first Commissioners of Irish fisheries, 1821 to 1830;
Reports of the second Commissioners of Irish fisheries 1842 to 1846;
Reports of the Inspectors of Irish Fisheries 1869 to 1901;
Department of Agriculture and Technical Instruction Reports on the Irish Fisheries, 1902 to 1923;
Reports of the Congested Districts Board of Ireland 1893 to 1922;
Reports on the Development Grant (Ireland) Act, 1905 to 1908;
Reports of the Development Commissioners, 1912 to 1919;
Statements of Particulars regarding the Foreshore under the Crown Lands Act 1866 from 1872 to 1919.

Surveys and Reports

Bjuke, Carl G., Report on the Clark, Mary and Doran, Gráinne. Archives of the Dublin Mansion House Fund for the Relief of Distress in Ireland 1880: descriptive list. Dublin City Archives. www.dublincity.ie/.

Colleran, J., and D.P. Hanley., *Assessment of Piers, Harbours and Landing Places in County Galway.* Galway: Galway County Council, 2001. www.galwaycoco.ie.

Halpin, Sarah, and Gráinne O'Connor, Clare Coastal and Architectural Heritage Survey. Ennis, County Clare: Clare County Council. Available on the website www.clarelibrary.ie.

Maps

Larkin, W., Map of the County of Galway in the Province of Connaught in Ireland from Actual Survey by William Larkin. Engraved by S.I. Neale and Son, London 1819.

Monteguy, MS chart, *Carte Hydrographique de la Baie de Galloüay, 1690*. Bibliotethèque Nationale de France, Departement des Cartes et Plans, Portefeuille 27 du Service hydrographique, Division

7, Pieces 2 D and 2[1] D. Reproduced and discussed in Jane Conroy, *Journal of the Galway Archaeological and Historical Society,* 49 (1997), pp. 36–48.
Nimmo, A., Map of the Harbours of Galway Bay surveyed for the Commissioners for the Irish Fisheries, 1822, by Alexander Nimmo C.E., R.I.A.
—, A. Costello and Greatmans Bays surveyed for the Commissioners of Irish fisheries by Alexander Nimmo civil engineer, FRSE, MRIA, 1823.
—, The Harbours of Roundstone and Birterbuy with Part of the Coast of Galway from Slyne Head to Mynish Island, surveyed for the Commissioners of Irish Fisheries by Alexander Nimmo civil engineer, MRIA, 1823.
—, Cleggan Harbour Co. Galway. Coloured map. NLI, maps, 16 H. 5 (5).
Ordnance Survey of Ireland, Six-inch Series, First Edition. Co. Galway sheets.
Ordnance Survey of Ireland, Six-inch Series, Second Edition. Co. Galway sheets.
Ordnance Survey of Ireland, Six-inch Series, Second Edition. Co. Clare sheets.
Ordnance Survey of Ireland, Discovery Series of Maps. Numbers 37, 44, 45, 46, 51, 52.
Robinson, T., Oileán Arann: A map of the Aran Islands with a companion to the map. Roundstone, Co. Galway: Folding Landscapes, 1996.
—, Connemara. Part 1: Introduction and gazetteer. Part 2: A one-inch map. Roundstone Co. Galway: Folding Landscapes, 1990.
—, The Burren. A map of the uplands of North-West Clare, Éire. Roundstone, Co. Galway: Folding Landscapes, 1999.

Miscellaneous

Fisheries Series. Correspondence, July 1848 to January 1847, relative to the Measures adopted for the Relief of Distress in Ireland and Scotland. BPP 1847 [765].
Placenames (Ceantair Ghaeltachta) Order. S.I. No 599/2011. Dublin: Iris Oifigiúil, 25 November 2011.

Files on Piers in the National Archives of Ireland

OPW/8/ and OPW/5/ files: These are files of the Office of Public Works conserved in the National Archives of Ireland. Some files contain only a single item, usually a memorial for a pier or quay; others contain many items, both printed and manuscript. A complete listing of all the OPW/8 files is available online at www.nationalarchives.ie. Entries here are in the form given in the National Archives web list, but with dates added in brackets. Names are as given on the files, even where misspelt (e.g. 'Farrea Kinvarra Bay. OPW/8/143 (1846)', which concerns Tarrea). There can be more than one file for any given pier and the contents of some files may contain material relevant to others.

Killeaney, Aran Islands OPW/8/11; Aughrusbeg Breakwater OPW/8/15 (1886); Ballinacourty Pier Co. Galway OPW/8/17 (1824–1827); Ballinakill Pier OPW/8/19 (1846–1848) [Keelkyle]; Ballyvaughan Pier (Co. Clare) OPW/8/29 (1848); Barna Pier OPW/8/35 (1842–1843); Barnaderg Bay OPW/8/36 (1841–1854); Bealandangan OPW/8/37 (1836); Blackhead Pier (Co. Clare) OPW/8/42 (1825); Black Rock (Co. Galway) OPW/8/44 (1827–1828); Bundoughlas OPW/8/53 (1884); Bunowen Pier. OPW/8/56 (1839–1854); Bush Harbour OPW/8/58 (1883); Carna OPW/8/59 (1883); Cashel OPW/8/68 (1884); Cashla Beg OPW/8/69 (1883); Calla OPW/8/73 (1883); Canowen OPW/8/74 (1884) [Canower]; Claddagh Pier OPW/8/79 (1827–1854); Claren

OPW/8/81 (1842) [Clarinbridge]; Cleggan Pier OPW/8/83 (1822–1884); Clifden Pier OPW/8/84 (1825–1846); Derrynea Harbour OPW/8/106 (1883); Doohulla OPW/8/113 (1846); Dunloughan Pier OPW/8/133 (1883); Dynish Island OPW/8/136 (1884); Ennislacken Island OPW/8/138 (1847); Errislanan Harbour OPW/8/139 (1846–1849); Farrea Kinvarra Bay OPW/8/143 (1846) [Tarrea]; Furbough Harbour OPW/8/148 (1842–1846) [Furbo]; Galway Harbour sub-series OPW/8/150 (1821-1850>); Lettermullen Island. OPW/8/151 (1887) [Kiggaul]; Garumna Island OPW/8/153 (1822); Glaninagh (Co. Clare) OPW/8/158 (1846); Gortdromagh and Fekeragh OPW/8/164 (1846); Greatman's Bay Pier OPW/8/166 (1837); Gurey, Owey Island, Aughrus Bay OPW/8/171 (1843–1883); Inishark OPW/8/180 (1885); Inisheer OPW/8/181 (1885); Inishmaan OPW/8/182 (1883); Inverin Harbour OPW/8/186 (1822–1834); Island Eddy, Galway Bay OPW/8/191 (1846); Kilkieran Harbour OPW/8/201 (1846–1849); Killery Harbour, Co. Mayo OPW/8/206 (1817–1846); Coolackey, Ballinahinch OPW/8/214 (1846); Kinvarra Harbour OPW/8/216 (1845–1848); Lackancasha Harbour, Ballinahinch OPW/8/221 (1846–1847); Laughawn Lea Pier or Errislannan Pier OPW/8/223 (1846–1848); Letterfrack OPW/8/23 (1884); Lettermullen, Moycullen OPW/8/234 (1846); Mason Island, Ard Bay, Roundstone OPW/8/244 (1846); Maumeen Pier, Greatman's Bay OPW/8/245 (1837); General files OPW/8/246; Meenish Island OPW/8/248 (1846) [Mweenish]; Moynish Island, Galway Bay OPW/8/258 (1846); Mynish Island, Ballinahinch OPW/8/265 (1846) [Mweenish]; New Harbour (Co. Clare) OPW/8/267 (1827); Old Quay (Co. Clare) OPW/8/273 (1883); Oranmore Harbour OPW/8/274 (1807–1847); Partha OPW/8/276 (1885) [Moyrus]; Portagh Quay,Moyrus OPW/8/281 (1887); Rosroe Pier OPW/8/325 (1846–1848); Roundstone Harbour OPW/8/326 (1825–1846); Spiddal Harbour OPW/8/341 (1823–1847); Tarrea Pier, Kinvara OPW/8/349 (1846–1849); Rossmuck Pier, Moycullen OPW/8/372 (1846); Cloghballymore OPW/8/383 (1845).

Files in the Archive, Galway County Library

GC/CSO/2/56; GC/CSO/2/58.

Files in Dublin City Archives

Ch1/21; Ch1/70; Ch1/71.

Secondary Sources

Blake, John A., *The History and Position of the Sea Fisheries of Ireland and how they may be made to afford Increased Food and Employment* (Waterford: J.H. McGrath, 'The Citizen' office, 1866. Reprinted by General Books LLC, Memphis, USA, 2012).

—, *The Irish Salmon Fisheries. Replies to Arguments advanced against the Bill now before Parliament, for assimilating the Fishery Laws of Ireland to those of England* (London: Robert Hardwicke, 1863).

Brady, Thomas F., Proposed Home or Refuge for Lost and Starving Dogs in Dublin (Dublin: the Author, 1878).

Breathnach, Caoilte, *A Word in your Ear. Folklore from the Kinvara area* (Kinvara: Comhairle Phobail Chinn Mhara, N.D. [*c.* 1990]).

Breathnach, Ciara, *The Congested Districts Board of Ireland 1891–1923* (Dublin: Four Courts Press, 2005).

Brennan, Agnes T., *Notes on some common Irish Seaweeds* (Dublin: Institute of Industrial Research and Standards. Published by the Stationery Office, 1950).

Colleran, J., and D.P. Hanley, *Assessment of Piers, Harbours and Landing Places in County Galway* [The Ryan Hanley Survey] (Galway: Galway County Council, 2001). 4 vols. The survey data, including maps and pictures, are in the public domain and can be accessed on the internet as follows: http://data-galwaycoco.opendata.arcgis.com/datasets/. Click on 'Piers and Harbours of Co. Galway' in 'Recent Databases' on that site. (Latest accessed 2 March 2017.)

Collins, Timothy (Ed.), *Decoding the Landscape* (Galway: Centre for Landscape Studies, 2003),3rd edition.

Conner, H.D., *Manual of Fisheries (Ireland) Acts* (Dublin: HMSO, 1904).

Conroy, Jane, 'Galway Bay, Louis XIV's Navy and the "Little Bougard"'. *Journal of the Galway Archaeological and Historical Society,* vol. 49 (1997), pp. 36–48.

Coombes, James, *Utopia in Glandore* (Butlerstown: Muintir na Tíre, 1970).

Cullen, L.M., 'The Galway Smuggling Trade in the 1730s', *Journal of the Galway Archaeological and Historical Society,* vol. 30 (1962–63), pp. 7–40.

—, 'H.M.S. Spy off the Galway Coast in the 1730s: The Politics and Economics of Wool Smuggling.' *Journal of the Galway Archaeological and Historical Society,* vol. 65 (2013), pp. 27–47.

—, 'Five Letters relating to Galway Smuggling in 1737.' *Journal of the Galway Archaeological and Historical Society,* vol. 65 (2013), pp. 27–47.

—, *Life in Ireland* (London: Batsford, 1968).

Davis, K., *Oranmore in Days of Yore* (Galway: Comhairle Phobail Chinn Mhara, ND [*c.* 2005]). (NUI Galway Hardiman Library, SCRR 941.74 Dav).

Duffy, P., *Galway History on a Postcard* (Dublin: Currach Press, 2013).

Evans, E.E., *Mourne Country* (Dundalk: Dundalgan Press, 1967), 2nd edition.

Gosling, P., B. MacMahon and C. Roden, 'Nausts, púcáns and "mallúirs". *Archaeology Ireland,* Vol. 24, no. 3 (Autumn 2010), pp. 30–34.

Griffith, Sir John Purser, and J.W. Griffith, *Western Harbours of Ireland. An Enquiry into the Suitability of various Bays and Estuaries on the West Coast of Ireland for the formation of a Commercial Port and Naval Base, March 1918.* Privately printed. (Available in NUI Galway Hardiman Library, SPCOL 387.109415Gri.)

Griffiths, A.R.G., *The Irish Board of Works, 1831–1878* (New York: Garland Publishing Inc., 1987).

Hall, S.C. and A.M., *Ireland, its Scenery, character etc.* (London: 1845).

Hardiman, J., *A History of the Town and the County of the Town of Galway* (Galway: Reprinted by Galway Tribune Ltd., 1994).

Hargrave, F., 'A Treatise in Three Parts from a Manuscript of Lord Chief Justice Hale.' *Collection of Tracts relative to the Law of England,* vol. 1. (Dublin: E. Lynch and others, 1787).

Heaney, M.H., *Sacerdotes Tuamenses.* Unpublished document. Information supplied by Fr Kieran Waldron.

Hoctor, D., *The Department's Story – A History of the Department of Agriculture* (Dublin: Institute of Public Administration, 1971).

Jackson, Gordon, *The History and Archaeology of Ports* (Tadworth, Surrey: The Windmill Press, 1983).

Jarvis, Adrian, 'Dock and Harbour Provision for the Fishing Industry since the Eighteenth Century' in D.J. Starkey, C. Reid, and N. Ashcroft (eds), *England's Sea Fisheries* (London: Chatham Publishing, 2000).

Johnson, Samuel, *A Dictionary of the English Language* (London: Henry Bohn, 1828).

Joynt, W.L., *The Salmon Fishery and Fishery Laws of Ireland. A Paper read before the Dublin Statistical Society, the 21st of January 1861* (Dublin: E. Ponsonby, 1861).

Kean, N. (ed.), *South and West Coasts of Ireland Sailing Directions* (Dublin: Irish Cruising Club, 2008), Twelfth Edition.

King, C., and C. McNamara (eds), *The West of Ireland. New Perspectives on the Nineteenth Century* (Dublin: The History Press, 2011).
Korff, A., *The Book of Aran* (Kinvara: Tír Eolas Press, 1994).
Lafferty, S., P. Commins and J.A. Walsh, *Irish Agriculture in Transition. A Census Atlas of Agriculture in the Republic of Ireland* (Dublin: Teagasc, 1999).
Lee, J.J., *The Modernisation of Irish Society 1848 to 1918* (Dublin: Gill and Macmillan, 1973).
Lewis, S., *Topographical Dictionary of Ireland* (London: S. Lewis, 1837).
Lohan, Rena, *Guide to the Archives of the Office of Public Works* (Dublin: The Stationery Office, 1994).
Lynam, S., *Humanity Dick. A Biography of Richard Martin, M.P. 1754–1834* (London: Hamish Hamilton, 1975).
Mac Cárthaigh, C. (ed), *Traditional Boats of Ireland. History, Folklore and Construction* (Cork: The Collins Press, 2008).
MacLochlainn, Alf., 'The Claddagh Piscatory School', in Anon., *Two Galway Schools* (Galway: Labour History Group, 1993). ISBN 0951737112.
Mainer, Geo., 'Hamilton Hartley Killaly' in G. Mainer (ed.), *Dictionary of Canadian Biography, vol. 10* (Toronto: University of Toronto/Université Laval, 1972).
McNamara, Conor, 'This Wretched People' in C. King and C. McNamara (eds), *The West of Ireland. New Perspectives on the Nineteenth Century* (Dublin: The History Press, 2011).
Melvin, P., *Estates and Landed Society in Galway* (Blackrock Co. Dublin: Edmund Burke, 2012).
Micks, W.L., *History of the Congested Districts Board for Ireland 1891 to 1923* (Dublin: Eason and Sons, 1925).
Morrissey, J. (compiler and editor), *On the Verge of Want. The base line reports of the Inspectors of the Congested Districts Board, 1892* (Dublin: Crannóg Books, 2001).
— (compiler and editor), *Inishbofin and Inishark, Connemara* (Dublin: Crannóg Books, 2012).
Mullins, M.B., 'Memoir of Barry Duncan Gibbons, late vice-president of the Institute of Civil Engineers of Ireland.' *Trans. Inst. Civ. Engs. of Ireland* 7, p. 168 ff.
Neff, J., *Irish Coastal Habitats: A Study of impacts on designated Conservation areas.* Data taken from *Coastal management a case for action.* Published by EOLAS on behalf of the County and City Engineers' Association.
Ní Chinnéide, S., 'Coquebert's Impressions of Galway City and County in the Year 1791', *Journal of the Galway Archaeological and Historical Society,* vol. 25 (1952–53), pp. 1–14.
Ó Dónaill, N., *Foclóir Gaeilge-Béarla [Irish-English Dictionary]* (Dublin: The Stationery Office, 1977).
O'Donoghue, B., *The Irish County Surveyors 1834–1944. A Biographical Dictionary* (Dublin: Four Courts Press, 2007).
O'Dowd, Anne, 'Resources and Life: Aspects of Working and Fishing on the Aran Islands'. In J. Waddell, J.W. O'Connell and A. Korff (eds), *The Book of Aran* (Kinvara, Co. Galway: Tír Eolas Press, 1999).
O'Dowd, P., *Down by the Claddagh* (Galway: Kenny's Bookshop, 1993).
O'Sullivan, J.J., *Breaking Ground. The Story of William T. Mulvany* (Cork: Mercier Press, N.D. [*c.* 2004]).
O'Sullivan, M.D., 'Glimpses of the Life of Galway Merchants and Mariners in the Early Seventeenth Century', *Journal of the Galway Archaeological and Historical Society,* 15, pp. 129–40.
O'Toole, P., *Aran to Africa.* (N.P.: Nuascéalta Teo., 2013).
O'Tuarisg, L., 'Stair Chois Fharraige.' In *Biseach: Iris Chumann Forbartha Chois Fharraige* (Galway: Chumann Forbartha Chois Fharraige, 1999).
Pethica, J.L., and J.C. Roy (eds), *To the Land of the Free from this Island of Slaves. Henry Stratford Persse's Letters from Galway to America, 1821–1832* (Cork: Cork University Press, 1998).

Prunty, J., and P. Walsh, *Galway/Gaillimh. Irish Historic Towns Atlas No. 28* (Dublin: Royal Irish Academy, 2016).

Roberts, Callum, *The Unnatural History of the Sea* (London: Octopus Publishing Group, 2007).

Robins, J., *Custom House People* (Dublin: Institute of Public Administration, 1993).

Robinson, R., 'The Line and Trawl Fisheries in the Age of Sail' in DJ. Starkey, C. Reid, and N. Ashcroft (eds), *England's Sea Fisheries* (London: Chatham Publishing, 2000).

Robinson, T., *Stones of Aran. Pilgrimage* (Dublin: The Lilliput Press, 1986).

—, *Stones of Aran. Labyrinth* (Dublin: The Lilliput Press, 1995).

—, *Connemara. Listening to the Wind* (Dublin: Penguin, 2006).

—, *Connemara. The Last Pool of Darkness* (Dublin: Penguin, 2008).

—, *Connemara. A Little Gaelic Kingdom* (Dublin: Penguin, 2011).

Roney, Sir C., *How to Spend a Month in Ireland.* (London: John Camden Hotten, 1872) (first published 1866).

Saorstát Éireann, *Report of Department of Fisheries for the years 1923 to 1925* (Dublin: The Stationery Office, 1927).

Selden, J., *Of the Dominion or Ownership of the Sea,* translated by M. Nedham (London: William Du Gard, 1652; reprint edition by Arno Press, 1972).

Semple, M., *Reflections on Lough Corrib* (Galway: the author, 1974).

Starkey, D.J., C. Reid, and N. Ashcroft (eds), *England's Sea Fisheries* (London: Chatham Publishing, 2000).

Synge, J.M., *In Wicklow, West Kerry and Connemara* (Dublin: Maunsel and Co., 1919).

Villiers-Tuthill, K., *Alexander Nimmo and the Western District* (Clifden, Co. Galway: Connemara Girl Publications, 2006).

Waddell, J., J.W. O'Connell and A. Korff (eds), *The Book of Aran* (Kinvara, Co. Galway: Tír Eolas Press, 2nd printing 1999).

Wainwright, Mr, *The first section of the Galway Estates of the late Thomas Barnewell Martin Esq. [The Martin Estates Connemara] to be sold in August 1849 by Auction.*

Went, Arthur E.J., *Foreign Fleets off the Irish Coast* (JCHAS home).

West, T., *Horace Plunkett: Co-Operation and Politics, an Irish biography* (Gerard's Cross: Colin Smythe, 1986).

Wilkins, N.P., *Ponds, Passes and Parcs. Aquaculture in Victorian Ireland* (Dublin: The Glendale Press, 1989).

—, *Squires, Spalpeens and Spats: Oysters and Oystering in Galway Bay* (Galway: the Author, 2001).

—, *Alexander Nimmo, Master Engineer 1783–1832* (Dublin: Irish Academic Press, 2009).

Woodman, K., *Safe and Commodious. The Annals of the Galway Harbour Commissioners, 1830–1997* (Galway: The Galway Harbour Company, 2000).

INDEX

Note: Page locators in bold refer to tables; those in italics to figures and illustrations.